Gene Therapy and Tissue Engineering in Orthopaedic and Sports Medicine

Johnny Huard
Freddie H. Fu
Editors

With 51 Figures

Birkhäuser
Boston • Basel • Berlin

Johnny Huard
Growth and Development Laboratory
Department of Orthopaedic Surgery
  and Molecular Genetics and
  Biochemistry
Children's Hospital of Pittsburgh and
  University of Pittsburgh
Pittsburgh, PA 15213-2583
USA

Freddie H. Fu
Department of Orthopaedic Surgery
Musculoskeletal Research Center
Center for Sports Medicine
University of Pittsburgh
Pittsburgh, PA 15213-2583
USA

RD
732
.G46
2000

Library of Congress Cataloging-in-Publication Data
Gene therapy and tissue engineering in orthopaedic and sports medicine / Johnny Huard,
editor; Freddie H. Fu, co-editor.
     p.   cm. — (Methods in bioengineering)
   Includes bibliographical references and index.
   ISBN 0-8176-4071-1 (hc.:alk. paper)
     1. Musculoskeletal system—Diseases—Gene therapy.   2. Orthopedics.   3. Gene
therapy.   4. Sports medicine.   I. Huard, Johnny.   II. Fu, Freddie H.   III. Series.
RD732.G46   2000
616.7'042—dc21   99-058338

Printed on acid-free paper.

© 2000 Birkhäuser Boston   ***Birkhäuser*** Ⓡ

All rights reserved. This work may not be translated or copied in whole or in part without the
written permission of the publisher (Birkhäuser Boston, c/o Springer-Verlag New York, Inc.,
175 Fifth Avenue, New York, NY 10010, USA), except for brief excerpts in connection with
reviews or scholarly analysis. Use in connection with any form of information storage and
retrieval, electronic adaptation, computer software, or by similar or dissimilar methodology
now known or hereafter developed is forbidden.
The use of general descriptive names, trade names, trademarks, etc., in this publication, even
if the former are not especially identified, is not to be taken as a sign that such names, as understood by the Trade Marks and Merchandise Marks Act, may accordingly be used freely by
anyone.

ISBN 0-8176-4071-1
ISBN 3-7643-4071-1   SPIN 10753794

Typeset by Best-set Typesetter Ltd., Hong Kong.
Printed and bound by Maple-Vail Book Manufacturing Group, York, PA.
Printed in the United States of America.

9 8 7 6 5 4 3 2 1

Methods in Bioengineering

Series Editor

Daniel L. Farkas
Center for Light Microscope Imaging and Biotechnology
Carnegie Mellon University
4400 Fifth Avenue
Pittsburgh, PA 15213, USA

This is a new series of books that will review and describe important methods in bioengineering and biomedical engineering, focusing on those areas that best define these fields and operate at the forefront of technology and relevance. The scope of the series is wide, going from molecular, genetic, and cellular techniques up to more macroscopic ones such as biomechanical measurement, bioimaging, and prosthetic development. The books will provide guides to "cutting-edge" and more established techniques that will be helpful to students and researchers who may well have differing backgrounds in biology, medicine, and engineering. It is also hoped that the series will enhance public awareness of bioengineering and of its exciting potential.

To my wife, Marcelle, who is not only my source of inspiration, but also the mother of my son as well as the technician in chief of my laboratory. Thank you for your outstanding support, encouragement, and love.

Johnny Huard

Contents

Part I: Background

Part II: Gene Therapy Applications for the Musculoskeletal System

Part III: Tissue Engineering in Orthopaedics and Sports Medicine

Part IV: Hurdles and Development of New Vectors for Gene Therapy and Tissue Engineering

Preface

This book has been written in response to the many physicians and scientists working on the development of biological approaches to providing therapies for many orthopaedic disorders as well as to improving the healing of many tissues of the musculoskeletal system. The first goal of this book is to make the language compatible between the bench scientist and the clinician working in orthopaedic and sports medicine in order to cover specific areas of the orthopaedic discipline where the treatment can be improved and/or changed by the advancements in molecular medicine.

Advancements in molecular biology, which encompass the study of the genetic basis of disease, have produced new diagnostic methods and drug therapies for genetic diseases and acquired disorders. The growth in the understanding of human genetics has also led to the initiation of many human gene therapy experiments. Although many approved therapeutic clinical trials using this new technology have been performed in the last ten years, the first clinical trial using this technology in the area of orthopaedics was performed at the University of Pittsburgh.

Gene therapy, by definition, is the insertion of a functioning gene into the cells of a patient to correct an inborn error of metabolism or to provide a new function of the cells. This broad definition includes the potential treatment of essentially all types of human diseases, including all genetic disorders, cancer, infectious diseases, cardiovascular disorders, and auto-immune disorders through the genetic modification of cells of the human body to prevent or eliminate the disease.

With the development of improved methods of transferring genes into cells, new treatment options are becoming available in orthopaedics and sports medicine to aid in the healing capacity of musculoskeletal tissues. In this book, current techniques in gene therapy and tissue engineering will be described in addition to their potential application in improving the healing of the musculoskeletal system. This is the first book to comprehensively cover the new methods of gene therapy and tissue engineering with fifteen chapters by the leading practitioners of the applications of this technology to many tissues of the musculoskeletal system.

The book is divided into four parts. The first provides a scientific background for the concepts involved in gene therapy and tissue engineering for the musculoskeletal system. The basic application of gene therapy in orthopaedics is addressed in Chapter 1, and the use of different vectors and strategies for gene transfer will be discussed in Chapter 2.

The second part covers applications of gene therapy in orthopaedics and sports medicine, including the use of this technology for hard tissues (Chapter 3), for spinal disorders (Chapter 4), for improving the healing following orthopaedic-related muscle injuries (Chapter 5), for the repair of osteochondral articular defects (Chapter 6), for Osteogenesis imperfecta (Chapter 7), and in, combination with nitric oxide, for sports injuries, (Chapter 8).

The third part covers the spectacular combination of gene therapy and tissue engineering for the musculoskeletal system, including chapters on the use of stem cells in tissue engineering (Chapter 9), tissue engineering of ligament healing (Chapter 10), muscle-based tissue engineering and orthopaedic applications (Chapter 11), and the application of tissue engineering to cartilage repair (Chapter 12).

The final part of the book covers the hurdles facing the application of gene therapy and tissue engineering for the musculoskeletal system as well as the development of strategies to overcome these limitations. We discuss immune reaction following cell and gene therapy (Chapter 13) and the development of new-generation vectors for gene therapy (Chapters 14 and 15).

We truly hope that this book will provide a valuable resource for clinicians and scientists and a source of inspiration that stimulates the field of molecular medicine toward the development of new therapeutics for orthopaedic and sports medicine for the new millennium.

Pittsburgh, Pennsylvania

Johnny Huard
Freddie H. Fu

Acknowledgments

We would like to thank Miss Dana Och for her outstanding editorial assistance in the completion of this book as well as the following individuals for their thoughtful review of many chapters: Morey S. Moreland, Vladimir Martinek, Romain Seil, Kazumasu Fukushima, Chang Woo Lee, Douglas S. Musgrave, Christian Lattermann, and Dalip Pelinkovich. We want to also thank Mr. Ryan Pruchnic, Marcelle Pellerin, and Arvydas Usas for their technical help on the tissue processing and imaging of many illustrations included in this book.

Pittsburgh, Pennsylvania

Johnny Huard
Freddie H. Fu

Contributors

David Amiel, Connective Tissue Biochemistry, University of California School of Medicine, Department of Orthopaedics 0630, La Jolla, CA 92093-0630, USA

Scott D. Boden, The Emory Spine Center, Decatur, Georgia 30033, USA

Patrick P. Bosch, Growth and Development Laboratory, Department of Orthopaedic Surgery, Children's Hospital of Pittsburgh with University of Pittsburgh, Pittsburgh, PA, 15213-2583, USA

Charles S. Day, Growth and Development Laboratory, Department of Orthopaedic Surgery, Children's Hospital of Pittsburgh with University of Pittsburgh, Pittsburgh, PA, 15213-2583, USA

Christopher H. Evans, Department of Molecular Genetics and Biochemistry and Orthopaedic Surgery, University of Pittsburgh School of Medicine, and Megabios Corporation (Burlingame, CA), Pittsburgh, PA 15261, USA

Freddie H. Fu, Department of Orthopaedic Surgery, Musculoskeletal Research Center, Center for Sports Medicine, University of Pittsburgh, Pittsburgh, PA, 15213, USA

Lars G. Gilbertson, Musculoskeletal Research Center, Department of Orthopaedic Surgery, University of Pittsburgh Medical Center, E1641 Biomedical Science Tower Pittsburgh, PA 15213, USA

Steven C. Ghivizzani, Department of Molecular Genetics and Biochemistry and Megabios Corporation, Burlingame, CA; University of Pittsburgh, Pittsburgh, PA 15261, USA

Joseph C. Glorioso, Department of Molecular Genetics and Biochemistry, University of Pittsburgh, and Megabios Corporation, Burlingame, CA; Pittsburgh, PA 15261, USA

Randal S. Goomer, University of California San Diego School of Medicine, La Jolla, CA 92093-0630, USA

Joel Greenberger, University of Pittsburgh Medical Center, Department of Radiation Oncology, Pittsburgh, PA 15213, USA

Kevin A. Hildebrand, Department of Orthopaedic Surgery, University of Pittsburgh, E1641 Biomedical Science Tower, Pittsburgh, PA, 15213, USA

Johnny Huard, Growth and Development Laboratory, Department of Orthopaedic Surgery and Molecular Genetics and Biochemistry, Children's Hospital of Pittsburgh and University of Pittsburgh, Pittsburgh, PA, 15213-2583, USA

Kazutaka Izawa, The Laboratory for Soft Tissue Research and Sports Medicine Service, The Hospital for Special Surgery affiliated with Cornell University Medical College, New York, NY, 10021, USA

James D. Kang, Musculoskeletal Research Center, Department of Orthopaedic Surgery, University of Pittsburgh Medical Center, E1641 Biomedical Science Tower Pittsburgh, PA 15213, USA

Channarong Kasemkijwattana, Department of Orthopaedic Surgery, Vajira Hospital, Faculty of Medicine, Srinakhrinwirot University, 681 Samsen Road, Dusit, Bangkok 10300, Thailand

Christian Lattermann, Department of Orthopaedic Surgery, University of Pittsburgh, E1641 Biomedical Science Tower, Pittsburgh, PA, 15213, USA

Juan Li, Department of Molecular Genetics and Biochemistry, Room 1213, Biomedical Science Tower, University of Pittsburgh, Pittsburgh, PA, 15261, USA

Jay Lieberman, Department of Orthopaedic Surgery, UCLA School of Medicine, Center for the Health Sciences, Los Angeles, CA 90095-6902, USA

Jacques Ménétrey, Clinique D'Othopedie et de Chirurgie de L'Appareil Moteur, Hôpital Universitaire de Geneve, 24 Rue Micheli-du-Crest, CH-1211 Geneve 14, Switzerland

Morey S. Moreland, Department of Orthopaedic Surgery, University of Pittsburgh School of Medicine; William and Jean Donaldson Professor, Division of Pediatric Orthopaedics, Children's Hospital of Pittsburgh, Pittsburgh, PA, 15213, USA

Douglas S. Musgrave, Growth and Development Laboratory, Department of Orthopaedic Surgery, Children's Hospital of Pittsburgh with University of Pittsburgh, Pittsburgh, PA, 15213-2583, USA

Kotaro Nishida, Musculoskeletal Research Center, Department of Orthopaedic Surgery, University of Pittsburgh Medical Center, E1641 BST, Pittsburgh, PA 15213, USA

Christopher Niyibizi, University of Pittsburgh, 986 Scaife Hall, Pittsburgh, PA, 15261, USA

Ryan Pruchnic, Growth and Development Laboratory, Department of Orthopaedic Surgery, Musculoskeletal Research Center, Division of Sports Medicine, University of Pittsburgh, Pittsburgh, PA, 15213-2583, USA

Scott A. Rodeo, The Laboratory for Soft Tissue Research and Sports Medicine Service, The Hospital for Special Surgery affiliated with Cornell University Medical College, New York, New York, 10021, USA

Paul D. Robbins, Department of Molecular Genetics and Biochemistry, University of Pittsburgh School of Medicine, and Megabios Corporation (Burlingame, CA), Pittsburgh, PA 15261, USA

Patrick N. Smith, University of Pittsburgh, 986 Scaife Hall, Pittsburgh, PA, 15261, USA

Jun-Kyo Suh, Department of Biomedical Engineering, Tulane University; New Orleans, LA 70118, USA

Jacques P. Tremblay, Unité de Génétique Humaine Centre, Hospitalier de l'Université Laval, Quebec, Canada G1V4G2.

Nobuyoshi Watanabe, Department of Orthopaedic Surgery, University of Pittsburgh, E1641 Biomedical Science Tower, Pittsburgh, PA, 15213, USA

Savio L-Y. Woo, Musculoskeletal Research Center, Department of Orthopaedic Surgery, University of Pittsburgh, E1641 Biomedical Science Tower, Pittsburgh, PA, 15213, USA

Xiao Xiao, Department of Molecular Genetics and Biochemistry, Room W1213 Biomedical Science Tower, Pittsburgh, PA, 15213, USA

Henry E. Young, Division of Basic Medical Science, Mercer University School of Medicine, Macon, Georgia 31207, USA

Part I
Background

1
Gene Therapy in Orthopaedics

CHRISTIAN LATTERMANN and FREDDIE H. FU

Orthopaedics has seen several breakthrough innovations in the past that changed the quality and possibilities for treatment of orthopaedic diseases and conditions. Orthopaedics was initially a nonoperative discipline until after several wars the need for more and more efficient surgical techniques for amputations became obvious. Primary stabilization and repair of injured bones and soft tissues has then been integrated into the field of orthopaedics. Plate osteosynthesis and refinement of soft tissue techniques built the basis for the development of minimally or noninvasive treatment options such as arthroscopic joint surgery or percutaneous plating and intramedullary nailing of fractures. Biomechanical analysis and testing have been of great importance for the development of these techniques. Toward the end of this century, these techniques have been further refined and again build the ground for the integration of yet another scientific discipline into the realm of orthopaedic surgery, biology. The following chapters in this book aim to provide a profound insight into gene therapy, one of the newest and most exciting aspects of applied biological principles in orthopaedics.

Throughout the last 20 years, it has become clear that mechanical stabilization and biomechanical analysis of operative techniques alone do not guarantee surgical success. Only recently have we realized that the biology of soft tissue and bone healing is paramount to the outcome and success of our practice. Soft tissue and bone healing as well as chronic orthopaedic conditions such as rheumatoid arthritis (RA) are highly dependent on, or even maintained by, the action of growth factors, cytokines, antigen presentation, and protein interactions in the widest sense.

As we investigate the healing responses and the onset of RA or osteoarthritis (OA), we find that molecular biological, immunological, and biochemical aspects seem to play a key role. Additionally, in genetic diseases of primarily orthopaedic interest such as osteogenesis imperfecta (OI) or Duchenne muscular dystrophy (DMD), today's treatment options are purely symptomatic. In order to provide a curative treatment approach, we have to understand the genetic defect and its biological effect.

As we learn more about the is action of growth factors and cytokines, this knowledge is used to improve the treatment of chronic diseases and accelerate the healing response after soft tissue injury. The direct application of proteins such as growth factors or antiinflammatory cytokines has shown that these factors can be potentially beneficial (Rodeo et al. 1995; Einhorn and Trippel 1997; Scherping et al. 1997; Hildebrand et al. 1998) for the healing of soft tissues. The direct application of these proteins, however, suffers from several shortcomings. A very short biological halflife makes the multiple application of growth factors necessary. Alternatively, very high initial dosages are needed to achieve a sufficiently prolonged biological activity. The application of very high dosages, however, may impair the healing response or may even be toxic. In genetic defects such as OI, the direct application of the missing protein is not feasible. The introduction of the gene for this protein would enable the cell to continuously produce the missing protein.

Gene therapy is one of the most elegant techniques to overcome these problems. It comes out of the field of molecular biology and is a technique that allows for the sustained, local delivery of therapeutic proteins such as growth factors or cytokines. Using viral or nonviral delivery vehicles (vectors) the genetic information, usually a complementary deoxyribonucleic acid (cDNA), encoding the protein of interest is inserted into a living cell (Figure 1.1). The genetically modified cell has the potential to express the protein encoded by the transferred DNA in a sustained manner, making it a valuable vehicle for the targeted, long-term delivery of a protein of interest. Using this technique, growth factors can be locally applied to tissues with low healing capacity without the necessity of repeated injections or systemic administration.

Although this gene therapy technique can be used in acquired or genetic diseases. The purposes and structure of gene therapy, protocols in these two groups of orthopaedic diseases, are fundamentally different.

Gene Therapy for Genetic or Acquired Diseases

Based upon studies in the early 1970s on the mechanisms of viral infection and on the upsurge of recombinant DNA technology in the late 1970s the concept of gene therapy has become a reality. The initial target of gene therapy was the correction of single gene defects responsible for inheritable diseases, such as the adenosine deaminase deficiency syndrome which leads to a severe combined immunodeficiency. While genetic diseases may result from the deletion or alteration of just one gene, the effect of the gene product in different tissues may be complex. Additionally, gene expression may be problematic on the translational and posttranslational levels and must be coordinated with another gene product. The complexity of genetic diseases remains the reason that only few of the

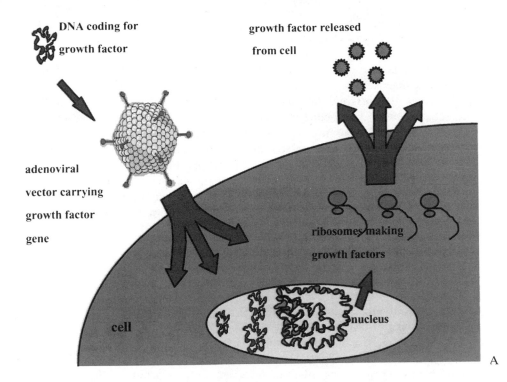

DNA coding for
growth factor

growth factor released
from cell

adenoviral
vector carrying
growth factor
gene

ribosomes making
growth factors

cell

nucleus

A

GENE EXPRESSION PATHWAY

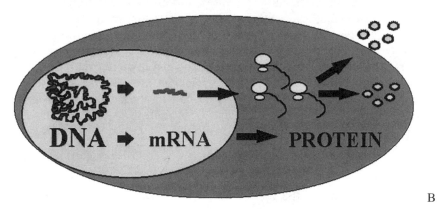

DNA → mRNA → PROTEIN

B

FIGURE 1.1. The DNA coding for a protein is inserted into a vector (**A**). This vector can be viral or nonviral in nature. The vector attaches itself to the cell surface and inserts its gene into the nucleus of the host cell. The genetic information for the protein then uses the cell's transcription mechanism to translate into a messenger ribonucleic (mRNA) (**B**) which leaves the nucleus and is translated into aminoacid sequences in the ribosomal machinery of the cell. The aminoacid sequence arranges itself to form the protein which then is released through the endoplasmatic reticulum into the cells' surrounding. The gene therapy approach, thus turns a host cell into a "factory" for the protein of interest which could be a growth factor or a cytokine.

approved gene therapy protocols in humans are designed for the treatment of genetic diseases.

Whereas the original notion of gene therapy addressed single gene defects, this view has been superceded by the possibility of using genes for delivery of therapeutic agents directly into somatic tissue cells, an approach that opens up an extraordinarily diverse set of clinical possibilities. This somatic gene therapy can be performed locally or systemically. Whereas systemic gene therapy may be advantageous in cardiac or vascular diseases, orthopaedic injuries or diseases are usually more local in their distribution. Thus a local gene delivery into the diseased or injured tissue is desirable.

Applications of Somatic Gene Therapy in Orthopaedics

Soft tissues in the musculoskeletal system exhibit different capacities to heal following injuries and, with the exception of bone, there is no restitutio ad integrum. The healing response in soft tissue always leads to the formation of a scar with a lower mechanical strength than the original tissue. This scar tissue can be modified in different ways according to the location and properties of the surrounding uninjured tissue.

Growth factors have been found to be essential for the onset and progression of the initial healing responses, and are believed to be the key to accelerated healing and tissue regeneration (Einhorn and Trippel 1997; Trippel 1997). Gene therapy provides an excellent technique to deliver these growth factors locally into the site of injury. While only one application is necessary, the protein will be released in a sustained manner for days or weeks, depending on the vector used. Hence, the advantageous continuous delivery of specific growth factors can be achieved using the gene therapy approach. The gene for a specific growth factor can be introduced into the injured tissue directly or indirectly. In direct gene delivery, the delivery vehicle (vector) is directly injected into the tissue and inserts its gene into surrounding cells. The indirect approach uses cells that have been taken from the injured tissue through a muscle biopsy or skin biopsy. These cells are cultured and the vector, viral or nonviral in nature, is given onto the cell culture. The process is referred to as transfection when nonviral vectors are used, and transduction when viral vectors are used. Transfection and transduction include the attachment of the vector to the cell membrane and insertion of the encoded gene into the cellular genome (Evans and Robbins 1995).

How Can Gene Therapy Help the Orthopaedic Surgeon?

The most common problems seen in the day-to-day practice of orthopaedic surgeons are early or progressive osteoarthritis, fractures, ligamentous and tendinous injuries, cartilage defects, and other soft tissue injuries. Thus, a major driving force behind the development of new treatment techniques for orthopaedic injuries has been the reduction of recovery time. The introduction of minimally invasive surgery, intramedullary nailing, and arthroscopic techniques have reduced postoperative morbidity. Additionally, modern rehabilitation schemes have accelerated recovery after injury. The key for the treatment of any of these conditions, however, remains supporting or reestablishing a healing response. Fractures have to be stabilized, tendons and ligaments repaired or reconstructed, and cartilage stimulated in order to establish the ground for an undisturbed healing response. As we have seen from animal experiments, a healing response can be improved by the local targeted application of certain growth factors into the healing injury site (Rodeo et al. 1995; Hildebrand et al. 1998; Trippel 1997; Ménétrey, Kasemkijwattana, Fu, Moreland and Huard 1999), which point to gene therapy's major role in orthopaedic surgery. The direct introduction of growth factors and cytokines is not sufficient due to their above-mentioned short biological half-life in the magnitude of minutes to hours. Using the gene therapy approach, however, we can locally deliver these growth factors in a sustained manner for a specific amount of time, thereby approximating their use as a pharmacological substance. This approach offers the opportunity for a single time application that yields a prolonged effect during tissue healing.

In Which Diseases or Injuries Can Gene Therapy Help the Orthopaedic Surgeon?

An understanding of where gene therapy can be an asset to the orthopaedic surgeon must be based on an understanding of different tissues and their injury patterns.

Fracture Healing

Approximately 5.6 million fractures occur annually in the United States, 10% of which result in delayed or nonunions (Praemer, Furner, and Rice 1992). These delayed or nonunions constitute a significant cost for the health system. Current treatment options have not evolved significantly and usually involve surgical procedures or prolonged conservative treatment rationales (Muller and Thomas 1979). The application of growth factors, particularly bone morphogenic proteins (BMPs), has shown promising results for the acceleration of the healing of segmental defects in animal experiments (Gerhart et al. 1993; Zegzula, Buck, Brekke, Wozney, and Hollinger 1997; Lieberman et al. 1998). The one time delivery of a growth factor may not be sufficient to stimulate bone formation, particularly in complex settings such as compound fractures, grade 3 open fractures, or atrophic nonunions. Using a gene therapy approach, factors such as BMPs or angiogenetic factors can be delivered more efficiently to these severe injuries and may thus improve the healing response of these severe conditions without risking further soft tissue damage

Ligament Healing

With the general increase in athletic activity level of the average population, the incidence of ligament and capsular injuries around the knee joint and the shoulder has risen enormously. The rupture of an anterior cruciate ligament (ACL), for example, leads to a functional knee instability that sidelines recreational and professional athletes for months. Additionally, because the healing response of an intraarticular ligament such as the ACL is greatly impaired, a torn ACL is not repaired but reconstructed using ACL grafts. ACL grafts undergo remodeling but never regenerate toward a regular ACL (Arnoszky, Tarvin, and Marshall 1982). The application of growth factors using the direct or indirect gene therapy approach opens a whole new perspective on the idea of regenerating an ACL from a tendon graft. Restructuring may occur in response to the local production of growth factors within the graft.

Another problematic area in sports medicine is the fixation of tendon in bone tunnel grafts, since even after several years post injury the tendon does not integrate into the bone tunnel (L'Insalata, Klatt, Fu, and Harner 1997). In order to remodel or recreate a regular ligamentous insertion of a tendon graft, the reestablishment of a 4 zonal direct ligament insertion is necessary. However, certain embryological growth factors may be capable of achieving this task provided that they can be targeted and applied to the ligament insertion site (Wolfman et al. 1997). Gene therapy offers an excellent technique to achieve this goal.

Cartilage Healing

One of the most challenging clinical problems associated with joint diseases is the degeneration or loss of the articular surface. Articular cartilage is a highly organized but avascular tissue. Hence, its healing capacity is low. Many techniques have been developed in the last 2 decades that aim to restore the articular surface. Cartilage debridement and resurfacing, drilling of the subchondral bone plate (Insall 1974; Blevins 1998), transplantation of cartilage plugs (Hangody et al. 1998), and cartilage cell transplantation (Brittberg et al. 1994) are promising attempts to restore the articular surface. Unfortunately, none of these techniques has so far been capable of restoring articular cartilage, since the transplanted cartilage or chondrocytes seem to dedifferentiate and produce collagen type I rather than collagen type II (Breinan et al. 1997). In vitro studies have shown that chondrocytes can be influenced to stay differentiated if the gene for certain growth factors has been transfected into the genome of the chondrocytes

(Goto et al. 1999). Gene therapy could, therefore, be of the greatest importance in the exciting and novel field of cartilage regeneration.

Muscle

Although muscle contusions and strains are among the most common soft tissue injuries, little progress has been made toward the healing of these injuries. Particularly in the field of sports medicine, muscle injuries are a major problem sidelining numerous high-level professional athletes. Although muscle is a very well perfused tissue, the healing of muscle injuries results in a fibrous scar, which potentially impairs muscle function. Recent studies show that the application of growth factors can potentially improve muscle healing and significantly reduce scar formation, leading to a better functional performance after injury (Ménétrey et al. 1998). The application of gene therapy as a therapeutic tool offers exciting perspectives for the development of treatment options after muscle injuries. The main interest of muscle tissue for gene therapy researchers, however, focuse on the utilization of muscle cells as ex vivo delivery vehicles. Immature muscle cells (myoblasts) can be easily transfected with viral or nonviral delivery vehicles (Dhawan et al. 1991; Huard, Acsadi, Jani, Massie, and Karpati 1994). Protein synthesis of myoblasts is more efficient than that of any other cell in the human body. This renders myoblasts particularly attractive as growth factor producing cells after they have been transfected with a vector. Furthermore, myoblasts can fuse with other myoblasts and create myotubes that again, in contrast to all other cells in the human body, continue producing large amounts of proteins even in the postmitotic stage. This technical aspect makes the myoblast a potentially highly efficient cell type for the delivery of growth factors and cytokines in the ex vivo gene therapy approach.

Osteoarthritis

One of the most common orthopaedic diseases is osteoarthritis. Usually the result of wear and tear due to repetitive overuse or altered joint kinematics, osteoarthritis develops as a progressive disease that currently cannot be prevented or cured. The general belief has been that mechanical factors play the predominant role in the onset of the disease. New data suggest that the progress of the disease is influenced and controlled by an imbalance of intraarticular cytokines. The delicate balance between anabolic and catabolic enzymes on the one hand, and pro- and anti-inflammatory cytokines on the other, is disturbed in osteoarthritic joints and leads to a progressive destruction of the articular surface (Lohmander, Roos, Dahlberg, Hoerrner, and Lark 1994; Lohmander 1995; Cameron, Fu, Paessler, Schneider, and Evans 1994; Cameron 1997). Direct application of antiinflammatory proteins may be problematic due to their rapid intraarticular clearing and short half-life of the protein. The administration of genetically manipulated synoviocytes expressing antiinflammatory proteins, however, has shown promising results toward reducing the amount of inflammation and degradation in acutely arthritic joints (Ghivizzani et al. 1998). Gene therapy has been shown to be a potentially powerful tool for the improvement and possible cure of progressive joint destruction due to osteoarthritis.

Rheumatoid Arthritis

Rheumatoid arthritis is a chronically aggressive autoimmune disease resulting in total joint destruction. Orthopaedic surgeons usually see RA patients at the final stage when a total joint replacement becomes necessary. Total joint replacements in these patients are more difficult to perform due to the great amount of inflammation and bony erosion. More disturbingly, RA patients need total joint replacement at a very young age. A significant amount of basic research has been performed to investigate the causes and mechanisms of RA. Some valuable approaches have been pursued within the last 10 years. For example the attempt to block the primary inflammation has been promising in animal experiments. The blockage of primary inflammation mediators, such as interleukin 1 (IL1), or redirection of primary immune cells, such as Th1 and Th2 lymphocytes, has been able to alleviate the severity of RA in animal models (Wooley et al. 1993; Whalen 1999). In addition, the application of these observations to a gene therapy protocol has proven to be extremely successful in preventing onset and severity of RA when used as a prophylactic measure (Whalen 1999). These progressive data have led to the first human clinical trial of gene therapy in orthopaedics (Evans et al. 1996). Designed as a feasibility and safety study, this trial has been successfully completed and opens the road to further gene therapy trials in this area of research.

Prospects of Gene Therapy in Orthopaedics

Gene therapy is not yet an established therapeutic technique for the treatment of orthopaedic diseases. However, we believe that it has great potential for clinical application in the treatment of injured tissues. These techniques may be particularly helpful in tissues with low healing capacity such as ligaments, menisci, and articular cartilage. In particular, the transfer of growth factor genes to sites of tissue injury may influence the healing process by encouraging migration of cells to the injured site, increasing the rate of cell differentiation and proliferation, and upregulating matrix synthesis. Gene therapy will also be a great therapeutic asset in situations such as meniscal tears, where the surrounding environment impairs the healing process or where there are comminuted fractures with severe soft tissue damage as well as established bony nonunions. Chronic diseases, such as

OA and RA, will greatly benefit from the possibility of delivering antiinflammatory and cartilage anabolic cytokines as well as growth factors to the joints.

The greatest potential of gene therapy lies in the continuous local delivery of growth factors and cytokines to the affected tissue for a prolonged period of time. Systemic side effects of the protein can be avoided and therapeutic levels sustained. The feasibility of gene transfer into soft tissues has been studied in recent years and has provided promising preliminary results. Transfer of marker genes to muscles, ligaments, menisci, articular cartilage, and bone has been demonstrated (Huard et al. 1994; Gerich, Kang, Fu, Robbins, and Evans 1997; Kang et al. 1997; Baltzer et al. 1999; Goto et al. 1999). In recent studies, the direct, adenovirally mediated transfer of a marker gene to the rabbit patella tendon, synovium, intervertebral disk, and bone has been shown (see chapter 4; Gerich et al. 1997; Ghivizzani et al. 1997; Baltzer et al. 1999). Expression of the transferred gene could be detected for up to 6 weeks in each tissue. Using an ex vivo approach, the transfer of a therapeutic gene to a segmental defect in a rat has been shown (Lieberman et al. 1998), with the healing response in a rat femoral fracture model as considerably accelerated. The transfer of marker genes and therapeutic genes to the site of a non-union has been shown in animal models for spinal fusions (Boden et al. 1998). Direct adenoviral gene transfer to atrophic nonunions and into the healing bone ligament interface after ligament replacement has also been demonstrated (Lattermann et al. 1999). The feasibility of gene transfer to different tissues and disease sites has recently been shown. The task for the future will be to investigate growth factors, cytokines, and molecular mechanisms in order to fully understand and use gene transfer technology for the benefit of our patients.

Conclusion

Our understanding of the biochemical properties and molecular mechanisms of connective tissue injuries and diseases is still very limited. The available data indicate that complicated mechanisms involving cellular, enzymatic, and humoral processes regulate the release and distribution of growth factors and cytokines in orthopaedic diseases and injuries. As our understanding of these mechanisms and of tissue homeostasis grows, our ability to understand and influence or control changes during connective tissue disease and healing will also expand. New, exciting therapeutic fields have been opened with the concept of gene therapy. There is now a great potential for the advancement of treatment modalities in orthopaedic surgery. Gene therapy is one of the most advanced and exciting concepts on the road to better and less invasive treatment options for our patients, and we will have to adapt to these new technological tools. It is now our responsibility and that of future generations of orthopaedic surgeons to understand and advance these new biological concepts so that they may become an integral part of our thinking and therapeutic regimen.

References

Arnozsky, S.P., Tarvin, G.B., and Marshall, J.L. 1982. Anterior cruciate ligament replacement using patellar tendon. An evaluation of graft revascularization in the dog. *J Bone Joint Surg* 64-A, No.: 2:217–24.

Baltzer, A.W., Lattermann, C., Whalen, J.D., Braunstein, S., and Robbins, P.D. 1999. Evans A gene therapy approach to accelerating bone healing. Evaluation of gene expression in a New Zealand white rabbit model. *Knee Surg Sports Traumatol Arthrosc* 7(3):197–202.

Blevins, F.T., Steadman, J.R., Rodrigo, J.J., and Silliman, J. 1998. Treatment of articular cartilage defects in athletes: an analysis of functional outcome and lesion appearance. *Orthopedics* 21(7):761–7.

Boden, S.D., Titus, L., Hair, G., Liu, Y., Viggeswarapu, M., Nanes, M.S., and Baranowski, C. 1998. Lumbar spine fusion by local gene therapy with a cDNA encoding a novel osteoinductive protein LMP-1. *Spine* (1):23:2486–92.

Breinan, H.A., Minas, T., Hsu, H.P., Nehrer, S., Sledge, C.B., and Spector, M. 1997. Effect of cultured autologous chondrocytes on repair of chondral defects in a canine model. *J Bone Joint Surg Am* 79(10):1439–51.

Brittberg, M., Lindahl, A., Nilsson, A., Ohlsson, C., Isaksson, O., and Peterson, L. 1994. Treatment of deep cartilage defects in the knee with autologous chondrocyte transplantation. *N Engl J Med* 6;331(14):889–95.

Cameron, M., Buchgraber, A., Passler, H., Vogt, M., Thonar, E., Fu, F., and Evans, C.H. 1997. The natural history of the anterior cruciate ligament-deficient knee. Changes in synovial fluid cytokine and keratan sulfate concentrations. *Am J Sports Med* 25(6):751–4.

Cameron, M.L., Fu, F.H., Paessler, H.H., Schneider, M., and Evans, C.H. 1994. Synovial fluid cytokine concentrations as possible prognostic indicators in the ACL-deficient knee. *Knee Surg Sports Traumatol Arthrosc* 2(1):38–44.

Dhawan, J., Pan, L.C., Pavlath, G.K., Travis, M.A., Lanctot, A.M., and Blau, H.M. 1991. Systemic delivery of human growth hormone by injection of genetically engineered myoblasts. *Science* 6;254(5037):1509–12.

Einhorn, T.A. and Trippel, S.B. 1997. Growth factor treatment of fractures. *Instr Course Lect* 46:483–6.

Evans, C.H. and Robbins, P.D. 1995. Possible orthopaedic applications of gene therapy. *J Bone Joint Surg Am* 77(7):1103–14.

Evans, C.H., Robbins, P.D., Ghivizzani, S.C., Herndon, J.H., Kang, R., Bahnson, A.B., Barranger, J.A., Elders, E.M., Gay, S., Tomaino, M.M., Wasko, M.C., Watkins, S.C., Whiteside, T.L., Glorioso, J.C., Lotze, M.T., and Wright, T.M. 1996. Clinical trial to assess the safety, feasibility, and efficacy of transferring a potentially anti-arthritic cytokine gene to human joints with rheumatoid arthritis. *Hum Gene Ther* 20;7(10):1261–80.

Gerhart, T.N., Kirker-Head, C.A., Kriz, M.J., Holtrop, M.E., Hennig, G.E., Hipp, J., Schelling, S.H., and Wang, E. 1993. Healing segmental femoral defects in sheep using recombinant human bone morphogenetic protein. *Clin Orthop* 293: 317–26.

Gerich, T.G., Kang, R., Fu, F.H., Robbins, P.D., and Evans, C.H. 1997. Gene transfer to the patellar tendon. *Knee Surg Sports Traumatol Arthrosc* 5(2): 118–23.

Ghivizzani, S.C., Lechman, E.R., Kang, R., Tio, C., Kolls, J., Evans, C.H., and Robbins, P.D. 1998. Direct adenovirus-mediated gene transfer of interleukin 1 and tumor necrosis factor alpha soluble receptors to rabbit knees with experimental arthritis has local and distal anti-arthritic effects. *Proc Natl Acad Sci USA* 14;95(8):4613–8.

Ghivizzani, S.C., Lechman, E.R., Tio, C., Mule, K.M., Chada, S., McCormack, J.E., Evans, C.H., and Robbins, P.D. 1997. Direct retrovirus-mediated gene transfer to the synovium of the rabbit knee: implications for arthritis gene therapy. *Gene Ther* 4(9):977–82.

Goto, H., Schuler, F.D., Lamsam, C., Moeller, H.D., Niyibizi, C., Fu, F.H., Robbins, P.D., and Evans, C.H. 1999. Transfer of lacI marker gene to the meniscus. *J Bone Joint Surg Am* Jul;81(1):918–25.

Hangody, L., Kish, G., Karpati, Z., Udvarhelyi, I., Szigeti, I., and Bely, M. 1998. Mosaicplasty for the treatment of articular cartilage defects: application in clinical practice. *Orthopedics* 21(7):751–6.

Hildebrand, K.A., Woo, S.L., Smith, D.W., Allen, C.R., Deie, M., Taylor, B.J., and Schmidt, C.C. 1998. The effects of platelet-derived growth factor-BB on healing of the rabbit medial collateral ligament. An in vivo study. *Am J Sports Med* 26:549–54.

Huard, J., Acsadi, G., Jani, A., Massie, B., and Karpati, G. 1994. Gene transfer into skeletal muscles by isogenic myoblasts. *Hum Gene Ther* 5(8):949–58.

Insall, J. 1974. The Pridie debridement operation for osteoarthritis of the knee. *Clin Orthop* 101(01):61–7.

Kang, R., Marui, T., Ghivizzani, S.C., Nita, I.M., Georgescu, H.I., Suh, J.K., Robbins, P.D., and Evans, C.H. 1997. Ex vivo gene transfer to chondrocytes in full-thickness articular cartilage defects: a feasibility study. *Osteoarthritis Cartilage* 5(2):139–43.

L'Insalata, J.C., Klatt, B., Fu, F.H., and Harner, C.D. 1997. Tunnel expansion following anterior cruciate ligament reconstruction: a comparison of hamstring and patellar tendon autografts. *Knee Surg Sports Traumatol Arthrosc* 5(4):234–8.

Lattermann, C., Clatworthy, M., Weiss, K.R., Robbins, P.D., Fu, F.H., and Evans, C.H. 1999. Targeting of gene therapy in ligament insertions: Bone or tendon? *Trans Orthop Orth Res Soc 45th Meeting. Anaheim* p:1066.

Lieberman, J.R., Le, L.Q., Wu, L., Finerman, G.A., Berk, A., Witte, O.N., and Stevenson, S. 1998. Regional gene therapy with a BMP-2-producing murine stromal cell line induces heterotopic and orthotopic bone formation in rodents. *J Orthop Res* 16(3):330–9.

Lohmander, L.S. 1995. The release of aggrecan fragments into synovial fluid after joint injury and in osteoarthritis. *J Rheumatol Suppl* 43:75–7.

Lohmander, L.S., Roos, H., Dahlberg, L., Hoerrner, L.A., and Lark, M.W. 1994. Temporal patterns of stromelysin-1, tissue inhibitor, and proteoglycan fragments in human knee joint fluid after injury to the cruciate ligament or meniscus. *J Orthop Res* 12(1):21–8.

Ménétrey, J., Kasemkijwattana, C., Fu, F.H., Moreland, M.S., and Huard, J. 1999. Suturing versus immobilization of a muscle laceration: a morphological and functional study in a mouse model. *A J Sports Med* Mar–Apr;27(2):229–9.

Muller, M.E. and Thomas, R.J. 1979. Treatment of non-union in fractures of long bones. *Clin Orthop* 138:141–53.

Praemer, A., Furner, S., and Rice, D. 1992. *Musculoskeletal Conditions in the United States* 85–124, Park Ridge, Illinois, The American Academy of Orthopaedic Surgeons.

Rodeo, S.A., Hannafin, J.A., Suzuki, K., Deng, X., Hidaka, C., Arnoczky, S.P., and Warren, R.F. 1995. Bone morphogenetic protein enhances early healing in a bone tunnel. *Trans Orth Res Soc* 20:288.

Scherping, S.C. Jr., Schmidt, C.C., Georgescu, H.I., Kwoh, C.K., Evans, C.H., and Woo, S.L. 1997. Effect of growth factors on the proliferation of ligament fibroblasts from skeletally mature rabbits. *Connect Tissue Res* 36(1):1–8.

Trippel, S.B. 1997. Growth factors as therapeutic agents. *Instr Course Lect* 46:473–6.

Wolfman, N.M., Hattersley, G., Cox, K., Celeste, A.J., Nelson, R., Yamaji, N., Dube, J.L., DiBlasio-Smith, E., Nove, J., Song, J.J., Wozney, J.M., and Rosen, V. 1997. Ectopic induction of tendon and ligament in rats by growth and differentiation factors 5, 6, and 7, members of the TGF-beta gene family. *J Clin Invest* 15;100(2):321–30.

Wooley, P.H., Whalen, J.D., Chapman, D.L., Berger, A.E., Richard, K.A., Aspar, D.G., and Staite, N.D. 1993. The effect of an interleukin-1 receptor antagonist protein on type II collagen-induced arthritis and antigen-induced arthritis in mice. *Arthritis Rheum* 36(9):1305–14.

Whalen, J.D., Lechman, E.L., Carlos, T. et al. 1999. Adenoviral transfer of the viral IL-10 gene periarticularly to mouse paws suppresses development of collagen-induced arthritis in both injected and noninjected paws. *J Immunol* Mar 15;162(6):3625–32.

Zegzula, H.D., Buck, D.C., Brekke, J., Wozney, J.M., and Hollinger, J.O. 1997. Bone formation with use of rhBMP-2 (recombinant human bone morphogenetic protein-2). *J Bone Joint Surg Am* 79(12):1778–90.

2
The Use of Different Vectors and Strategies for Gene Transfer to the Musculoskeletal System

JOHNNY HUARD

Introduction

With the development of improved methods of transferring genes into cells, new treatment options are becoming available in orthopaedics and sports medicine to aid in the healing capacity of musculoskeletal tissues. In this chapter, current techniques in gene therapy and tissue engineering will be described in addition to their potential application in improving the healing of the musculoskeletal system, including articular cartilage, muscle injuries, and damages to intraarticular structures such as ligament, cartilage, and meniscus.

Gene therapy, a novel form of molecular medicine, could have a major impact on human healthcare in the future. Two different forms of gene therapy are possible: germ line and somatic cell gene therapy. Germ line gene therapy is greatly limited by technical complexities and ethical issues. Somatic gene therapy, which exclusively benefits the individual and cannot be passed to the next generation, arouses much interest in the scientific community.

A number of somatic tissues have already been investigated for gene therapy approaches: bone marrow, keratinocytes, hepatocytes, endothelial cells, neuron/glia, cardiac myocytes, fibroblasts, and myoblasts. Even though gene therapy approaches have been designed for correcting genetic disorders, including cystic fibrosis, hemophilia, familial hypercholesterolemia, and Duchenne muscular dystrophy (DMD), this technology may also be amenable to treating acquired diseases and conditions of the musculoskeletal system and helping to improve the healing of various tissues (bone, ligament, cartilage, meniscus, and muscle).

In this chapter, we will summarize current knowledge and up to date achievements pertaining to gene therapy techniques for the musculoskeletal system. This chapter will be divided into three different sections covering (1) vectors for gene therapy, (2) strategies for gene transfer into muscle, and (3) targeted therapeutic genes and potential applications of these technologies for orthopaedic and sports medicine. Although the main topic of this chapter is the application of gene therapy to the musculoskeletal system, we will focus on the development of gene therapy approaches to skeletal muscle. The different vectors and gene delivery strategies developed for skeletal muscle are similar and, consequently, directly applicable to other tissues of the musculoskeletal system. It is clear that therapeutic genes and potential applications of gene therapy for skeletal muscle differ from those for other tissues; this concept will be discussed at the end of the chapter.

Vectors for Gene Therapy

Somatic gene therapy, the transferring of a functional gene into a particular tissue to alleviate a biochemical deficiency, has emerged as a novel and exciting form of molecular medicine. Due to a number of properties, muscle tissue has been suggested as a promising target for gene therapy. First, since skeletal muscle is composed of multinucleated, postmitotic myofibers, it may facilitate high and long term expression of introduced genes. Second, the mononucleated myogenic precursor cells (satellite cells), which are located between the extracellular matrix and the plasma membrane of myofibers, can be relatively easily isolated and cultivated in vitro and efficiently transduced using either viral or nonviral vectors. In addition, the ability of the myogenic cells to remove fuse with, or into, myofibers in vivo has established them as promising gene delivery vehicles. Finally, the high level of vascularization of muscle tissue may facilitate the systemic delivery of potentially therapeutic muscle and nonmuscle proteins, such as growth factors, factor IX, or erythropoietin (see Blau and Springer 1995; Svensson, Tripathy, and Leiden 1996).

To date, most of the studies for gene transfer into muscle cells have been aimed at therapy for DMD. This common X linked recessive muscular dystrophy is caused by a deficiency of dystrophin, an important component of the plasma membrane cytoskeleton of muscle fibers (Hoffman, Brown, and Kunkel 1987; Arahata et al. 1988; Bonilla et al. 1988; Watkins, Hoffman, Slayter, and Kunkel 1988; Zubryzcka–Gaarn et al. 1988). Dystrophin deficiency leads to a continuous loss of muscle fibers with progressive muscle weakness and the early death of patients in their second decade of life. Effective gene therapy for DMD will require the transfer of adequate copies of either the full length dystrophin gene or dystrophin minigenes (Acsadi, Dickson et al. 1991; Dunckley, Davies, Walsh, Morris, and Dickson 1992; Dunckley, Wells, Walsh, and Dickson 1993; Ragot et al. 1993; Vincent et al. 1993; Acsadi, Lochmueller et al. 1996) in affected muscle to restore adequate production of dystrophin for alleviation of the muscle weaknes. The following sections will describe five different vectors that have been extensively used for gene transfer to skeletal muscle and nonmuscle tissue.

Table 2.1 summarizes the different vectors used for gene transfer to skeletal muscle and describes their general characteristics.

Plasmid DNA

Naked functional DNA can be taken into mammalian cells by a number of physical methods, including coprecipitation with calcium phosphate, use of polycation or lipid to complex with DNA, or by encapsulation of DNA into liposome. The intramuscular injection of plasmid DNA (nonviral vector) to deliver genes to muscle has shown the advantages of low toxicity and immunogenicity (Wolff et al. 1992). An initial disadvantage has been the relatively low transfection efficiency, despite the use of large amounts of

TABLE 2.1. Characteristics of different vectors used for gene transfer to skeletal muscle.

Vector	Integrating	Muscle Cells Infection				General
		In Vitro		In Vivo		
		myoblasts	myotubes	newborn muscle fibers	adult muscle fibers	
Plasmid DNA and Liposomes	No	+	+	+	+	very low efficiency of gene delivery low immunogenicity/ cytotoxicity
Retrovirus	YES	+++	−	−	−	low toxicity/immunogenicity infects mitotically active cells only low gene insert capacity
Adenovirus	NO	+++	−	+++	−	infects mitotic/postmitotic cells low cytotoxicity immune rejection problems new generation vectors (less immunogenic; under investigation)
Herpes Simplex Virus (type 1)	NO	+++	+++	+++	−	large insert capacity infects mitotic/postmitotic cells immune rejection problems new generation vectors (less cytotoxic; under development)
Adeno-associated Virus	YES	+++	+++	+++	+++	low immunogenicity/ cytoxicity high persistence of gene transfer low gene insert capacity

plasmid DNA (Acsadi et al.1991; Danko et al. 1993), which is probably due to the high tissue density of muscle. However, recent studies have shown improved transfection efficiencies in muscle with an increased accessibility of DNA to the myofibers following: pretreatment with a myonecrotic agent, such as cardiotoxin; preinjection of large volumes of hypertonic sucrose; or injection at the myotendinous junction (Davis, Demeneix et al. 1993, Davis, Whalen, and Demeneix 1993; Doh, Vahlsing, Hartikka, Liang, and Manthorpe 1997). Alternatively, several studies show improved plasmid transfection efficiencies in muscle using nontargeted liposomes and/or polylysine condensed plasmid DNA (Vitiello, Chonn, Wasserman, Duff, and Worton 1996; see also Chapter 6). In particular, the recent development of ligand (e.g., transferrin)-directed DNA–liposome complexes capable of transducing myogenic cells in a receptor-dependent manner (Feero et al. 1997) is promising.

Although the delivering of nucleic acids into mammalian cells can be made more efficient by the physical method, recent research has been extensively focused on viral vectors capable of infecting virtually every cell in a target population. Strategies of delivering genes to skeletal muscle using viral vectors based on retrovirus, adenovirus, herpes simplex virus, or adeno–associated virus have therefore evolved rapidly.

Retrovirus

Retrovirus vectors are relatively safe and can infect dividing myoblasts (progenitors of skeletal muscle) with a high efficiency (Barr and Leiden 1991; Dhawan et al. 1991; Dunckley et al. 1992; Dunckley et al. 1993; Salvatori et al. 1993). In addition, the ability of retroviruses to become stably integrated into the host cell genome can provide long term, stable expression of the delivered gene. However, retroviruses remain incapable of infecting postmitotic cells (Miller, D.G., Adam, and Miller, A.D. 1990). Muscle cells become postmitotic very early in their development, and mature muscle tissue has no actively dividing cells unless injured or affected by a genetic disorder (e.g., DMD). Therefore, muscle and other tissues of the musculoskeletal system show a dramatic loss of retroviral transduction during maturation. Other limitations to the use of retroviruses are their small gene insert capacity (less than 7 kb), relatively low titers (10^5–10^6 pfu/ml), and risk of insertional mutagenesis (see Table 2.1).

Adenovirus

Adenoviral vectors (Ad) can infect both mitotic myoblasts and postmitotic immature myofibers (Quantin, Perricaudet, Tajbakhsh, and Mandell 1992; Ragot et al. 1993; Vincent et al. 1993; Acsadi, Jani, Massie et al. 1994; Acsadi et al. 1996; Huard, Lochmuller, Acsadi et al. 1995; Huard, Lochmuller, Jani et al. 1995), and they can be prepared at high titers (10^9–10^{11} pfu/ml).

However, the stability and long term expression of transgenes delivered to skeletal muscle using first generation Ad have been limited by immune rejection and maturation dependent adenoviral transduction of myofibers (Acsadi, Jani, Huard et al. 1994; Yang et al. 1994; van Deutekom, Floyd et al. 1998; van Deutekom, Hoffman and Huard et al. 1998). However, the overall application of Ad gene transfer to skeletal muscle has been hindered by the immune rejection problem and then inability to transduce mature myofiber. The low gene insert capacity (<8 kb) of first generation Ad has recently been overcome by the development of new mutant Ad lacking all viral genes and having an expanded insert capacity of 28 kb (Haecker et al. 1996; Kochanek et al. 1996; Kumar-Singh and Chamberlain 1996). These Ad, which were developed specifically for the delivery of large dystrophin of up to 28 kba of exogeneous DNA, are the first to deliver full length dystrophin to skeletal muscle. Moreover, they promise to dramatically reduce immunogenicity.

Herpes Simplex Virus

Viral vectors derived from the herpes simplex virus type 1 (HSV–1) are naturally capable of carrying large DNA fragments, such as the 14 kb dystrophin cDNA, and they have recently been studied for their ability to transduce muscle cells (Huard, Goins, and Glorioso 1995; Huard et al. 1996; Huard, Akkaraju, Watkins, Cavalcoli, and Glorioso; Huard, Krisky et al. 1997). Herpes simplex viral vectors, which can persist in the host cell in a nonintegrated state and be prepared at adequately high titers (10^7–10^9 pfu/ml), have shown efficient transduction of myoblasts, myotubes, and immature myofibers. However, their relatively high cytotoxicity and immunogenecity, which hamper long term transgene expression, have been identified as a major disadvantage of the first generation herpes simplex viral vectors. In recent attempts to overcome this hurdle, deletion of viral immediate–early (IE) genes from the mutant vectors has been shown to reduce cytotoxicity in many cell types and tissues including skeletal muscle (Marconi et al. 1996; Huard, Krisky et al. 1997).

Adeno–Associated Virus

Recombinant adeno–associated viral vectors (rAAV) have recently been used as gene delivery vehicles for muscle cells. Although long term transgene expression (up to 18 months) and a high efficiency of muscle cells at different maturities (see Chapter 15) have been observed in mouse skeletal muscle (Xiao and Samulski 1996; Fisher et al. 1997; Reed–Clark, Sferra, and Johnson 1997), the application of adeno-associated viral vectors for gene therapy purposes may be limited by their restrictive gene insert capacity (<5 kb). For the musculoskeletal system, where some specific growth factors have been identified (see the following section), the small genes of the growth factor are directly amenable to the insert capacity of rAAV, making this vector a potential delivery vehicle for the musculoskeletal system.

Strategies for Gene Transfer into Muscle

Various approaches can be used to achieve gene transfer to the musculoskeletal system, including cell therapy, gene therapy, and a combination of both techniques in tissue engineering approaches. We will describe the different approaches that have been used to deliver growth factors into injured skeletal muscle. Different biological techniques can be used to enhance tissue healing following injuries: the delivery of specific growth factor proteins, the use of cell therapy (myoblast transplantation), and gene therapy based on viral and nonviral vectors. In fact, many growth factors have already been shown to be capable of stimulating the growth and protein secretion of many musculoskeletal cells (Trippel et al. 1996). This section will review the application to enhance muscle healing of these different biological approaches that are also directly amenable to other tissues of the musculoskeletal system.

Direct Injection of Human Recombinant Growth Factor Proteins

During muscle regeneration, it is presumed that trophic substances released by the injured muscle activate the satellite cells (Schultz, Jaryszak, and Valliere 1985; Allamedine, Dehaupas, and Fardeau 1989; Schultz 1989; Hurme and Kalimo 1992; Bischoff, 1994; Rao and Kohtz 1995). During growth and development, specific growth factors have been shown to be capable of eliciting variable responses from skeletal muscle. Preliminary data has further suggested that individual growth factors play a specific role during muscle regeneration (Jennische and Hansson 1987; Jennische 1989; Anderson, Liu, and Kardami 1991; Grounds 1991; Lyles, Amin, Bock, and Weil 1993; McFarland, Pesall, and Gilkerson 1993; Johnson and Allen 1995; Lefaucheur and Sebille 1995; Chambers and McDermott 1996). It has been reported that the terminal differentiation of skeletal myoblasts is under negative control from several mitogenic growth factors, including basic fibroblast growth factors (bFGF), platelet derived growth factor BB (PDGF–BB), and other uncharacterized serum mitogens (Gospodarowitz, Weseman, Moran, and Lindstrom 1976; Linkhart, Clegg, and Hauschka 1981; Florini, Roberts et al. 1986; Olson, Sterrnberg, Hu, Spizz, and Wilcox 1986; Allen and Boxhorn 1989; Jin, Rahm, Claesson–Welsh, Helden, and Sejerson 1990; Yablonka–Reuvini, Balestri, and Bowen–Pope 1990; Harrington, Daub, Song, Stasek, and Garcia 1992).

Another group of growth factors, the insulin–like growth factors (IGF–1 and 2), have been found to be mitogenic for myoblasts (Florini, Ewton, Falen, and Van Wyk 1986; Engert, Berlund, and Rosenthal 1996; Quinn and Haugk 1996; Damon, Haugk, Birnbaum, and Quinn 1998). IGF–1 is critical in mediating the growth of muscle and other tissues (Florini, Ewton, and Coolican 1996). Systemic administration of IGF–1 results in increased muscle protein content and reduced protein degradation (Zdanowicz et al. 1995). More important, overexpression of IGF–1 has been correlated with muscle hypertrophy in transgenic mouse lines (Coleman et al. 1995). With aging, there is a decrease in the production and activity of the growth hormone/IGF–1 axis, which leads to an increase in catabolic processes (Lamberts, van den Beld, and van der Lely 1997) and is exhibited by the age related loss of muscle mass and strength. The prevention of muscle mass loss has been achieved in healthy individuals by growth hormone administration, and is associated with increases in IGF–1 levels (Papadakis et al. 1996).

We have characterized the efficiency of muscle recovery following common muscle injuries (contusions, lacerations, and strains) and observed massive muscle regeneration occurring in the early steps of healing that eventually becomes impaired by the development of scar tissue in the injured muscle. We have also identified specific growth factors (IGF–1, bFGF, and nerve growth factor (NGF)) capable of enhancing myoblast proliferation and differentiation in vitro, and demonstrated the delivery of these growth factors into the injured muscle to improve muscle healing following orthopaedic-related injuries (see Chapter 5).

Although direct injection of the recombinant proteins bFGF, NGF, and IGF–1 displays some beneficial effect on muscle healing following injuries (improvement of muscle regeneration and muscle strength), very high dosages and repeated injections are often required because of their short biological half-life and rapid dissemination by the bloodstream. Different gene transfer systems based on cell and gene therapy are being investigated for delivering high levels and persistent expression of growth factor to enhance muscle regeneration and improve muscle healing following injuries (see "Targeted Therapeutic Genes . . . for Orthopaedic and Sports Medicine," page 27).

Cell Therapy

Myoblast transplantation consists of the implantation of normal myoblast precursors (satellite cells) into the dystrophic muscle in order to create a reservoir of normal myoblasts that can regenerate the dystrophic muscle. The transplantation of normal myoblasts into dystrophin-deficient muscle to create a reservoir of dystrophin producing myoblasts capable of fusing with dystrophic myofibers has been studied extensively in both mdx mice (an animal model for DMD) and DMD patients (for review, see Partridge

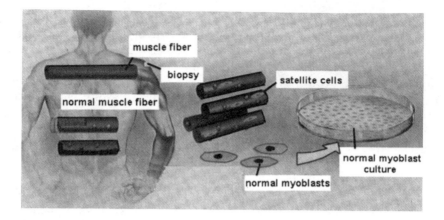

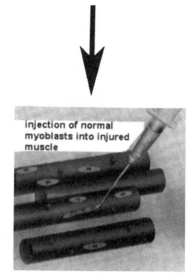

FIGURE 2.1. Schematic representation of cell therapy based on myoblasts.

1991). Myoblast transplantation has been found to be capable of delivering the missing protein (dystrophin) and increasing muscle strength in DMD muscles (see Figure 2.1 for schematic representation).

Similarly, we have observed that the injection of primary myoblasts in notexin-injured muscle (snake venom) enhances muscle regeneration in the injected muscle (Huard, Verreault, Roy, Tremblay, M., and Tremblay, J.P. 1994). Numerous transduced myofibers have been observed, showing that myoblast transplantation can enhance regeneration in injured muscle. This observation suggests that endogenous myoblasts transplanted into the injured muscle will be capable of regenerating and releasing various growth factors to further improve muscle healing (contusion, laceration, and strain). The transplantation of myoblasts has recently been shown to be capable of improving the muscle strength of injured muscle following laceration, contusion, and strain (Kasemkijwattana, Ménétrey, Somogyi et al. 1998; Kasemkijwattana, Ménétrey, Day et al. 1998; see also Chapter 5).

Although these studies showed transient restoration of dystrophin in dystrophic muscle and improved strength in injured muscle, immune rejection (see Chapter 14) as well as the poor survival and spread of injected myoblasts posttransplantation have greatly limited the success of myoblast transplantation (Morgan, Watt, Slopper, and Partridge et al. 1988; Morgan, Hoffman, and Partridge 1990; Morgan, Pagel, Sherrat, and Partridge 1993; Karpati et al. 1989; Karpati and Worton 1992; Partridge, Morgan, Coulton, Hoffman, and Kunkel 1989; Huard, Labrecque, Dansereau, Robitaille, and Tremblay 1991; Huard, Bouchard et al. 1991; Gussoni et al. 1992; Gussoni, Blau, and Kunkel 1997; Huard, Bouchard et al. 1992; Huard, Roy et al. 1992; Tremblay et al. 1993; Beauchamp, Morgan, Pagel, and Partridge 1994; Huard, Guerette et al. 1994; Huard, Verreault et al. 1994; Huard, Acsadi, Jani, Massie, and Karpati 1994; Kinoshita et al. 1994; Mendell et al. 1995; Vilquin, Wagner, Kinoshita, Roy, and Tremblay 1995; Fan, Maley, Beilharz, and Grounds 1996). Although approaches are being developed to improve the efficiency of myoblast transplantation in muscular dystrophy and muscle injuries/repair (Guerette, Asselin, Skuk, Entman, and Tremblay 1997; Qu et al. 1998), recent research has increasingly focused on the application of gene therapy using viral or nonviral vectors to deliver genes to skeletal muscle.

Tissue Engineering Based on the ex vivo Gene Transfer Approach

The ex vivo procedure, which combines myoblast transplantation and gene therapy using an autologous myoblast transfer (AMT), has also been utilized in delivering genes to skeletal muscle. This approach was originally used as an alternative to verify the success of myoblast transfer without immunological problems, since an AMT is performed. The idea behind this approach involves the establishment of a primary myoblast cell culture from dystrophic and injured muscles which is engineered by adequate transfection or transduction with a vector encoding dystrophin and growth factors cDNA in vitro (see Figure 2.2). These engineered myoblasts are then injected into the same host in order to bypass immunological problems against the injected myoblasts (see Figure 2.2). This method has been performed using adenovirus (Ad), retrovirus (RV), and HSV-1 carrying reporter genes, and has shown that the transduced myoblasts (isogenic myoblasts) fused and reintroduced the reporter genes into the injected muscle (Salvatori et al. 1993; Huard, Acsadi et al. 1994; Booth et al. 1997). Moreover, we have recently observed that the ex vivo approach may be

Ex-vivo Gene Transfer Approach

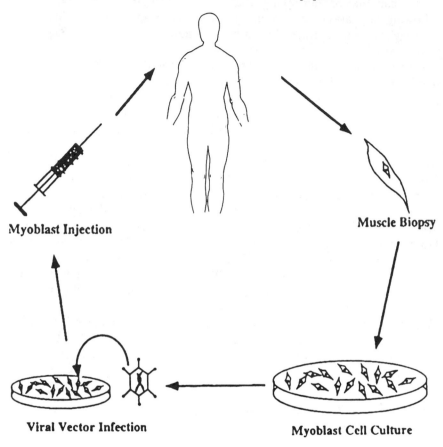

FIGURE 2.2. Schematic representation of the *ex vivo* gene transfer approach.

used to deliver full-length dystrophin into dystrophic muscle, and the efficiency of viral transduction using the ex vivo approach has been found to
be higher than that of direct injection of the same amount of virus (Floyd
et al. 1998).

   We have also investigated the use of ex vivo gene transfer based on autologous myoblast to deliver genes into injured muscle (lacerated, contused,
and strain injured muscle; see Chapter 5). In fact, primary myoblasts have
been isolated, engineered following transduction with an Ad carrying the
β–galactosidase (LacZ) reporter gene, and injected into the injured muscle.
The LacZ transduced myofibers have been found capable of persisting in
injected muscle for at least 35 days postinjection, suggesting that the ex vivo

gene transfer of autologous myoblast transfer is feasible and leads to a persistent expression of marker gene in injured muscle (Kasemkijwattana, Ménétrey, Somogyi et al. 1998; Kasemkijwattana, Ménétrey, Day et al. 1998; see also Chapter 5). We have recently shown that the myoblast mediated ex vivo gene transfer of Ad, carrying the expression of IGF–1, significantly improves muscle healing following orthopaedic-related injuries.

This ex vivo technique, which combines cell and gene therapy, may be advantageous for the musculoskeletal system because the cells can be used as a reservoir of secreting molecules as well as a source as of exogeneous cells capable of participating in the healing process. In fact, the use of pluripotent muscle derived stem cells in our ex vivo approach using muscle-based tissue engineering (see review in Chapter 11) may become very attractive for the healing of the musculoskeletal system.

Direct Gene Transfer Approach

Direct gene therapy, which involves injecting viral vectors directly into the muscle without using myoblasts as intermediates, is another approach to mediate gene delivery to skeletal muscle (see Figure 2.3). It has been shown that muscle cells can be successfully transduced in vitro and in vivo using intramuscular inoculation of replication defective Ad, RV and HSV carrying luciferase and β–galactosidase as reporter genes (Acsadi et al. 1994a,b; van Deutekom, Floyd et al. 1998; van Deutekom, Hoffman et al. 1998). However, a differential transducibility has been observed using these viral vectors throughout skeletal muscle development, representing a major limitation associated with AV as a gene delivery vector to skeletal muscle (van Deutekom, Floyd et al. 1998; van Deutekom, Hoffman et al. 1998).

We have observed that direct gene transfer of recombinant Ad carrying the β–galactosidase reporter gene is capable of highly transducing injured muscle (laceration, contusion, and strain). Many LacZ expressing myofibers have been found in the injured site of the contused, lacerated, and strained muscle at 5 days following the direct gene transfer approach (Kasemkijwattana, Ménétrey, Somogyi et al. 1998; Kasemkijwattana, Ménétrey, Day et al. 1998; see also Chapter 5). Although a transient expression of the transgene has been observed that is likely due to immune rejection problems, the use of new generation Ad (Haecker et al. 1996; Kochanek et al. 1996; Kumar Singh and Chamberlain 1996) as well as the adeno–associated virus (AAV) (Xiao et al. 1996) will likely reduce the possibility of immune rejection and consequently allow a persistent expression of the transgene into the injected muscle. In fact, AAV has been found to be capable of efficiently transducing injured muscle, and the transduced myofibers capable of long term persistence without cytotoxic and immunogenic problems (see Chapter 15).

Gene Therapy

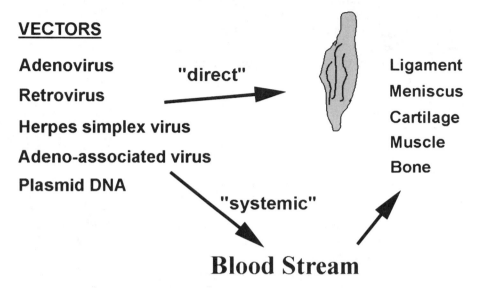

VECTORS

Adenovirus

Retrovirus

Herpes simplex virus

Adeno-associated virus

Plasmid DNA

"direct"

"systemic"

Blood Stream

Ligament

Meniscus

Cartilage

Muscle

Bone

FIGURE 2.3. Schematic representation of the direct and systemic gene therapy approaches.

Systemic Gene Therapy

Systemic delivery of the viral vector is very attractive especially when the host tissue is difficult to target by direct injection (see Figure 2.3). In addition, systemic delivery may achieve a better distribution of therapeutic genes in the tissues in comparison to direct injection, which often leads to a transgene expression localized only at the injection site. Systemic delivery of viral vector has been performed to deliver genes in muscle and nonmuscle tissue (Huard, Lochmuller, Acsadi et al. 1995), but this technique has been hindered by the high number of viral particles required, the lack of specificity of transduction in the targeted tissue, and the loss of viral particles in the lung and liver. Although approaches are being designed to improve systemic delivery of viral and nonviral vectors, tissue displaying a poor vascularity, such as meniscus and cartilage, may not be amenable to this gene therapy approach. The aforementioned approaches of gene transfer to skeletal muscle and nonmuscle tissue represent important considerations for different applications of gene therapy to the musculoskeletal system. In fact, the application of gene therapy and tissue engineering for arthritis (Chapter 14), hard tissue (Chapter 3), spinal disorders (Chapter 4), muscle injuries (Chapter 5), articular defect (Chapter 6), osteogenesis imperfecta (Chapter 7), and cartilage, meniscus, and ligament repair (Chapters 10, 11, and 12) will be presented. The development of approaches capable of enhancing tissue regeneration following injury by delivering substances through cell and gene therapy may promote efficient healing and complete functional recovery following injury.

Targeted Therapeutic Genes and Potential Applications of Human Gene Therapy for Orthopaedic and Sports Medicine

Although many of the following chapters will discuss the application of gene therapy to specific tissues of the musculoskeletal system, we will briefly summarize potential applications of gene therapy to several tissues of the musculoskeletal system: gene therapy to treat intraarticular diseases and conditions (arthritis, ligament, meniscus, cartilage), muscle injuries and repair, and bone defects.

Intraarticular Disorders and Conditions

Arthritis

Gene therapy, historically focused on the treatment of heritable diseases, is being intensely researched for novel approaches for treating acquired disease states. A specific example from the University of Pittsburgh illustrates the applicability of gene therapy to nonfatal musculoskeletal conditions. A human clinical trial has been initiated by Dr. Chris Evans's research group to confirm the safety of gene therapy for rheumatoid arthritis. In this trial, synovial cell mediated gene therapy is being used to deliver interleukin–1 receptor antagonist protein (IRAP) to inhibit the progression of the inflammatory response (Evans and Robbins 1994, 1995a, 1995b). IRAP has already been successfully delivered to the joint to exert a clinical response in both animal models (rabbit) and human patients (Evans and Robbins 1994, 1995a, 1995b). This therapeutic model highlights two of the main advantages of gene therapy over conventional treatment: (1) incorporating these genes encoding for a desired protein into the anatomical site supplies a constant source of therapeutic effect; and (2) by exerting its effect at the cellular level, gene therapy confers the ability to intervene early in the disease process.

Experience has also shown the potential importance of cell selection for the delivery system. In synovial cells, the expression of the therapeutic IRAP protein is initially high but declines with time, becoming undetectable around 4 to 6 weeks (Evans and Robbins 1994, 1995a, 1995b). The reason for this decline is unknown; cell death due to natural cellular turnover is a possible explanation. We have alternatively investigated whether cells from peripheral tissues, such as skeletal muscle, can be used as a gene delivery vehicle to the joint. We have observed that myoblasts can differentiate into myotubes and myofibers in many intraarticular structures of the knee; we are investigating whether this muscle-based gene therapy approach will allow high and long-term transgene expression in the joint (Day et al. 1997; see also Chapter 11).

Ligament

The incidence of ligament injuries is estimated at 178 injuries per 100,000 people per year in the general population (Miyasaka, Daniel, Stone, and Hirschman 1991). By extrapolation of these results, the number of anterior cruciate ligament tears can be estimated at 100,000 a year.

The anterior cruciate ligament (ACL) is an intraarticular structure of fiber bundles that twist along the axis of the ligament (Woo, Suh, Parson, Wang, and Watanabe 1998). Histologic studies in rabbit models have shown that the ACL contains type I and type III collagen as well as a high concentration of glycoaminoglycans and reducible collagen cross links (Amiel, Frank, Harwood, Fronek, and Akeson 1984). The ACL has a low healing capacity, which might be due to the encasement of the ligament in a synovial sheath and to the fact that it is bathed in synovial fluid. Complete tears of the ACL do not demonstrate a natural capacity for spontaneous healing. Thus, restoration of the function of a torn ACL requires surgical reconstruction using autograft or allograft tissue (O'Donaghue et al. 1971; Hawkins, Misamore, and Merrit 1986). Arnoczky, Rarvin, and Marshall (1982) have shown that the ACL graft undergoes a ligamentization process consisting of four phases: ischemic necrosis, revascularization, cell proliferation, and eventually collagen remodeling. Rougraff, Shelbourne, Gerth, and Warner (1993) have reported that in a human study this ligamentization process can take up to 3 years to reach completion.

Recently, several studies have addressed the effect of growth factors on the metabolism of ACL fibroblasts. Using in vitro models, different authors have demonstrated the stimulating effect of platelet derived growth factor-AB PDGF–AB, epidermal growth factor (EGF) and bFGF on rabbit and canine ACL fibroblast proliferation (Schmidt et al. 1995; DesRosiers, Yahia, and Rivard 1996; Scherping, Jr., et al. 1997), and transforming growth factor-beta (TGF–β) on fibroblasts from sheep explants (Spindler, Imro, Mayes, and Davidson 1996). In vitro, the synthesis of collagen and noncollagenous protein has been found to be increased by TGF–β and increased in a dose-dependent manner by EGF (Marui et al. 1997). Furthermore, in vivo studies have recently shown that PDGF, TGF–β, and EGF promoted the healing of the medial collateral ligament (Conti and Dahners 1993; Batten, Hansen, and Dahners 1996; Marui et al. 1997; Hildebrand et al. 1998). The data suggest that these specific growth factors may improve the healing of the

ACL and ACL graft ligamentization. Therefore, exogenous growth factor administration is a promising approach to improve ligament healing. However, the successful clinical implementation of this technique is currently limited by the problem of maintaining an adequate concentration of growth factor in the lesion site or the target tissue. The short half-life of growth factors and systemic lavage may lead to a rapid clearance of the substances from the desired site. To address these issues, gene therapy may be an interesting delivery system to the ligament. Studies of gene therapy and tissue engineering will demonstrate the feasibility of gene transfer to ligament using a LacZ marker gene. We will present results in Chapter 11 regarding the use of direct, myoblast, and fibroblast mediated ex vivo gene transfer in the anterior cruciate ligament.

Cartilage

Four different growth factors influencing cartilage metabolism and chondrocyte differentiation might also be used to improve the healing of cartilage following injury: TGF–β, IGF–1, EGF and bFGF. A number of bone morphogenetic proteins (BMPs) also stimulate the synthesis of proteoglycan and collagen type II in human chondrocyte cultures (Sellers, Peluso, and Morris 1997). Two other members of the transforming growth factor β superfamily were identified and designated as cartilage derived morphogenetic protein–1,2 (CDMP–1,2) (Chang et al. 1994; Luyten 1995). Moreover, it has recently been reported (Kang et al. 1997) that successful transfer of genes (marker genes) to chondrocytes opens new opportunities in treatments to improve cartilage healing. We will present new approaches based on gene therapy and tissue engineering to deliver marker genes and, eventually, growth factor into cartilage (see Chapters 11 and 12).

Meniscus

The meniscus has an important role in the functioning of the knee joint, including load transmission at the tibiofemoral articulation, shock absorption, lubrification, and stabilization of the entire knee joint. Meniscal lesions and menisectomy have been found to predispose to the development of early osteoarthritis. The healing ability of the injured meniscus remains limited since it requires the proliferation of meniscal fibrochondrocytes either from an intrinsic source at the site of injury or via an extrinsic source from the blood supply or synovium. The development of approaches to improve meniscal healing postinjury has become paramount. Some growth factors have been shown to be capable of improving meniscal healing (Webber, Harris, and Hough, Jr. 1985; Hashimoto, Kurosaka, Yoshiya, and Horohata 1992; Spindler et al. 1995; McAndrews and Arnoczky 1996; Natsu-Ume et al. 1997; Lamsam, Fu, Robbins, and Evans 1998), but the appropriate doses of growth factor to enhance healing have not yet been defined.

Although TGF–β, PDGF, and hepatocyte growth factor (HGF) have been shown to be capable of stimulating proliferation of meniscal cells in vitro, we have further characterized the effects of 9 growth factors on the meniscal fibrochondrocyte proliferation and collagen/noncollagen synthesis and identified EGF, transforming growth factor alpha (TGF-α), bFGF and PDGF–A,B as candidate molecules to improve meniscal healing. The direct administration of the recombinant growth factor protein is likely to be hindered by the limited biological half-life of these proteins and the rapid clearance of the injected proteins. We have, then, evaluated the ability to deliver marker genes into the meniscus and found that direct and myoblast mediated ex vivo gene transfer could eventually be used to deliver high levels and persistent expression of these growth factors into the injured meniscus. The development of an efficient gene transfer procedure to deliver those specific factors may eventually improve meniscal healing after injuries (see Chapter 11).

Bone Defect

The success of gene therapy has prompted consideration of numerous other acquired musculoskeletal conditions that may be amenable to this approach, including such relatively common problems in the orthopaedic field as segmental bone defects and nonunions. The current protocol for proper union of a bone defect usually requires an operative procedure for bone grafting that often leads to a lengthy recovery period for the patient, with an unsatisfactory end result. Osteogenic proteins, such as BMP–2, can promote bone healing in segmental bone defects. Since the injection of the recombinant protein is hindered by the limited half-life of these products and their rapid clearance by the bloodstream, large quantities of the proteins are needed to enhance bone healing potential. Cell mediated gene therapy in osseous defects may be an alternative approach to persistently deliver these osseous proteins (see Chapter 3). Osteoblast is the logical cell to be considered as a gene delivery vehicle to bone, but its use is greatly hindered by many limitations. Alternatively, muscle cells may offer at least 4 advantages as a gene delivery vehicle to the bone: (1) a muscle tissue harvest is less invasive than bone marrow aspiration and can potentially be obtained during initial debridement procedures; (2) myoblasts offer substantially higher transgene expression levels due, in part, to their ability to fuse into myotubes and myofibers; (3) their ability to differentiate into post-mitotic myotubes may provide greater potential for long term expression than stromal cells; and (4) recent data have indicated that muscle tissue may also contain inducible osteoprogenitor cells, negating the theoretical benefit of using bone stromal cells. The use of gene therapy and tissue engineering based on muscle cells to deliver recombinant human bone morphogenic protein 2 (rhBMP–2) to enhance fracture healing in segmental defects will be discussed in Chapter 11.

Muscle Injuries and Repair

Muscle injuries are a challenging problem in traumatology and the most frequent occurrences in sports medicine. Muscle damage occurs through a variety of mechanisms ranging from direct mechanical deformation (contusions, lacerations, and strains) to indirect mechanisms related to ischemia and neurological injuries. Contraction injuries (strains), muscle lacerations, and contusions are among the most common muscle injuries. Although these injuries are capable of healing, recovery remains very slow and occasionally only an incomplete functional recovery occurs. Following laceration, contusion, and strain, the injured muscle undergoes a rapid process of skeletal muscle regeneration that is hindered by the development of scar tissue formation in the injured area. We will discuss biological approaches

TABLE 2.2. Effects of growth factors on the musculoskeletal system.

Growth Factors		Tissue			
	Skeletal Muscle	Hyaline Cartilage	Meniscus	Ligament	Bone
Insulin Growth Factor (IGF–1)	+	+	–	–	+
Basic Fibroblast Growth Factor (bFGF)	+	+	+	+	+
Nerve Growth Factor (NGF)	+	ND	–	–	ND
Acidic Fibroblast Growth Factor (a–FGF)	+/–	ND	–	–	ND
Platelet-Derived AA	–	ND	–	+	ND
Growth Factor AB	–	ND	+	+	ND
(PDGF) BB	–	ND	ND	+/–	ND
Epidermal Growth Factor (EGF)	–	+	+	+	ND
Transforming Growth Factor-alpha (TGF–α)	–	ND	+	–	ND
Transforming Growth Factor-beta (TGF–β)	+/–	+	+	+	+
Bone Morphogenetic Protein (BMP–2)	ND	+	ND	ND	+
Chondrocyte Morphogenetic Protein CDMP (1,2)	ND	+	ND	ND	ND
Hepatocyte Growth Factor (HGF)	ND	ND	+	ND	ND

+ = positive effect.
– = no effect.
ND = not determinate.

to improve muscle healing following injury through the enhancement of muscle regeneration and the consequential minimization of fibrosis development. Based on the identification of growth factors (IGF–1, bFGF, and NGF) that are capable of improving muscle regeneration, we will present different approaches to using gene delivery in injured muscle. Different gene delivery systems to improve muscle healing following laceration, contusion, and strain—direct injection of the selected substances, direct and ex vivo gene transfer of recombinant adenovirus, and myoblast transfer will be discussed in Chapter 5. These studies should further understanding of the factors in muscle responsible for enhancing the proliferation and differentiation of muscle cells, as well as help in the development of new and innovative strategies to promote efficient muscle healing and complete functional recovery following the most common muscle injuries.

Summary and Conclusion

Gene therapy and tissue engineering are novel approaches with various potential applications in almost every medical field. Although still in their infancy, they are the subject of much research, and almost all studies related to the orthopaedic and sports medicine fields have used the E. coli β–galactosidase reporter gene in animal models. Two problems with gene therapy, however, are transient expression in transduced cells and our incomplete knowledge regarding the regulation of gene expression. For example, gene expression has been observed to persist up to 6 weeks in muscle, ligament, and meniscus. Even with such a limited period, this technique holds particular promise in the field of sports medicine. It may be advantageous in the treatment of acute injuries, because long time gene expression may not absolutely be required. Finally, the feasibility of gene therapy and the identification of potential growth factors to stimulate the growth of muscle, meniscus, bone, cartilage, and ligaments (see Table 2.2) provide promising new concepts for future treatment in orthopaedic and sports medicine related injuries and repair.

References

Acsadi, G., Dickson, G., Love, D.L., Jani, A., Walsh, F.S., Gurusinghe, A., Wolff, J.A., and Davies, K.E. 1991. Human dystrophin expression in mdx mice after intramuscular injection of DNA constructs. *Nature* 352:815–18.

Acsadi, G., Jani, A., Huard, J., Blaschuk, K., Massie, B., Holland, P., Lochmueller, H., and Karpati, G. 1994b. Cultured human myoblasts and myotubes show markedly different transducibility by replication-defective adenovirus recombinant. *Gene Ther* 1:338–40.

Acsadi, G., Jani, A., Massie, B., Simoneau, M., Holland, P., Blaschuk, K., and Karpati, G.A. 1994a. Differential efficiency of adenovirus-mediated in vivo gene transfer into skeletal muscle cells at different maturity. *Hum Mol Gen* 3:579–84.

Acsadi, G., Lochmueller, H., Jani, A., Huard, J., Massie, B., Prescott, S., Simoneau, M., Petrof, B., and Karpati, G. 1996. Dystrophin expression in muscles of mdx mice after adenovirus-mediated in vivo gene transfer. *Hum Gene Ther* 7:129–40.

Allamedine, H.S., Dehaupas, M., and Fardeau, M. 1989. Regeneration of skeletal muscle fiber from autologous satellite cells multiplied in-vitro. *Muscle Nerve* 12:544–55.

Allen, R.E. and Boxhorn, L.K. 1989. Regulation of skeletal muscle satellite cell proliferation and differentiation by transforming growth factor-beta, insulin-like growth factor-1, and fibroblast growth factor. *J Cell Physiol* 138:311–15.

Anderson, J.E., Liu, L., and Kardami, E. 1991. Distinctive patterns of basic fibroblast growth factor (bFGF) distribution in degenerative and regenerating areas of dystrophic (mdx) striated muscle. *Dev Biol* 147:96–109.

Amiel, D., Frank, C., Harwood, F., Fronek, J., and Akeson, W. 1984. Tendons and ligaments: a morphological and biochemical comparison. *J Orthop Res* 1:257–65.

Arahata, K., Ishiura, S., Ishiguro, T., Tsukahara, T., Suhara, Y., Eguchi, C., Ishihara, T., Nonaka, I., Ozawa, E., and Sugita, H. 1988. Immunostaining of skeletal and cardiac muscle surface membrane with antibody against Duchenne muscular dystrophy peptide. *Nature* 333:861–3.

Arnoczky, S.P., Rarvin, G.B., and Marshall, J.L. 1982. Anterior cruciate ligament replacement using patellar tendon. *J Bone Joint Surg* 64-A:217–24.

Barr, E. and Leiden, J.M. 1991. Systemic delivery of recombinant proteins by genetically modified myoblasts. *Science* 254:1507–9.

Batten, M.L., Hansen, J.C., and Dahners, L.E. 1996. Influence of dosage and timing of application of platelet-derived growth factor on early healing of the rat medial collateral ligament. *J Orthop Res* 14:736–41.

Beauchamps, J.R., Morgan, J.E., Pagel, C.N., and Partridge, T.A. 1994. Quantitative studies of the efficacy of myoblast transplantation. *Muscle Nerve* 18:S261.

Bischoff, R. 1994. The satellite cell and muscle regeneration. In *Myology*. 2nd Edition. 97–118. New York: McGraw-Hill.

Blau, H.M. and Springer, M.L. 1995. Molecular medicine: muscle based gene therapy. *N Engl J Med* 333:1554–6.

Bonilla, E.C.E., Samitt, A.F., Miranda, A.P., Hays, G., Salviati, S., Dimauro, S., Kunkel, L.M., Hoffman, E.P., and Rowland, L.P. 1988. Duchenne muscular dystrophy: deficiency of dystrophin at the muscle cell surface. *Cell* 54:447–52.

Booth, D.K., Floyd, S.S., Day, C.S., Glorioso, J.C., Kovesdi, I., and Huard, J. 1997. Myoblast mediated ex vivo gene transfer to mature muscle. *J Tissue Eng* 3(2):125–33.

Chambers, R.L. and McDermott, J.C. 1996. Molecular basis of skeletal muscle regeneration. *Can J Appl Physiol* 21(3):155–84.

Chang, S.C., Hoang, B., Thomas, T., Vukicevic, S., Luyten, F.P., Ryba, N.J.P., Kozak, A.H., Reddi, A.H., and Moos, M. 1994. Cartilage-derived morphogenetic proteins. *Biological Chem* 269(45):28227–34.

Coleman, M.E., DeMayo, F., Yin, K.C., Lee, H.M., Geske, R., Montgomery, C., and Schwartz, R.J. 1995. Myogenic vector expression of insulin like growth factor I stimulates muscle cell differentiation and myofiber hypertrophy in transgenic mice. *J Biol Chem* 270(20):12109–16.

Conti, N.A. and Dahners, L.E. 1993. The effect of exogenous growth factors on the healing of ligaments. *Trans Orthop Res Soc* 18:60.

Damon, S.E., Haugk, K.L., Birnbaum, R.S., and Quinn, L.S. 1998. Retrovirally mediated over-expression of insulin–like growth factor is required for skeletal muscle differentiation. *J Cell Physiol* 175:109–20.

Danko, I., Fritz, J.D., Latendresse, J.J., Herweijer, H., Schultz, E., and Wolff, J.A. 1993. Dystrophin expression improves myofiber survival in mdx muscle following intramuscular plasmid DNA injection. *Hum Mol Genet* 2(12):2055–61.

Davis, H., Demeneix, B.A., Quantin, B., Coulombe, J., and Whalen, R.G. 1993. Plasmid DNA is superior to viral vectors for direct gene transfer into adult mouse skeletal muscle. *Hum Gene Ther* 4:733–40.

Davis, H.L., Whalen, R.G., and Demeneix, B.A. 1993. Direct gene transfer into skeletal muscle in vivo: factors affecting efficiency of transfer and stability of expression. *Hum Gene Ther* 4:151–9.

Day, C.S., Kasemkijwattana, C., Menetrey, J., Floyd, S.S., Booth, D.K., Moreland, M.S., Fu, F.H., and Huard, J. 1997. Myoblast mediated gene transfer to the joint. *J Orth Res* 15:894–903.

DesRosiers, E.A., Yahia, L., and Rivard, C.H. 1996. Proliferative and matrix synthesis response of canine anterior cruciate ligament fibroblasts submitted to combined growth factors. *J Orthop Res* 14:200–8.

Dhawan, J., Pan, L.C., Pavlath, G.K., Travis, M.A., Lanctot, A.M., and Blau, H.M. 1991. Systemic delivery of human growth hormone by injection of genetically engineered myoblasts. *Science* 254:1509–12.

Doh, S.G., Vahlsing, J., Hartikka, J., Liang, X., and Manthorpe, M. 1997. Spatial-temporal patterns of gene expression in mouse skeletal muscle after injection of LacZ plasmid DNA. *Gene Ther* 4:648–63.

Dunckley, M.G., Davies, K.E., Walsh, F.S., Morris, G.E., and Dickson, G. 1992. Retroviral mediated transfer of a dystrophin minigene into mdx mouse myoblasts in vitro. *Febs Letter* 296(2):128–34.

Dunckley, M.G., Wells, D.G., Walsh, F.S., and Dickson, G. 1993. Direct retroviral-mediated transfer of dystrophin minigene into mdx mouse muscle in vivo. *Hum Mol Genet* 2:717–23.

Engert, J.C., Berlund, E.B., and Rosenthal, N. 1996. Proliferation precedes differentiation in IGF-1 stimulated myogenesis. *J Cell Biol* 125L:431–40.

Evans, C. and Robbins, P.D. 1994. Prospects for treating arthritis by gene therapy. *J Rheumatol* 21:779–82.

Evans, C. and Robbins, P.D. 1995a. Current concepts review. Possible orthopaedic applications of gene therapy. *J Bone Joint Surg Am* 77:1103–14.

Evans, C. and Robbins, P.D. 1995b. Progress toward the treatment of arthritis by gene therapy. *Ann Med* 27:543–6.

Fan, Y., Maley, M., Beilharz, M., and Grounds, M. 1996. Rapid death of injected myoblasts in myoblast transfer therapy. *Muscle Nerve* 19:853–60.

Feero, W.G., Li, S., Rosenblatt, J.D., Sirianni, N., Morgan, J.E., Partridge, T.A., Huang, L., and Hoffman, E.P. 1997. Selection and use of ligands for receptor-mediated gene delivery to myogenic cells. *Gene Ther* 4:664–74.

Fisher, K.J., Jooss K., Alston, J., Yang, Y., Haectier, S.E., High, K., Pathak, R., Raper, S.E., and Wilson, J.M. 1997. Recombinant adeno-associated virus for muscle directed gene therapy. *Nat Med* 3:306–12.

Florini, J.R., Ewton, D.Z., and Coolican, S.A. 1996. Growth hormone and the insulin-like growth factor system in myogenesis. *Endocrine Review* 17(5):481–517.

Florini, J.R., Ewton, D.Z., Falen, S.Z., and Van Wyk, J.J. 1986. Biphasic concentration dependency of the stimulation of myoblast differentiation by somatomedins. *Am J Physiol* 250:C771–8.

Florini, J.R., Roberts, A.B., Ewton, D.Z., Falen, S.B., Flanders, K.C., and Sporn, M.B. 1986. Transforming growth factor-b. A very potent inhibitor of myoblast differentiation, identical to the differentiation inhibitor secreted by Buffalo rat liver cells. *J Biol Chem* 261:16509–13.

Floyd, S.S., Clemens, P.R., Ontell, M.R., Kochanek, S., Day, C.S., Yang, J., Hauschka, S.D., Balkir, L., Morgan, J.E., Moreland, M.S., Feero, W.G., Epperly, M., and Huard, J. 1998. Ex vivo gene transfer using adenovirus mediated full length dystrophin delivery to dystrophic muscles. *Gene Ther* 5:19–30.

Gospodarowitz, D., Weseman, J., Moran, J.S., and Lindstrom, J. 1976. Effect of fibroblast growth factor on the division and fusion of bovine myoblasts. *J Biol Chem* 70:395–405.

Grounds, M.D. 1991. Towards understanding skeletal muscle regeneration. *Path Res Pract* 187:1–22.

Guerette, B., Asselin, I., Skuk, D., Entman, M., and Tremblay, J.P. 1997. Control of inflammatory damage by anti–LFA–1: Increase success of myoblast transplantation. *Cell Trans* 6(2):101–7.

Gussoni, E., Pavlath, P.K., Lanctot, A.M., Sharma, K., Miller, R.G., Steinman, L., and Blau, H.M. 1992. Normal dystrophin transcripts detected in DMD patients after myoblast transplantation. *Nature* 356:435–8.

Gussoni, E., Blau, H.M., and Kunkel, L.M. 1997. The fate of individual myoblasts after transplantation into muscles of DMD patients. *Natl Med* 3:970–7.

Haecker, S.E., Stedman, H.H., Balice-Gordon, R.J., Smith, D.B., Greelish, J.P., Mitchell, M.A., Wells, A., Sweeney, H.I., and Wilson, J.M. 1996. In vivo expression of full length human dystrophin from adenoviral vectors deleted of all viral genes. *Hum Gene Ther* 7:1907–14.

Harrington, M.A., Daub, R., Song, A., Stasek, J., and Garcia, J.G.N. 1992. Interleukin-1a mediated inhibition of myogenic terminal differentiation: Increased sensitivity of Ha–*ras* transformed cultures. *Cell Growth Diff* 3:241–8.

Hashimoto, J., Kurosaka, M., Yoshiya, S., and Horohata, K. 1992. Meniscal repair using fibrin sealant and endothelial cell growth factor: an experimental study in dogs. *Am J Sports Med* 20(5):537–41.

Hawkins, R.J., Misamore, G.W., and Merrit, T.R. 1986. Follow–up of acute nonoperated isolated anterior cruciate ligament tears. *Am J Sports Med* 14:205–10.

Hildebrand, K.A., Woo, S.L.Y., Smith, D.W., Allen, C.R., Deie, M., Taylor, B.J., and Schmidt, C.C. 1998. The effects of platelet-derived growth factor–BB on healing of the rabbit medial collateral ligament. An in vivo study. *Am J Sports Med* 26:549–54.

Hoffman, E.P., Brown, J., and Kunkel, L.M. 1987. Dystrophin: the protein product of the duchenne muscular dystrophy locus. *Cell* 51:919–28.

Huard, J., Acsadi, G., Jani, A., Massie, B., and Karpati, G. 1994. Gene transfer into skeletal muscles by isogenic myoblasts. *Hum Gene Ther* 5:949–58.

Huard, J., Akkaraju, G., Watkins, S.C., Cavalcoli, M.P., and Glorioso, J.C. 1997. LacZ gene transfer to skeletal muscle using a replication defective herpes simplex virus type 1 mutant vector. *Hum Gene Ther* 8:439–52.

Huard, J., Bouchard, J.P., Roy, R., Malouin, F., Dansereau, G., Labrecque, C., Albert, N., Richards, C.L., Lemieux, B., and Tremblay, J.P. 1992. Human myoblast transplantation: preliminary results of 4 cases. *Muscle Nerve* 15:550–60.

Huard, J., Bouchard, J.P., Roy, R., Malouin, F., Labrecque, C., Dansereau, G., Lemieux, B., and Tremblay, J.P. 1991. Myoblast transplantation produced dystrophin positive muscle fibers in a 16–year–old patient with Duchenne muscular dystrophy. *Clin Science* 81:287–8.

Huard, J., Feero, W.G., Watkins, S.C., Hoffman, E.P., Rosenblatt, D.J., and Glorioso, J.C. 1996. The basal lamina is a physical barrier to HSV mediated gene delivery to mature muscle fibers. *J Virol* 70(11):8117–23.

Huard, J., Goins, B., and Glorioso, J.C. 1995. Herpes Simplex virus type 1 vector mediated gene transfer to muscle. *Gene Ther* 2:1–9.

Huard, J., Guerette, B., Verreault, S., Tremblay, G., Roy, R., Lille, S., and Tremblay, J.P. 1994. Human myoblast transplantation in immunodeficient and immunosuppressed mice: Evidence of rejection. *Muscle Nerve* 17:224–34.

Huard, J., Krisky, D., Oligino, T., Marconi, P., Day, C.S., Watkins, S.C., and Glorioso, J.C. 1997. Gene transfer to muscle using herpes simplex virus-based vectors. *Neuromusc Disord* 299–313.

Huard, J., Labrecque, C., Dansereau, G., Robitaille, L., and Tremblay, J.P. 1991. Dystrophin expression in myotubes formed by the fusion of normal and dystrophic myoblasts. *Muscle Nerve* 14:178–82.

Huard, J., Lochmuller, H., Acsadi, G., Jani, A., Massie, B., and Karpati, G. 1995. The route of administration is a major determinant of the transduction efficiency of rat tissues by adenoviral recombinants. *Gene Ther* 2:107–15.

Huard, J., Lochmuller, H., Jani, A., Holland, P., Guerin, C., Massie, B., and Karpati, G. 1995. Differential short term transduction efficiency of adult versus newborn mouse tissues by adenoviral recombinants. *Exp Mol Pathol* 62:131–43.

Huard, J., Roy, R., Bouchard, J.P., Malouin, F., Richards, C.L., and Tremblay, J.P. 1992. Human Myoblast transplantation between immunohistocompatible donors and recipients produces immune reactions. *Transpl Proc* 24(6):3049–51.

Huard, J., Verreault, S., Roy, R., Tremblay, M., and Tremblay, J.P. 1994. High efficiency of muscle regeneration following human myoblast clone transplantation in SCID mice. *J Clin Invest* 93:586–99.

Hurme, T. and Kalimo, H. 1992. Activation of myogenic precursor cells after muscle injury. *Med Sci Sports Exerc* 24(2):197–205.

Jennische, E. 1989. Sequential immunohistochemical expression of IGF–1 and the transferin receptor in regenerating rat muscle in vivo. *Acta Endocrinol* 121:733–8.

Jennische, E. and Hansson, H.A. 1987. Regenerating skeletal muscle cells express insulin–like growth factor 1. *Acta Physiol Scand* 130:327–32.

Jin, P., Rahm, M., Claesson-Welsh, L., Heldin, C.H., and Sejersen, T. 1990. Expression of PDGF A–chain and b–receptor genes during rat myoblast differentiation. *J Cell Biol* 110:1665–72.

Johnson, S.E. and Allen, R.E. 1995. Activation of skeletal muscle satellite cells and the role of fibroblast growth factor receptors. *Exper Cell Res* 2119:449–53.

Kang, R., Marui, T., Ghivizzani, S.C., Nita, I.M., Georgescum, H.I., Suhm, J.K., Robbins, P.D., and Evans, C.H. 1997. Ex vivo gene transfer to chondrocytes in full-thickness articular cartilage defects: a feasibility study. *Osteoarthritis Cartilage* 5:139–43.

Karpati, G., Pouliot, Y., Zubrzycka-Gaarn, E.E., Carpenter, S., Ray, P.N., Worton, R.G., and Holland, P. 1989. Dystrophin is expressed in mdx skeletal muscle fibers after normal myoblast implantation. *Am J Pathol* 135:27–32.

Karpati, G. and Worton, R.G. 1992. Myoblast transfer in DMD: problems and interpretation of efficiency. *Muscle Nerve* 15:1209.

Kasemkijwattana, C., Ménétrey, J., Somogyi, G., Moreland, M.S., Fu, F.H., Buranapanitkit, B., Watkins, S.C., and Huard, J. 1998. Development of approaches to improve the healing following muscle contusion. *Cell Transpl* 7(6):585–98.

Kasemkijwattana, C., Ménétrey, J., Day, C.S., Bosch, P., Buranapitkit, B., Moreland, M.S., Fu, F.H., Watkins, S.C., and Huard, J. 1998. Biologic intervention in muscle healing and regeneration. *Sports Med Arthrosc Rev* 6:95–102.

Kinoshita, I., Vilquin, J.T., Guerette, B., Asselin, I., Roy, R., and Tremblay, J.P. 1994. Very efficient myoblast allotransplantation in mice under FK506 immunosuppression. *Muscle Nerve* 17:1407–15.

Kochanek, S., Clemens, P.R., Mitani, K., Chen, H.H., Chan, S., and Caskey, C.T. 1996. A new adenoviral vector: replacement of all viral coding sequences with 28 kb of DNA independently expressing both full length dystrophin and beta–galactosidase. *Proc Natl Acad Sci USA* 93:5731–6.

Kumar-Singh, R. and Chamberlain, J.S. 1996. Encapsilated adenovirus minichromosomes allow delivery and expression of a 14 kb dystrophin cDNA to muscle cells. *Hum Mol Genet* 5:913–21.

Lamberts, S.W.J., van den Beld, A.W., and van der Lely, A.J. 1997. The endocrinology of aging, *Science* 278:419–24.

Lamsam, C., Fu, F.H., Robbins, P.D., and Evans, C.H. 1998. Gene transfer to meniscal fibrochondrocytes (address). New Orleans, Lousiana. *American Academy of Orthopaedic Surgeons.*

Lefaucheur, J.P. and Sebille, A. 1995. Muscle regeneration following injury can be modified in vivo by immmune neutralization of basic fibroblast growth factor, transforming growth factor β–1 or insulin–like growth factor 1. *J Neuroimmunology* 57:85–91.

Linkhart, T.A., Clegg, C.H., and Hauschka, S.D. 1981. Myogenic differentiation in permanent clonal mouse myoblast cell lines: regulation by macromolecular growth factors in the culture medium. *Dev Biol* 86:19–30.

Luyten, F.P. 1995. Cartilage-derived morphogenetic proteins: key regulators in chondrocyte differentiation. *Acta Orth Scand* 66:51–4.

Lyles, J.M., Amin, W., Bock, E., and Weil, C.L. 1993. Regulation of NCAM by growth factors in serum free myotubes cultures. *J Neurosci Res* 34(3):273–86.

Marconi, P., Krisky, D., Oligino, T., Poliani, P.L., Ramakrishnan, R., Goins, R.G., Fink, D.J., and Glorioso, J.C. 1996. Replication defective HSV vectors for gene transfer in vivo. *Proc Natl Acad Sci USA* 93:11319–20.

Marui, T., Niyibizi, C., Georgescu, H.I., Cao, M., Kavalkovich, K.W., Levine, R.E., and Woo, S.L.Y. 1997. The effect of growth factors on matrix synthesis by ligament fibroblasts. *J Orthop Res* 15:18–23.

McAndrews, P.T. and Arnoczky, S.P. 1996. Meniscal repair enhancement techniques. *Clinics in Sports Medicine* 15(3):499–510.

McFarland, D.C., Pesall, J.E., and Gilkerson, K.K. 1993. The influence of growth factors on turkey embryonic myoblasts and satellite cells in vitro. *General and Comparative Endocrinology* 89(3):415–24.

Mendell, J.R., Kissel, J.T., Amato, A.A., King, W., Signore, L., Prior, T.W., Sahenk, Z., Benson, S., McAndrew, P.E., and Rice, R. 1995. Myoblast transfer in the treatment of Duchenne's muscular dystrophy. *N Eng J Med* 333:832–8.

Miller, D.G., Adam, M.A., and Miller, A.D. 1990. Gene transfer by retrovirus vectors occurs only in cells that are actively replicating at the time of infection. *Mol Cell Biol* 10:4239–42.

Miyasaka, K.C., Daniel, D.M., Stone, M.L., and Hirschman, P. 1991. The incidence of knee ligament injuries in the general population. *Am J Knee Surg* 4:3–8.

Morgan, J.E., Hoffman, E.P., and Partridge, T.A. 1990. Normal myogenic cells from newborn mice restore normal histology to degenerating muscle of the mdx mouse. *J Cell Biol* 111:2437–49.

Morgan, J.E., Pagel, C.N., Sherrat, T., and Partridge, T.A. 1993. Long-term persistence and migration of myogenic cells injected into pre-irradiated muscles of mdx mice. *J Neurol Sci* 115:191–200.

Morgan, J.E., Watt, D.J., Slopper, J.C., and Partridge, T.A. 1988. Partial correction of an inherited defect of skeletal muscle by graft of normal muscle precursor cells. *J Neurol Sci* 86:137–47.

Natsu-Ume, T., Nakamura, N., Shino, K., Matsumoto, N., Nakata, K., Horibe, S., Iwamoto, M., Matsumoto, N., Nakamura, T., and Ochi, T. 1997. The effect of hepatocyte growth factor/scatter factor (HGF/SF) on meniscal fibrochondrocytes. San Francisco, CA: 43rd Annual Meeting, Orthopaedic Research Society, Feb. 9–13, 1997, p. 421.

O'Donoghue, D.H., Frank, G.R., Jeter, G.L., Johnson, W., Zeiders, J.W., and Keynon, R. 1971. Repair and reconstruction of the anterior cruciate ligament in dogs. *J Bone Joint Surg* 53-A:710–18.

Olson, E.N., Sterrnberg, E., Hu, J.S., Spizz, G., and Wilcox, C. 1986. Regulation of myogenic differentiation by type beta transforming growth factor. *J Cell Biol* 103:1799–806.

Papadakis, M.A., Grady, D., Black, D., Tierney, M.J., Gooding, A.W., Schambelan, M., and Grunfield, C. 1996. Growth hormone replacement in healthy older men improves body composition but not functional ability. *Ann Intern Med* 124:708–16.

Partridge, T.A. 1991. Myoblast transfer: a possible therapy for inherited myopathies. *Muscle Nerve* 14:197–212.

Partridge, T.A., Morgan, J.E., Coulton, G.R., Hoffman, E.P., and Kunkel, L.M. 1989. Conversion of mdx myofibers from dystrophin negative to positive by injection of normal myoblasts. *Nature* 337:176–9.

Qu, Z., Balkir, L., van Deutekom, J.C.T., Robbins, P.R., Priuchnic, R., and Huard, J. 1998. Development of approaches to improve cell survival in Myoblast Transfer Therapy. *J Cell Biol* 142(5):1257–67.

Quantin, B., Perricaudet, L.D., Tajbakhsh, S., and Mandell, J.L. 1992. Adenovirus as an expression vector in muscle cells in vivo. *Proc Natl Acad Sci USA* 89: 2581–4.

Quinn, L.S. and Haugk, K.L. 1996. Over expression of the type–1 insulin–like growth factor receptor increases ligand dependent proliferation and differentiation in bovine skeletal myogenic cultures. *J Cell Physiol* 168:34–41.

Ragot, T., Vincent, M., Chafey, P., Vigne, E., Gilgenkrantz, H., Couton, B., Cartaud, J., Briand, B., Kaplan, J.C., Perricaudet, M., and Kahn, A. 1993. Efficient adenovirus mediated gene transfer of a human mini-dystrophin gene to skeletal muscle of mdx mice. *Nature* 361:647–50.

Rao, S.S. and Kohtz, D.S. 1995. Positive and negative regulation of D type cyclin expression in skeletal myoblasts by basic fibroblast growth factor and transforming growth factor b. A role for cyclin D1 in control of myoblast differentiation. *J Biol Chem* 270(8):4093–100.

Reed–Clark, K., Sferra, T.J., and Johnson, P.R. 1997. Recombinant adeno–associated viral vectors mediated long-term transgene expression in muscle. *Hum Gene Ther* 8:659–69.

Rougraff, B., Shelbourne, K.D., Gerth, P.K., and Warner, J. 1993. Arthroscopic and histologic analysis of human patellar tendon autografts used for anterior cruciate ligament reconstruction. *Am J Sports Med* 21:277–84.

Salvatori, G., Ferrari, G., Messogiorno, A., Servidel, S., Colette, M., Tonalli, P., Giarassi, R., Cosso, G., and Mavillo, F. 1993. Retroviral vector mediated gene transfer into human primary myogenic cells leads to expression in muscle fibers in vivo. *Hum Gene Ther* 4:713–23.

Scherping, Jr., S.C., Schmidt, C.C., Georgescu, H.I., Kwoh, C.K., Evans, C.H., and Woo, S.L.Y. 1997. Effects of growth factors on the proliferation of ligament fibroblasts from skeletally mature rabbits. *Connect Tissue Res* 36:1–8.

Schmidt, C.C., Georgescu, H.I., Kwoh, C.K., Blomstrom, G.L., Engle, C.P., Larkin, L.A., Evans, C.H., and Woo, S.L.Y. 1995. Effect of growth factors on the proliferation of fibroblasts from the medial collateral and anterior cruciate ligaments. *J Orthop Res* 13:184–90.

Schultz, E. 1989. Satellite cell behavior during skeletal muscle growth and regeneration. *Med Sci Sports Exerc* 21:181.

Schultz, E., Jaryszak, D.L., and Valliere, C.R. 1985. Response of satellite cells to focal skeletal muscle injury. *Muscle Nerve* 8:217.

Sellers, R.S., Peluso, D., and Morris, E.A. 1997. The effects of recombinant human bone morphogenetic protein–2 (rhBMP–2) on the healing of full thickness defects of articular cartilage. *J Bone Joint Surg* 79(10):1452–63.

Spindler, K.P., Mayes, C.E., Miller, R.R., Imro, A.K., and Davidson, J.M. 1995. Regional mitogenic response to the meniscus to platelet derived growth factor (PDGF-AB). *J Orthop Res* 13(2):201–7.

Spindler, K.P., Imro, A.K., Mayes, C.E., and Davidson, J.M. 1996. Patellar tendon and anterior cruciate ligament have different mitogenic responses to platelet derived growth factor and transforming growth factor. *J Orthop Res* 14:542–6.

Svensson, E.C., Tripathy, S.K., and Leiden, J.M. 1996. Muscle based gene therapy: realistic possibilities for the future. *Mol Med Today* 2:166–72.

Tremblay, J.P., Malouin, F., Roy, R., Huard, J., Bouchard, J.P., Satoh, A., and Richards, C.L. 1993. Results of a blind clinical study of myoblast transplantations without immunosuppressive treatment in young boys with Duchenne muscular dystrophy. *Cell Trans* 2:99–112.

Trippel, S.B., Coutts, R.D., and Einhorn, T. 1996. Growth factors as therapeutic agents. *J Bone Joint Surg* 78-A:1272–86.

van Deutekom, J.C.T., Floyd, S.S., Booth, D.K., Oligino, T., Krisky, D., Marconi, P., Glorioso, J., and Huard, J. 1998. The development of approaches to improve viral gene delivery to mature skeletal muscle. *Neuromus Disorders* 8:135–48.

van Deutekom, J.C.T., Hoffman, E.P., and Huard, J. 1998. Muscle maturation: implications for gene therapy. *Mol Med Today* 4(5):214–20.

Vitiello, L., Chonn, A., Wasserman, J.D., Duff, C., and Worton, R.G. 1996. Condensation of plasmid DNA with polylysine improves liposome mediated gene transfer into established and primary muscle cells. *Gene Ther* 3:396–404.

Vilquin, J.T., Wagner, E., Kinoshita, I., Roy, R., and Tremblay, J.T. 1995. Successful histocompatible myoblast transplantation in dystrophin deficient mdx dystrophin. *J Cell Biol* 131(4):975–88.

Vincent, M., Ragot, T., Gilgenkrantz, H., Couton, D., Chafey, P., Gregoire, A., Briand, P., Kaplan, J.C., Kahn, A., and Perricaudet, M. 1993. Long-term correction of mouse dystrophic degeneration by adenovirus mediated transfer of a minidystrophin gene. *Nature Genet* 5:130–4.

Watkins, S.C., Hoffman, E.P., Slayter, H.S., and Kunkel, L.M. 1988. Immunoelectron microscopic localization of dystrophin in myofibers. *Nature* 333:863–6.

Webber, R.J., Harris, M.G., and Hough, Jr, A.J. 1985. Cell culture of rabbit meniscal fibrochondrocytes: proliferative and synthetic response to growth factors and ascorbate. *J Orthop Res* 3(1):36–42.

Wolff, J.A., Ludtke, J.J., Assadi, G., Williams, P., and Jani, A. 1992. Long term persistence of plasmid DNA and foreign gene expression in mouse muscle. *Hum Mol Genet* 1:363–9.

Woo, S.L.Y., Suh, J.K., Parson, I.M., Wang, J.H., and Watanabe, N. 1998. Biologic intervention in ligament healing: effect of growth factors. *Sports Med Anthrosc Rev* 6:74–82.

Xiao, X., Li, J., and Samulski, R.J. 1996. Efficient long term gene transfer into muscle tissue of immunocompetent mice by adeno–associated virus vector. *J Virol* 70:8098–108.

Yablonka–Reuvini, Z., Balestri, T.M., and Bowen–Pope, D.F. 1990. Regulation of proliferation and differentiation of myoblasts derived from adult mouse skeletal muscle by specific isoforms of PDGF. *J Cell Biol* 111:1623–9.

Yang, Y., Nunes, F.A., Berencsi, K., Furth, E.E., Gonczol, E., and Wilson, J.M. 1994. Cellular immunity to viral antigens limits E1-deleted adenoviruses for gene therapy. *Proc Natl Acad Sci USA* 91:4407–11.

Zdanowicz, M.M., Moyse, J., Wingertzahn, M.A., O'Conner, M., Teichberg, S., and Slonim, A.E. 1995. Effect of insulin–like growth factor I in murine muscular dystrophy. *Endocrinology* 1336:4880–6.

Zubryzcka–Gaarn, E.E., Bulman, D.E., Karpati, G., Burghes, A.H.M., Belfall, B., Klamut, H.J., Talbot, J., Hodges, R.S., Ray, P.N., and Worton, R.G. 1988. The Duchenne muscular dystrophy gene is localized in the sarcolemma of human skeletal muscle. *Nature* 333:466–9.

Part II
Gene Therapy Applications for the Musculoskeletal System

3
Regional Gene Therapy for Hard Tissues

Jay R. Lieberman

Introduction

Advances in molecular biology now allow for the transfer of genes to target cells that can evoke specific biological responses with possible applications for the treatment of clinical orthopaedic problems. The purpose of this review is to discuss such potential uses of gene therapy in hard tissues, specifically in diseases related to bone and cartilage.

Clinical approaches to a number of different diseases associated with bone and cartilage could benefit from gene therapy. These potential applications for the treatment of fractures, fracture nonunions, osteoporosis, cartilage defects, and osteoarthritis will be the focus of this chapter. In addition, the management of diseases that affect the development of the skeleton, including osteogenesis imperfecta, Gaucher's disease, achondroplasia, multiple epiphyseal dysplasia, and spondyloepiphyseal dysplasia, may also be amenable to treatment with gene therapy (Evans and Robbins 1995). These possibilities will be discussed in Chapter 14 of this book.

The treatment of bone and cartilage disorders with gene therapy will require the development of technology to deliver the appropriate genes to either osteoprogenitor cells, chondroprogenitor cells, osteoblasts, chondrocytes, synovium, or muscle (Barr and Leiden 1991; Dai, Reiman, Naviaux, and Verma 1992; Gomez–Foix et al. 1992; Rosen and Thies 1992; Thaller, Dart, and Tesluk 1993; Setoguchi, Jaffe, Daniel, and Crystal 1994; Baragi et al. 1995; Evans and Robbins 1995; Arai et al. 1997; Engstrand, Daluiski, Finerman, and Lieberman 1998; Lieberman, Daluiski, et al. 1999; Lieberman, Le, et al. 1998). In both in vitro and in vivo models, this type of gene delivery has been successful. An understanding of the biology of these conditions is necessary to optimize gene therapy strategies.

Human gene transfer can be performed using either an ex vivo or in vivo strategy (Rosenfield et al. 1992; Crystal 1995; Evans and Robbins 1995; Wilson 1995; Fang et al. 1996; Wilson 1996). In ex vivo gene transfer, the cDNA is transferred to cells in tissue culture, and the genetically modified cells are subsequently administered to the patient. An alternative in vivo technique is to transfer the gene directly into a specific anatomic location. The type of gene therapy system developed will be influenced by the duration of protein production required. The treatment of fresh fractures, fracture nonunions, and cartilage defects should only require a limited (days to weeks) duration of protein production. In contrast, the treatment of chronic problems (i.e., osteoporosis) or genetic defects requires long term production. Therefore, the selection of a vector, either viral or nonviral, that is episomal may be most appropriate for a limited duration of protein production, unless multiple administrations of the vector are planned (Scaduto and Lieberman 1999).

The location, defect size, condition of the surrounding soft tissue, and number of growth factors necessary to treat a particular problem are also important in determining the type of gene therapy to be used. Vectors for gene delivery may be either viral or nonviral. There are potential advantages and disadvantages associated with both systems. A detailed discussion of these vector alternatives is beyond the scope of this chapter. However, vectors that transfer genetic information into the genome of target cells (i.e., retroviruses) and can permanently modify the target cell genotype have potential advantages when treating hereditary and chronic disorders such as osteogenesis imperfecta or osteoporosis. However, there are risks associated with retroviruses, including potential for toxicity associated with chronic overexpression or insertional mutagenesis. In addition, since only dividing cells can be infected, retroviruses are best employed in an ex vivo gene transfer strategy (Crystal 1995).

In contrast, adenoviral vectors are well suited for in vivo gene transfer because they can be produced in high titers and they can efficiently infect both nonreplicating and replicating cells. The genetic information transferred by an adenovirus remains episomal and, therefore, does not alter the cellular genotype not is it passed on to daughter cells (Jones and Shenk 1979; Brett, Haddara, Prevek, and Graham 1994; Crystal 1995; Wilson 1995, 1996). However, adenoviral vectors have been noted to evoke nonspecific inflammatory responses. Therefore, the episomal position of the cDNA and the inflammatory responses to the viral proteins often limit the duration of expression to periods ranging from days to weeks (Rosenfeld et al. 1992; Mitani, Graham, and Caskey 1994; Setoguchi et al. 1994; Crystal 1995; Wilson 1996). As a result, adenoviral vectors are most appropriate for short-term use, or they will require chronic administration. However, the development of second generation adenoviral vectors, which are deleted in all viral protein coding sequences, may provide a safer, less immunogenic carrier with significantly greater insert capacity that will allow for the insertion of multiple cDNAs in the vector (Mitani, Graham, Caskey, and Kochanek 1995; Morsy et al. 1998).

Recombinant adeno-associated virus (rAAV) is a DNA virus that also has potential use as a vector for regional gene therapy. The advantages of rAAV include: (1) the ability to transduce dividing and nondividing cells;

(2) site specific integration into the host genome, and (3) the vector does not elicit a significant immune response. The mayor drawback to rAAV is a limited insert capacity. Inserts larger than 4.7 kilobases can not be packaged by the vector (Schwarz, in press).

Naked DNA is another attractive strategy for cDNA delivery in treating bone problems. This technique is simple and does not require the use of viruses (Fang et al. 1996). However, there are concerns about the efficiency of this gene transfer strategy and these will be discussed later in this chapter.

A number of other gene transfer strategies could be applied to the treatment of bone and cartilage defects. These include adeno–associated viruses, herpes simplex viruses, liposomes, and the gene gun (Crystal 1995; Evans and Robbins 1995). A complete discussion of the advantages and disadvantages of these gene delivery systems is beyond the scope of this chapter.

Fracture Healing and Nonunion

Approximately 5% to 10% of fractures heal slowly (delayed union) or do not heal at all (nonunion) (Praemer, Furner, and Rice 1992). These problems with bone healing are associated with a number of risk factors including the degree of soft tissue injury, location of the fracture, amount of bone loss, and infection. The physiology of bone formation and repair is influenced by growth factors, or proteins that can regulate critical cellular functions including proliferation, differentiation, and matrix synthesis (Canalis and Lian 1988; Canalis, McCarthy, and Centrella 1989; Bourque, Gross, and Hall 1993; Sandberg, Aro, and Vuorio 1993; Linkhart, Mohan, and Baylink 1996). Therefore, the transfer of the genetic sequence or cDNA for one of these growth factors (i.e., bone morphogenetic protein) could be used to enhance fracture healing or treat nonunions (Lieberman, Le, et al. 1998).

A number of different growth factors have been noted to influence fracture healing, including bone morphogenetic proteins (BMP) (Urist 1965; Wozney et al. 1988; Wang et al. 1990; Yasko et al. 1992; Gerhart et al. 1993, Cook et al. 1994a, 1994b; Cook, Wolfe, Salkeld, and Rueger 1995; Linkhart et al. 1996; Riley, Lane, Urist, Lyons, and Lieberman 1996), transforming growth factor beta TGF–B) (Joyce, Jingushi, and Bolander 1990; Centrella, Horowitz, Wozney, and McCarthy 1994; Linkhart et al. 1996), platelet derived growth factor (PDGF) (Graves et al. 1989; Linkhart et al. 1996), fibroblast growth factor (FGF) (Aspenberg, Thorngren, and Lomande 1991; Kawaguchi et al. 1994), and insulin growth factor (IGF) (Bagi, Deleon, Bommage, Rosen, and Sonner 1983; Canalis, Centrella, Burch, and McCarthy 1989; Ebeling et al. 1993; Thaller et al. 1993; Prockop 1997). In in vitro experiments, these growth factors have been noted to enhance mineralization of bone marrow cells in culture, and in in vivo studies they have enhanced healing of segmental bone defects in a variety of animal models.

The bone morphogenetic proteins (BMPs) are osteoinductive proteins that were originally identified by their ability to demineralize bone and have been demonstrated to be effective in healing intermediate size bone defects in a variety of animal models including rats, rabbits, dogs, sheep, and nonhuman primates (Urist 1965; Yasko et al. 1992; Gerhart et al. 1993, Cook et al. 1994a, 1994b; Cook et al. 1995). In general, young animals are used, and the remaining bone is well vascularized with an intact soft tissue envelope. However, such models do not truly simulate the clinical situation because fracture nonunions are often associated with poor bone stock, compromised blood supply, and abundant fibrous tissue that can inhibit bone repair. In this milieu, a single exposure to an exogenous growth factor may be insufficient to stimulate an adequate osteoinductive response. However, the development of a biological cellular vehicle, such as BMP producing bone marrow cells created via gene transfer, may provide an appropriate period of sustained production of BMP that would enable the host to adequately respond to this osteoinductive stimulus and heal the bone defect (Lieberman, Daluiski et al. 1998; Lieberman, Le et al. 1998). The bone formed must not only bridge the defect, but it must attain sufficient biomechanical strength to support normal function.

The treatment of a nonunion does not require long term protein production. Therefore, adenovirus and direct transfer of plasmid genes are appropriate treatment options. In our laboratory we have chosen to employ an ex vivo gene transfer system using autologous marrow cells that have been genetically manipulated. We are using an adenovirus containing a cDNA for BMP–2 because recombinant BMP–2 protein has been demonstrated to be osteoinductive in in vivo studies (Lieberman, Daluiski et al. 1998; Lieberman, Le et al. 1998).

The basic strategy is to harvest bone marrow cells from the patient, grow them in tissue culture, and then infect the bone marrow cells with BMP–2 containing adenovirus. The adherent fraction of bone marrow cells would be used as a biological carrier because these cells are easy to harvest and expand, and have also been demonstrated to be inherently osteoinductive. An adenovirus was selected because bone marrow cells grow slowly and the adenovirus can infect both replicating and nonreplicating cells. We have previously demonstrated in the laboratory the ability to infect both rat and human bone marrow cells with the BMP–2 containing adenovirus. We have infected W–20 cells, a mouse stromal cell line, with the adenoviral vector containing the cDNA for BMP–2. These W–20 BMP–2 producing cells were used successfully to heal a critical-sized segmental femoral defect in a nude rat (Lieberman, Le et al. 1998) (Figure 3.1).

We subsequently used BMP-producing Lewis rat bone marrow cells to heal a critical sized femoral segmental defect in syngeneic rats 2 months after implantation of the BMP–2 producing bone marrow cells. BMP–2

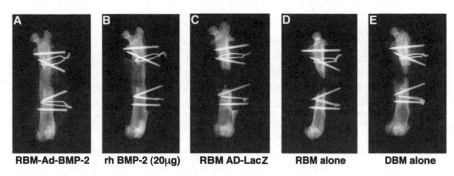

RBM-Ad-BMP-2   rh BMP-2 (20µg)   RBM AD-LacZ   RBM alone   DBM alone

FIGURE 3.1. Radiographs of specimens two month after the surgical procedure. Twenty milligrams of guanidium extracted demineralized bone matrix was used as a substrate in all 8 mm femoral defects. A: BMP-2-producing bone marrow cells created via adenovial gene transfer (Ad-BMP-2). B: rhBMP-2 (20 micrograms). C: *B* galactosidase producing (LacZ gene) rat bone marrow cells (Ad-lacZ, negative control). D: Uninfected rat bone marrow cells (negative control). E: Demineralized bone matrix alone (negative control). There was minimal bone formation in C,D, and E.

producing bone marrow cells were implanted in an 8 mm segmental femoral defect rat with guanidium extracted demineralized bone matrix that was used to hold the cells in place. Guanidium treatment eliminated the inherent osteoinductive activity of the demineralized bone matrix. Radiographic analysis revealed that 22 out of 24 bone defects healed at 2 months. Histomorphometric analysis revealed robust bone formation. There was no significant difference in biomechanical testing when assessing healed femurs that had been treated with recombinant human bone morphogenetic protein (rhBMP–2) (20 µg) and BMP–2-producing bone marrow cells with respect to failure torque, energy to failure, or torsional stiffness (Lieberman, et al. 1999).

In in vitro experiments assessing BMP–2 production over time, we have demonstrated that bone marrow cells can produce protein for approximately 28 days. The limited duration of transgene expression associated with a first generation adenoviral vector appears to be advantageous for this particular clinical application. Such a system of ex vivo gene transfer, consisting of the harvesting and subsequent infection of bone marrow cells with an adenovirus containing the BMP–cDNA, could be easily adapted for human use.

Baltzer (1999) has used an in vivo regional gene therapy strategy to heal femoral defects in a rabbit model. Adenoviral vectors containing the cDNA for either BMP–2 or TGF–B were diluted in a saline solution and then directly injected into a femoral defect (1.3 cm) in a rabbit. Thirteen weeks after injection 6 of 6 femurs treated with the BMP–2 containing adenoviral vector demonstrated radiographic evidence of healing. Sixteen weeks after injection 4 of 5 femurs injected with the TGF–B-containing adenovirus had healed. No biomechanical testing was performed. Although this strategy appears promising, an analysis of the immune response to direct adenoviral injection is necessary.

There has also been interest in using purified mesenchymal stem cells to enhance in vivo bone formation. These purified mesenchymal stem cells have been successfully used to heal a critical sized defect in the canine femur (Bruder, Kraus, Goldberg, and Kadiyala 1998). However, there are concerns that such mesenchymal stem cells may have limited osteoinductive potential in elderly individuals or in patients with osteoporosis or other metabolic bone diseases. Fortunately, the gene transfer of cDNAs for growth factors into these cells could enhance their osteoinductive potential.

Another potential alternative for gene therapy for fracture nonunions is to directly inject the adenovirus vector or plasmid DNA into a specific anatomic site. Recently, Fang et al. (1996) reported that a BMP–4 cDNA construct, which was delivered to a rat femoral segmental defect (5 mm) loaded on a collagen sponge (gene activated matrix or GAM), could provide sufficient BMP–4 to augment defect healing. Limited biomechanical testing was performed on these specimens. This same research group has used a gene-activated matrix containing a portion of the gene for parathyroid hormone (PTH 1–34) to enhance bone formation in a canine-femoral defect model. However, insufficient bone formation was induced to completely heal bone defects (Bonadio et al. 1999). The advantages of this delivery system using plasmid DNA are that it is relatively simple, no viruses are necessary, and it may be more cost effective than ex vivo gene transfer since the harvesting and reimplanting of cells is not necessary. One potential disadvantage of the technique is that it requires the plasmid to transfect local fibroblasts, and these fibroblasts will then secrete the protein. However, the surrounding soft tissue muscle and bone often have compromised vascularity that could limit the response to the transfected fibroblasts. Further investigation of this innovative treatment strategy remains necessary.

In summary, the treatment of a nonunion is quite appropriate for a gene therapy strategy. Our data suggests that using regional gene therapy to deliver BMP has significant osteoinductive potential in humans. However, other growth factors that are osteoinductive (i.e., TGF–B, PDGF, IGF) or a combination of growth factors that are both osteoinductive and/or angiogenic (i.e., FGF) may be advantageous. Further research is required to determine the efficacy of these techniques, but it is reasonable to consider adapting these strategies for human use.

Osteoporosis

Osteoporosis is characterized by progressive loss of bone mass and an increase in the fragility of bone that can predispose to fracture. In general, the treatment for osteoporosis at this time is to use agents that will enhance bone formation. The formation and resorption of bone is tightly regulated and is partially modulated by systemic hormones. However, local growth factors or cytokines generated in the local bone cell microenvironment should have a significant influence on bone remodeling (Mundy 1997). The same growth factors that could potentially be used to treat fracture nonunions and to enhance fracture healing are also candidates in the treatment of osteoporosis. These candidate growth factors include TGF–B, BMP, IGF, PDGF, and FGF (Mundy 1997). For example, IGF–1 has been used both systemically in an ovariectomized rat model and in subcutaneous injections in postmenopausal women (Bagi et al. 1983; Ebeling et al. 1993). In the rat model, administration of recombinant human IGF–1 promoted periosteal and endosteal bone formation. In postmenopausal women, subcutaneous injections of IGF–1 for 6 days produced dose dependent increases in the serum concentrations of type I procollagen carboxy–terminal propetide, which is an index of collagen synthesis (Ebeling et al. 1993).

A potential alternative to the delivery of recombinant growth factors would be to use genetically manipulated cells. For example, bone marrow cells could be genetically manipulated to produce one or several growth factors that could enhance bone formation. It would be helpful if the vector employed contained a promoter that allowed protein production to be turned on and off. These bone marrow cells could then be delivered directly into the systemic circulation or perhaps to specific bone sites such as the intertrochanteric region, lumbar spine, or distal radius where patients are likely to develop osteoporotic fractures. Clearly, we are a long way off from implementing this type of gene therapy, but it should be possible in the future.

Cartilage Defects

In contrast to bone, chondrocytes do not posses the inherent ability to regenerate themselves. Therefore, the repair of cartilage defects results in fibrocartilage tissue rather than normal cartilage. The prevention of the development of endstage arthritis is a significant clinical problem. Presently, treatment of cartilage defects includes autologous chondrocyte transplantation (Brittberg, Lindahl, Ohlsson, Isahsson, and Peterson 1994), drilling into subchrondral bone to facilitate migration of chondroprogenitor cells into the defect (Menche, Vangsness, Pitman, Gross, and Peterson 1998), and osteochondral allografts. However, none of these techniques has been consistently successful in producing normal hyaline articular cartilage.

Recently, attention has been focused on using growth factors to enhance and maintain the cartilage phenotype. TGF–B is known to enhance chondrogenesis and inhibit terminal differentiation of chondrocytes (Galera,

Redini et al. 1992a; Galera, Vivien et al. 1992). BMP–2 has been known to induce chondrogenesis in vitro and has been shown to enhance cartilage formation in a rabbit osteochondral defect model (Sellers, Peluso, and Morris 1997). However, no matter which method is used, healing the junction between new and host cartilage is critical for long-term durability of the repair cartilage. In addition, it is not clear that a single exposure to an exogenous growth factor will be sufficient to heal large cartilage defects. There are also concerns that the carriers used to deliver the growth factor may inhibit cartilage repair as they degrade. If biological degradation of the carrier is associated with an inflammatory response in the joint, healing of the cartilage defect could be inhibited.

Therefore, regional gene therapy may be a way to circumvent these problems. In vivo transduction of chrondrocytes using a viral—liposome suspension injected intraarticularly has already been performed (Tomita et al. 1997). However, efficient in vivo gene transfer may be limited secondary to the extracellular matrix surrounding the chondrocytes, and there is also concern that this gene transfer could result in transduction of neighboring cells.

It has already been demonstrated that autologus chondrocytes can be successfully harvested, grown in tissue culture, and reimplanted into an articular defect (Brittberg et al. 1994). This treatment leads to the growth of tissue in the defect that resembles hyaline articular cartilage. Mesenchymal stem cells have been used in animals models to heal articular cartilage defects (Wakitani et al. 1994). Therefore, it is reasonable to propose a strategy of ex vivo gene transfer with autologous chondrocytes. The proteins secreted by transduced chondrocytes or chondroprogenitor cells could enhance cartilage repair. This treatment strategy would include harvesting and subsequent infection of the cells, and delivery of the transduced cells to a specific cartilage defect site. The obvious disadvantage of the ex vivo transfer strategy is that it requires a 2 stage procedure, harvesting the cells for in vitro transfection and later reimplantation of the transfected cells. In addition, there are concerns that overexpression of a pluripotential growth factor such as (TGF–B) or BMP–2 could be toxic to the surrounding cells or cause other unwanted local side effects.

Arai et al. (1997) transduced a human chondrocyte—like cell line with an adenovirus containing the cDNA for TGF–B. The transduced cells secreted TGF–B in tissue culture for 21 days and there was elevated expression of type II collagen (Arai et al. 1997). In our laboratory, we have demonstrated the ability to infect rabbit chondrocytes with a TGF–B containing retrovirus. The cells were cultured in a high-density pellet culture system for up to 8 weeks. TGF–B was still expressed 8 weeks after transduction, and enhanced expression of type II collagen and aggrecan, 2 cartilage specific markers, were noted. Histological examination of the chondrocytes overexpressing TGF–B demonstrated an abundant extracellular matrix (Engstrand et al. 1998). The successful transduction of chondrocytes with these 2 vectors suggests that a similar strategy may be adapted for human use. However, the successful use of this type of gene transfer strategy in an in vivo animal model is necessary. In addition, it is not clear that TGF–B is the appropriate growth factor because of its pluripotential effects.

Insulin growth factor has been demonstrated to enhance cartilage proteoglycan production in in vitro experiments (Fortier et al. 1999). In addition, recombinant IGF enhances cartilage defect healing in in vivo models. IGF could be delivered to the cartilage defect by transducing chondrocytes or synovial cells in an ex vivo approach or by direct injection of an adenovirus into the synovial lining (Fortier et al. 1999; Nixon et al. 1999). Clearly, the successful implementation of a gene therapy program will require a more comprehensive understanding of the biology of chondrocyte matrix synthesis and cartilage repair.

Osteoarthritis

Gene therapy may also play a role in the treatment of osteoarthritis. Gene delivery to either chondrocytes or to the surrounding synovial tissue could have therapeutic potential for the treatment of osteoarthritic joints. Genes have been successfully introduced locally to the synovial tissue of knee joints of rabbits in both ex vivo and in vivo methods (Roessler et al. 1993). However, the drawback to this strategy is that the proteins produced may not be able to penetrate the cartilage matrix and influence chondrocyte behavior.

It has also been demonstrated that chondrocytes can be transduced with either retroviral or adenoviral vectors containing the cDNA for TGF–B. Overexpression of TGF–B by these cells could enhance matrix synthesis. IGF-1 could also be used in a gene therapy strategy to enhance matrix synthesis.

The treatment of osteoarthritis may also require gene transfer to chondrocytes either to compensate for mutations in structural proteins of the cartilage matrix or to secrete chondroprotective agents (Ala–Kokko, Baldwin, Moskowitz, and Prockop 1990; Baragi et al. 1995). With the use of an ex vivo gene transfer strategy, human chondrocytes have been successfully transduced with an adenovirus containing the cDNA for interleukin receptor antagonist protein (IL–1ra) (Baragi et al. 1995). It is hypothesized that IL–1ra is chondroprotective since it may block extracellular matrix degradation associated with IL–1. Transduced chondrocytes that were seeded onto a cartilage organ culture protected the cartilage from IL–1 induced extracellular matrix degradation. These transduced cells produced a sustained source of the chondroprotective protein, IL–1ra.

Developing a gene therapy strategy to treat osteoarthritis is obviously more complicated than treating a focal cartilage defect. It must not only provide sustained release of growth factors that enhance matrix synthesis, but also produce cytokines that are chondroprotective and inhibit cartilage degradation. The development of in vivo animal models that can test the potential therapeutic success of these gene transfer strategies is necessary.

Summary

Although, there are multiple potential applications for gene therapy to treat clinical problems related to diseases of both bone and cartilage, a number of different questions need to be answered so that we can optimize treatment of these clinical problems. What growth factors or cytokines will enhance treatment? What is the duration of protein products needed? What type of vector should be used, and what is the most appropriate type of gene transfer strategy (in vivo or ex vivo)? Finally, we need to enhance an understanding of the signal transduction pathways that control bone and cartilage formation and repair. These questions must first be addressed via rigorous testing in clinically relevant animal models and then in randomized-controlled trials in humans. However, despite these obstacles, orthopaedic patients can benefit from gene therapy, and hopefully this technology will be transferred from the laboratory to the clinic in the early part of the 21st century.

References

Ala–Kokko, L., Baldwin, C.T., Moskowitz, R.W., and Prockop, D.J. 1990. Single base mutation in the type II procollagen gene (COL2A1) as a cause of primary osteoarthritis associated with a mild chondysplasia. *Proc Natl Acad Sci USA* 87:6567–8.

Arai, Y., Kubo, T., Kappei, K., Kazushige, T., Taliahaski, K., Iveda, T., Imanishik, J., Takigawa, M., and Hirasawa, Y. 1997. Adenovirus vector mediated gene transduction to chondrocytes: in vitro evaluation of therapeutic efficacy of transforming growth factor–B, and heat shock protein to gene transduction. *J Rheumatol* 24:1787–95.

Aspenberg, P., Thorngren, K.G., and Lomande, L.S. 1991. Dose dependent stimulation of bone induction by basic fibroblast growth factor in rats. *Acta Orthop Scand* 62(5):481–4.

Bagi, C.M., DeLeon, E., Bommage, R., Rosen, D., and Sonner, A. 1983. Treatment of the ovariectomized rats with the complex of rhIGF–I/IGFBP–3 increases cortical and cancellous bone mass and improves structure in the femoral neck. *Calcif Tissue Internat* 35:578–85.

Baltzer, A.W.A. 1999. Bone healing by adenoviral based gene therapy with BMP–2 and TGFB [abstr]. *Trans Orthop Res Soc* 308.

Baragi, V.M., Rehkiewicz, R.R., Jordan, H., Banadio, J., Hartman, J.W., and Roessler, B.J. 1995. Transplantation of transduced chondrocytes protects articular cartilage from interleukin–1 induced extracellular matrix degradation. *J Clin Invest* 96(5):2454–60.

Barr, W. and Leiden, J.M. 1991. Special delivery of recombinant proteins by genetically modified myoblasts. *Science* 254:1507–8.

Bonadio, J., Smiley, E., Patil, P., and Goldstein, S. 1999. Localized, direct plasmid gene delivery in vivo: prolonged therapy results in reproducible tissue regeneration. *Nat Med* 5(7):753–9.

Bourque, W.T., Gross, M., and Hall, B.K. 1993. Expression of four growth factors during fracture repair. *Intern Dev Biol* 37(4):573–9.

Brett, A.J., Haddara, W., Prevek, L., and Graham, F.L. 1994. An efficient and flexible system for construction of adenovirus vectors with insertions or deletions in early regions 1 and 3. *Proc Natl Acad Sci USA* 91:8802–6.

Brittberg, M., Lindahl, A., Ohlsson, C., Isahsson, O., and Peterson, L. 1994. Treatment of deep cartilage defects in the knee with autologous chondrocyte transplantation. *N Engl J Med* 331:889–95.

Bruder, S., Kraus, K.H., Goldberg, V.M., and Kadiyala, S. 1998. The effect of implants loaded with autologous mesenchymal stem cells on the healing of canine segmental bone defects. *J Bone Joint Surg* 80A:985–96.

Canalis, E., Centrella, M., Burch, W., and McCarthy, T.L. 1989. Insulin–like growth factor 1 mediates selective anabolic effects of parathyroid hormone in bone cultures. *J Clin Invest* 83:60–5.

Canalis, E. and Lian, J.B. 1988. Effects of bone associated growth factors on DNA, collagen and osteocalcin synthesis in cultured fetal rat calvariae. *Bone* 9:243–6.

Canalis, E., McCarthy, T.L., and Centrella, M. 1989. The regulation of bone formation by local growth factors. *J Bone Mineral Res* 6:27–56.

Centrella, M., Horowitz, M.C., Wozney, J.M., and McCarthy, T.L. 1994. Transforming growth factor beta–gene family member and bone. *Endocrine Rev* 15: 27–39.

Cook, S.D., Baffes, G.C., Wolfe, M.W., Sampath, T.K., Rueger, D.C., and Whitecloud, T.S. 1994a. The effect of recombinant human osteogenic protein–1 on healing of large segmental bone defects. *J Bone Joint Surg* 76-A:827–38.

Cook, S.D., Baffes, G.C., Wolfe, M.W., Sampath, T.K., and Rueger, D.C. 1994b. Recombinant human bone morphogenetic protein–7 induces healing in a canine long-bone segmental defect model. *Clin Orthop* 301:302–12.

Cook, S.D., Wolfe, M.W., Salkeld, S.L., and Rueger, D.C. 1995. Effect of recombinant human osteogenic protein–1 on healing of segmental defects in nonhuman primates. *J Bone Joint Surg* 77-A:734–50.

Crystal, R.G. 1995. Transfer of genes to humans: early lessons and obstacles to success. *Science* 404–10.

Dai, Y., Reiman, M., Naviaux, R.K., and Verma, I.M. 1992. Gene therapy via primary myoblasts: long term expression of Factor IX protein following transplantation in vivo. *Proc Natl Acad Sci USA* 89:10892–5.

Ebeling, P.R., Jones, J.D., O'Fallon, W.M. Janes, C.L., and Riggs, B.L. 1993. Short-term effects of recombinant human insulin–like growth factor 1 on bone turnover in normal women. *J Clin Endocrinol Metab* 77:1384–7.

Engstrand, T., Daluiski, A., Finerman, G.A.M., and Lieberman, J.R. 1998. Enhanced chondrocytes overexpressing growth factor TGF–B. *Trans Orthop Orth Res Soc* 23:998.

Evans, C.H. and Robbins, P.D. 1995. Possible orthopaedic applications of gene therapy. *J Bone Joint Surg* 77A:1103–14.

Fang, J., Zhu, Y.Y., Smiley, E., Bonadio, J., Rouleau, J.P., Goldstein, S.A., McCauley, L.K., Davidson, B.L., and Roessler, B.J. 1996. Stimulation of new bone formation by direct transfer of osteogenic plasmid genes. *Proc Natl Acad Sci USA* 93:5753–8.

Fortier, L.A., Lust, G., Mohammed, H.O., and Nixon, A.J. 1999. Coordinate upregulation of cartilage matrix synthesis in fibrin cultures supplemented with exogenous insulin–like growth factor-1. *J Orthop Res* 17:467–74.

Galera, P., Redini, F., Vivien, D., Bonaventure, J., Penfornis, H., Loyau, G., and Pujol, J.P. 1992. Effects of transforming growth factor–b1 (TGF–b1) on matrix synthesis by monolayer cultures of rabbit articular chondrocytes during the dedifferentiating process. *Exp Cell Res* 200:379–92.

Galera, P., Vivien, D., Pronost, S., Bonaventure, J., Redini, F., Loyau, G., and Pujol, J.P. 1992. Transforming growth factor–beta 1 up regulation of collagen type II in primary cultures of rabbit articular chondrocytes involves increased mRNA levels without affecting mRNA stability and procollagen processing. *J Cell Physiol* 153:596–606.

Gerhart, T.N., Kirker–Head, C.A., Kriz, J.J., Holtrop, M.E., Hennig, G.E., Hipp, J., Schelling, S.H., and Wang, E. 1993. Healing segmental femoral defects in sheep using recombinant human bone morphogenetic protein. *Clin Orthop* 293: 317–26.

Gomez–Foix, A.M., Coats, W.S., Baques, S., Alam, T., Gerard, R.D., and Newgard, C.B. 1992. Adenovirus mediate transfer of the muscle glycogen phosphorylase gene into hepatocytes confers altered regulation of glycogen metabolism. *J Biol Chem* 267:25129.

Graves, D.T., Valentin–Opran, A., Delgado, R., Valente, A.J., Mundy, G., and Pichie, J. 1989. The potential role of platelet derived growth factor as an autocrine or paracrine factor for human bone cells. *Connect Tissue Res* 23:209–18.

Jones, N. and Shenk, T. 1979. Isolation of adenovirus type 5 host range deletion mutants defective for transformation of rat embryo cells. *Cell* 17:683.

Joyce, M.C., Jingushi, S., and Bolander, M.C. 1990. Transforming growth factors–B in the regulation of fracture repair. *Orthop Clin No Am* 21:199–209.

Kawaguchi, H., Kurokawa, T., Hanada, K., Aiyaman, Y., Tamira, M., Ogata, E., and Matsumoto, T. 1994. Stimulation of fracture repair by recombinant human basic fibroblast growth factor in normal and streptozotocin diabetic rats. *Endocrinology* 135:774–81.

Lieberman, J.R., Daluiski, A., Stevenson, S., McAllister, P., Lee, Y., Wu, L., Kabo, J.M., Finerman, G.A.M., and Witte, O.N. 1999. Regional gene therapy with BMP producing bone marrow cells heals segmental femoral defects in rats. *J Bone Joint Surg* 81A:905–17.

Lieberman, J.R., Le, L., Wu, L., Finerman, G.A.M., Berk, A.J., Witte, O.N., and Stevenson, S. 1998. Regional gene therapy with a BMP–2 producing cell murine stromal cell line induces heterotopic and orthotopic bone formation in rodents. *J Orthop Res* 16:330–9.

Linkhart, T.A., Mohan, S., and Baylink, D.J. 1996. Growth factors for bone growth and repair: IGF, TGF–B and BMP. *Bone* 19(1 Suppl):IS–12S.

Menche, D.S., Vangsness, C.T., Pitman, M., Gross, A.E., and Peterson, L. 1998. The treatment of isolated articular cartilage lesions in the young individual. *Instr Course Lect* American Academy of Orthopaedic Surgeons 47:505–15.

Mitani, K., Graham, F.L., and Caskey, C.T. 1994. Transduction of human bone marrow by adenoviral vector. *Hum Gene Ther* 5:941–8.

Mitani, K., Graham, F.L., Caskey, C.T., and Kochanek, S.T. 1995. Rescue, propagation and partial purification of a helper virus dependent adenovirus vector. *Proc Natl Acad Sci USA* 92:3854–8.

Morsy, M.A., Gu, M.C., Motzel, S., Zhao, J., Lin, J., Su, Q., Allen, H., Franklin, L., Parks, R.J., Grahem, F.L., Kochanek, S., Brett, A.J., and Caskey, C.T. 1998. An adenoviral vector deleted for all viral coding sequences results in enhanced safety and extended expression of a leptin transgene. *Proc Natl Acad Sci USA* 95:7866–71.

Mundy, G.L. 1997. Growth factors as potential therapeutic agents in osteoporosis. *Instr Course Lect* American Academy of Orthopaedic Surgeons 46:495–8.

Nixon, A.J., Fortier, L.A., Williams, J., and Mohammed, H.O. 1999. Enhanced repair of extensive articular defects by insulin–like growth factor–1 ladlin fibrin composites. *J Orthop Res* 17:475–87.

Praemer, A., Furner, S., and Rice, D.P. 1992. *Musculoskeletal Conditions in the United States*. 83–124, Park Ridge, IL: American Academy of Orthopaedic Surgeons.

Prockop, D.J. 1997. Marrow stromal cells as stem cells for nonhematopoietic tissues. *Science* 276:71–4.

Riley, E.H., Lane, J.M., Urist, M.R., Lyons, K.M., and Lieberman, J.R. 1996. Bone morphogenetic protein–2. Biology and applications. *Clin Orthop* 324:39–46.

Roessler, B.J., Allen, E.D., Wilson, J.M., Hartman, J.W., and Davidson, B.L. 1993. Adenoviral mediated gene transfer to rabbit synovium in vivo. *J Clin Invest* 92:1085–92.

Rosen, V. and Thies, R.S. 1992. The BMP proteins in bone formation and repair. *Trends Gen* 8(3):97–102.

Rosenfeld, M.A., Yoshimura, K., Trapnell, B.C., Yoneyama, K., Rosenthal, E.R., Dalemans, W.M., Fukayama, J., Bargon, L.E., Stier, L., Stratford–Perricaudet, M., Guggino, W.B., Pavirani, A., Lecocq, J.P., and Crystal, R.G. 1992. In vivo transfer of the human cystic fibrosis transmembrane conductane gene to the airway epithelium. *Cell* 68:143–55.

Sandberg, M.M., Aro, H.T., and Vuorio, E.I. 1993. Gene expression during bone repair. *Clin Orthop* 289:292–312.

Scaduto, A.A. and Lieberman, J.R. 1999. Gene therapy for osteoinduction. *Orthop Clin No Am* 30(4):625–33.

Schwarz, E.M. (In press). The adeno-associated virus vector for orthopaedic gene therapy. *Clin Orthop*.

Sellers, R.S., Peluso, D., and Morris, E.A. 1997. The effect of recombinant human bone morphogenetic protein–2 (rhBMP–2) on the healing of full thickness defects of articular cartilage. *J Bone Joint Surg* 79A:1452–69.

Setoguchi, Y., Jaffe, H.A., Danel, C., and Crystal, R.G. 1994. Ex vivo and in vivo gene transfer to the skin using replication deficient adenovirus vectors. *J Invest Dermatol* 102:415–21.

Thaller, S.R., Dart, A., and Tesluk, H. 1993. The effects of insulin–like growth factor on critical size calvarial defects in Sprague–Dawley rats. *Ann Plast Surg* 31:429–33.

Tomita, T., Hashimoto, H., Tomita, N., Morishita, R., Lee, S.B., Hayashida, K., Nakamura, N., Yonenubo, K., Kaneda, Y., and Ochi, T. 1997. In vivo direct gene transfer into articular cartilage by intraarticular injection mediated by HVJ (Sender virus) and liposomes. *Arthritis Rheum* 40:901–6.

Urist, M.R. 1965. Bone formation by autoinduction. *Science* 150:893–9.

Wakitani, S., Goto, T., Pineda, S.J., Young, R.G., Mansour, J.M., Caplan, A.I., and Goldberg, V.M. 1994. Mesenchymal cell based repair of large, full thickness defects of articular cartilage. *J Bone Joint Surg* 76A:579–92.

Wang, E.A., Rosen, V., D'Alessandro, J.S., Bauduy, M., Cordes, P., Harada, T., Israel, P.I., Hewick, R.M., Kerns, K.M., LaPan, P., Luxenberg, D.P., McQuaid, D., Martsatsos, I.K., Nove, J., and Wozney, J.M. 1990. Recombinant human bone morphogenetic protein induces formation. *Proc Natl Acad Sci USA* 87:2220–4.

Werntz, J.R., Lane, J.M., Burstein, A.H., Justin, R., Klein, R., and Tomin, E. 1996. Qualitative and quantitative analysis of orthopaedic bone regeneration by marrow. *J Orthop Res* 14:85–93.

Wilson, J.M. 1995. Gene therapy for cystic fibrosis: challenges and future directions. *J Clin Invest* 96:2547–754.

Wilson, J.M. 1996. Adenoviruses as gene delivery vehicles. *N Engl J Med* 1185–7.

Wozney, J.M., Rosen, V., Celeste, A.J., Mitsock, L.M., Whitters, J.J., Kriz, R.W., Hewick, R.M., and Wang, E.A. 1988. Novel regulators of bone formation: molecular clones and activities. *Science* 1528–34.

Yasko, A.W., Lane, J.M., Fellinger, E.J., Rosen, V., Wozney, J.M., and Wang, E.A. 1992. The healing of segmental bone defects induced by recombinant human bone morphogenetic protein (rhBMP–2): a radiographic, histological, and biomechanical study in rats. *J Bone Joint Surg* 74A:659–70.

4
Potential Applications of Gene Therapy to the Treatment of Spinal Disorders

KOTARO NISHIDA, JAMES D. KANG, SCOTT D. BODEN,
LARS G. GILBERTSON, JUN-KYO SUH, PAUL D. ROBBINS, and
CHRISTOPHER H. EVANS

Introduction

Two Spine-Related Problems in Orthopaedics and Sports Medicine: Disc Degeneration and Nonunion/Delayed Union in Spinal Fusion

Spinal disorders remain a formidable problem in orthopaedics and sports medicine. The scope of clinical diagnoses ranges from acute traumatic injuries (e.g., fractures and subluxations) to more chronic degenerative conditions (e.g., degenerative disc herniations with radiculopathy or myelopathy). Although the specific conditions underlying most spinal disorders are largely unknown, biomechanical and biological changes associated with disc degeneration are thought to be of etiological significance in many, making the intervertebral disc a primary focus of diagnostic investigation and management interventions. Unfortunately, few approaches are available clinically for the treatment or prevention of disc degeneration, mandating investigation and development of novel approaches such as gene therapy.

In many spinal disorders, the technology available for treatment is generally highly invasive and often involves surgical fusion. In a spinal fusion, two or more contiguous vertebrae are artificially linked together temporarily with spinal instrumentation (either anteriorly or posteriorly or both) until the fusion mass consolidates (i.e., bony union between the vertebrae occurs). The intended effect of a fusion is therefore to mechanically immobilize the segments within the fusion. While fusion surgery has been reported to be successful in treating the original diagnosis which necessitated the surgery, there are concerns in both the short term (e.g., nonunion or delayed union) and the long term (e.g., accelerated adjacent segment degeneration; see Figure 4.1) (Lee 1988). Recent efforts to stabilize the spine while retaining its flexibility include the development of semirigid fixation devices and artificial discs (Kostuik 1997; Lemaire et al. 1997; Papp, Porter, Aspden, and Shepperd 1997). Despite these advances,

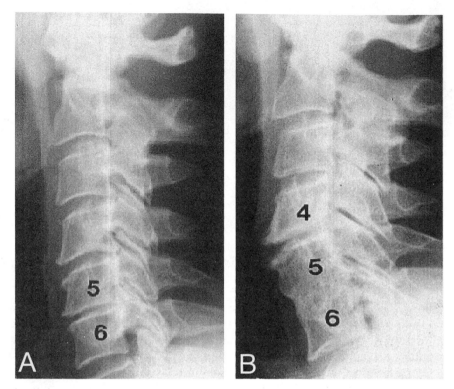

FIGURE 4.1. (A) Radiograph showing severe disc degeneration at the C5–6 level, necessitating removal of the disc and anterior fusion surgery. (B) Radiograph obtained $4\frac{1}{2}$ years after fusion surgery, demonstrating accelerated disc degeneration at the C4–5 level adjacent to the fusion.

effective treatment of spinal disorders may not be achievable based on manipulation of mechanical factors alone. One approach to the problem of nonunion in spinal fusion is biologic enhancement of the fusion through the application of growth factors or their encoding genes (i.e., gene therapy).

Here, then, are two challenging spinal problems—disc degeneration and nonunion/delayed union in spinal fusion—that are potential candidates for approaches based on gene therapy.

Gene Therapy

Although certain species of proteins, including growth factors, have promising therapeutic properties, *sustained* delivery of proteins to the spine—especially for chronic conditions—appears to be difficult to accomplish with present technology. An alternative possibility is to genetically modify cells of target organs in patients through gene therapy such that the cells

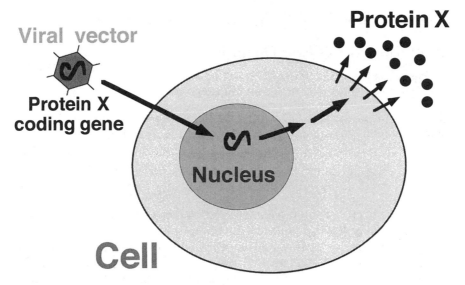

FIGURE 4.2. The idea of gene therapy is based on the fact that all proteins are synthesized using their genes. Through gene transfer, genetically modified cells can continuously manufacture desired proteins (such as growth factors) endogenously.

continuously manufacture the desired protein themselves. All proteins are synthesized in cells using their coding genes—the idea of gene therapy, therefore, is to deliver this encoding gene into the cell, with the result that the genetically modified cell continuously manufactures the desired protein endogenously (Figure 4.2). This is typically done either by introducing vectors containing the appropriate gene directly into the body or by removing target cells from the body, genetically altering them in vitro and then reimplanting them into the body. The former strategy is known as direct, or in vivo gene therapy, and the latter, as indirect, or ex vivo, gene therapy. The relative merits of these strategies depend upon the anatomy and physiology of the target organs, the pathophysiology of the disease, the vector of choice, safety considerations, and other variables (Evans and Robbins 1995). A further consideration is whether to attempt to deliver genes locally to the site of the disease (i.e., local gene therapy) or to attempt gene transfer distant from the disease site and use the circulatory system to deliver genetically modified cells and/or gene products (i.e., systemic gene therapy).

Although the target diseases for treatment with gene therapy originally were heritable, classic genetic disorders, recent advances have led to the potential use of gene therapy for acquired diseases—including disorders of the musculoskeletal system. This chapter focuses on potential applications of gene therapy to the treatment of spinal disorders, particularly those disorders associated with disc degeneration and spinal fusion.

Previous Studies of Gene Therapy in the Spine

Applications of Gene Therapy to the Intervertebral Disc

Adenovirus-Mediated Transfer of a Marker Gene to the
Intervertebral Disc

Our group has reported adenovirus mediated transfer of the LacZ marker
gene to rabbit intervertebral disc cells both in vitro and in vivo (Nishida et
al. 1998). The LacZ marker gene encodes for production of β–galactosidase,
detectable by X–Gal staining, and, thus, can provide evidence of both gene
transfer and gene expression. For the in vitro study, cell cultures were
established from the nucleus pulposus tissue of New Zealand white rabbits
and were infected with an adenovirus construct encoding the LacZ gene
(Ad–LacZ). For the in vivo study, the anterior aspects of lumbar intervertebral discs were surgically exposed, and Ad–LacZ in saline solution was
injected directly into the nucleus pulposus of selected discs. An equal
volume of saline only solution was injected into control discs. Expression
of the transferred gene was detected using X–Gal staining.

The results of the in vitro experiments demonstrated that nucleus pulposus cells were efficiently transduced by an adenoviral vector carrying
the LacZ gene. In vivo injection of Ad–LacZ into the nucleus pulposus
similarly resulted in the transduction of numerous cells. Marker gene
expression persisted in vivo at an apparently undiminished level for at least
12 weeks (Figure 4.3). No X–Gal staining was noted in control discs.

This demonstration of successful transfer of an exogenous marker gene
to the disc and sustained, long-term expression in an adult, immune-
competent animal model suggested that the adenoviral vector might be
suitable for delivery of *therapeutic* genes to the disc for the treatment of
spinal disorders, leading to the following study.

Adenovirus-Mediated Transfer of a Therapeutic Gene to the
Intervertebral Disc

In our next investigation, we performed an in vivo study, again using the
rabbit model, to determine the feasibility of adenovirus mediated transfer
of a therapeutic gene to the intervertebral disc (Nishida et al. 1999). We
used an adenovirus construct (Ad/CMV-hTGF–β1) containing the human
transforming growth factor–beta 1 (TGF–β1) encoding gene. TGF–β1 was
selected because it has a wide range of potentially therapeutic effects and
was previously found by Thompson, Oegema, Jr., and Bradford (1991) to
increase proteoglycan synthesis in cultured canine disc tissues. In our study,
the anterior aspects of lumbar intervertebral discs of 10 New Zealand white
rabbits were surgically exposed, and we directly injected 15 μl of saline with
adenovirus containing the therapeutic human TGF–β1 cDNA into the
nucleus pulposus of selected discs. The supraadjacent disc served as an

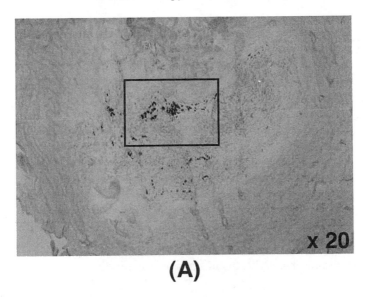

(A)

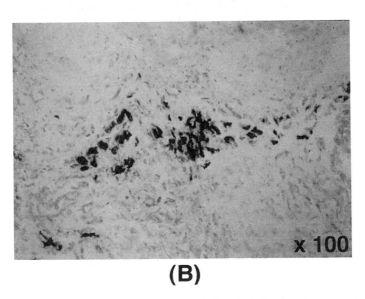

(B)

FIGURE 4.3. Representative section of an adult rabbit lumbar intervertebral disc stained with X–gal 12 weeks after injection of Ad–LacZ in vivo (A) original magnification: ×20, (B) original magnification: ×100. The dark stains indicate presence of β–galactosidase (the product of the LacZ gene), thus providing evidence of both successful gene transfer and gene expression in this in vivo animal model.

intact control for each rabbit. The rabbits were sacrificed 1 week later. Expression of the transferred gene was determined using enzyme-linked immunosorbent assay (ELISA), and proteoglycan synthesis was assessed by measurement of sulfate incorporation.

In vivo injection of Ad/CMV-hTGF–β1 into the nucleus pulposus was found to result in an approximately 5–fold increase in total (i.e., active and latent) TGF–β1 production over that of the intact control discs ($p < 0.05$). The discs of the therapeutic gene group exhibited a statistically significant 2–fold increase in proteoglycan synthesis compared to the intact control discs ($p < 0.05$) (Figure 4.4).

This study demonstrated the efficacy of adenovirus mediated transfer of a therapeutic gene to the intervertebral disc in vivo. The observation of a significant increase in proteoglycan synthesis secondary to gene transfer strongly suggests that gene therapy may have potential applications in altering the course of degenerative disc disease—a spinal disorder characterized in part by loss of proteoglycans in the nucleus pulposus.

Retrovirus-Mediated Gene Transfer to the Cartilaginous Endplate

Wehling and colleagues reported retrovirus mediated transfer of 2 exogenous genes to cultured chondrocytic cells from bovine intervertebral endplates (Wehling, Schulitz, Robbins, Evans, and Reinecke 1997). The exogenous genes were: (1) bacterial β–galactosidase (LacZ), and (2) cDNA of the human interleukin–1 receptor antagonist IL–1ra. β–galactosidase activity was determined by X–Gal staining, and IL–1ra was quantified by ELISA. Transfer of the LacZ marker gene resulted in approximately 1% β–galactosidase positive cells, while transfer of the IL–1ra encoding gene resulted in the production of $24\,ng/ml/10^6$ cells IL–1ra in 48 hours. Based on these results, the authors suggested that removal of endplate tissue from a degenerating disc (under arthroscopic or x–ray control) followed by transferal of therapeutic genes to the cells of the harvested tissue and reinjection of these cells into the disc (thereby turning the host tissues into sites for the synthesis of drugs) could open a new avenue for the treatment of degenerative diseases of the spine.

Applications of Gene Therapy to Spinal Fusion

Lumbar Spinal Fusion by Local Gene Therapy with a cDNA Encoding an Osteoinductive Protein (LMP–1)

Boden et al. (1998) have investigated spine fusion assisted by local ex vivo gene therapy using a cDNA encoding a novel intracellular osteoinductive protein (LMP–1). In their report, Boden and colleagues attempted single level posterior lumbar and thoracic arthrodesis in 14 athymic nude rats

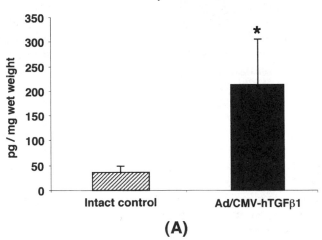

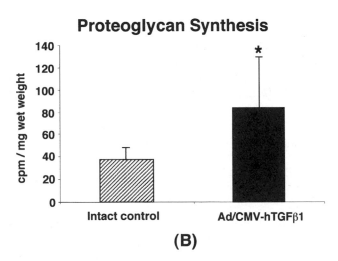

FIGURE 4.4. (A) Elevated TGF–β1 production and (B) elevated proteoglycan synthesis in rabbit nucleus pulposus tissue 1 week after in vivo injection of hTGF–β1 adenovirus construct into lumbar intervertebral discs, compared to intact control discs (n = 10 rabbits).
*Asterisk indicates a value significantly greater than that of intact control discs (p < 0.05)

as follows. After surgical exposure of the dorsal spine, graft material consisting of a devitalized bone matrix (no osteoinductive activity) was soaked with 0.75 to 1.5×10^6 bone marrow cells that had been transfected with the cDNA encoding the LMP–1 sequence. At control sites, marrow cells were transfected with the reverse copy of the cDNA that did not express any protein. Transfection of marrow cells for 2 hours ex vivo was accomplished using the mammalian expression vector pCMV2 and superfect transfection agent. The rats were euthanized after 4 weeks, and their spines were evaluated by manual palpation, radiographs, and noncalcified histology.

Successful spine fusion was obtained in 100% of the sites that received marrow cells transfected with active LMP–1 cDNA, and in 0% of the sites that received marrow cells transfected with reverse (inactive) LMP–1 cDNA ($p < 0.05$) (Figure 4.5). The new bone was normal membranous bone, with active trabeculae lined with osteoblasts. The devitalized bone matrix carrier was completely resorbed, and there was no evidence of any inflammatory response or ectopic bone formation outside the carrier. In specimens from sites that received the inactive form of LMP–1, no bone formation was seen, and much of the devitalized bone matrix carrier remained.

Boden et al. (1998) hypothesized that the LMP–1 protein activates a cascade of other growth factors (e.g., bone morphogenetic proteins (BMPs)) and results in sufficient signal amplification such that only a small percentage of cells needs to be transfected with the gene. These preliminary results suggest that local transient gene therapy with LMP–1 might provide a more promising method of osteoinduction than implantation of pharmacologic doses of recombinant or extracted osteoinductive proteins, and, hence, may be an alternative for bone regeneration and spinal fusion.

Bone Morphogenetic Protein Gene Therapy for the Induction of Spinal Arthrodesis

Alden, Hankins, Beres, Kallmes, and Helm (1998) have investigated spinal fusion assisted by local direct gene therapy using the bone morphogenetic protein 2 (BMP–2) encoding gene. In their report, 12 athymic nude rats were used. Recombinant replication defective type 5 adenovirus, with a universal promoter and BMP–2 gene (Ad–BMP–2), was used. A second adenovirus, constructed with a universal promoter and β–galactosidase (β–gal) gene (Ad–β–gal), was used as a control. Seven and one half microliters of virus was injected percutaneously and paraspinally at the lumbosacral junction in 3 groups (4 animals each): 1) Ad–BMP–2 bilaterally, 2) Ad–BMP–2 on the right, Ad–β–gal on the left, and 3) Ad–β–gal bilaterally. Computerized tomography (CT) scans of the lumbosacral spine were obtained at 3, 5, and 12 weeks. At 12 weeks, the animals were killed for histological inspection. Ectopic bone formation was seen both on 3–dimensional CT reconstruction and histologically in all the athymic nude

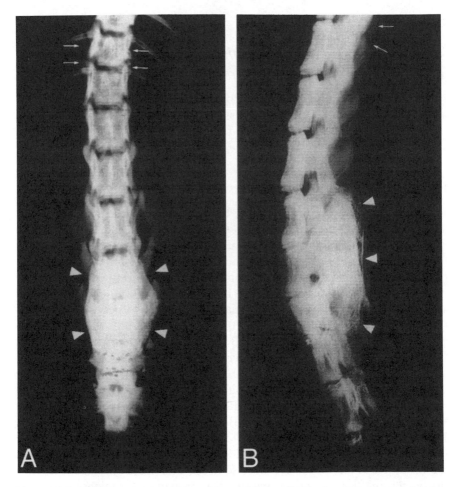

FIGURE 4.5. Anteroposterior (A) and lateral (B) radiographs of the thoracolumbar spine of a rat 4 weeks after implantation of devitalized bone matrix soaked with bone marrow cells that were transfected with the cDNA for LMP–1 in the correct orientation (lumbar spine, arrowheads) or in the reverse/inactive orientation (thoracic spine, arrows). Solid spine fusion was evident when the marrow cells expressed the LMP–1 protein (arrowheads), and no evidence of bone formation was seen where the marrow cells contained an inactive form of the LMP–1 cDNA (arrows).

rats at sites treated with Ad–BMP–2. Histological analysis revealed bone at different stages of maturity adjacent to the spinous processes, laminae, and transverse processes. This study clearly demonstrated that it is possible to produce in vivo endochondral bone formation by using direct adenoviral construct injection into the paraspinal musculature, which suggests that, in the future, gene therapy may be useful for achieving spinal fusion.

Discussion

Why Use Gene Therapy for the Treatment of Spinal Disorders?

In addressing this question, we will consider, first, the limitations of direct application of growth factors (necessitating transfer of their genes), and, second, potential clinical indications for gene therapy in the treatment of spinal disorders.

Direct Application of Growth Factors Versus Gene Therapy

Recent advancements in molecular biology such as recombinant DNA technology and the cloning of genes have enabled scientists to obtain sufficient amounts of pure gene products (i.e., RNA and proteins) to allow for clinical use. For example, granulocyte colony stimulating factor (G–CSF) is commercially available and has been found to be dramatically effective for treating neutropenia (Dale 1995). In the musculoskeletal system, many studies using recombinant proteins have been performed—for example, Hildebrand et al. (1998) have reported on the effectiveness of platelet derived growth factor–BB (PDGF–BB) in enhancing ligament healing in a rabbit model of medial cruciate ligament (MCL) injury.

In growth factor studies of spinal fusion, Boden and colleagues have demonstrated the usefulness of combining osteoinductive proteins with collagen carrier to accelerate spinal fusion in an animal model (Boden, Schimandle, Hutlon, and Chen 1995; Boden, Schimandle, and Hutlon 1995). In a growth factor study related to the intervertebral disc, Thompson et al. (1991) demonstrated that the addition of human transforming growth factor beta 1 (TGF–β1) to canine disc tissue in culture stimulated in vitro proteoglycan synthesis, suggesting that this growth factor might be used for the treatment of disc degeneration. However, so far, no successful treatment of disc degeneration in an animal model using recombinant proteins has been reported. This might be due to difficulties in delivering these proteins, especially for chronic conditions. For acute injuries and acute disease types, direct application of growth factor proteins or use of appropriate carriers combined with proteins seems to be effective and would appear to have certain advantages over gene therapy (such as ease in adjusting the dose and overall simplicity of the procedures). However, for chronic types of disease, such as disc degeneration, there is currently no practical method for producing sustained delivery of exogenous growth factors to the disc. Clearly, the relatively short half-life of the proteins and problems associated with diffusion in the tissue appear to limit the usefulness of direct delivery of growth factors for application to chronic diseases. Novel approaches such as gene transfer, therefore, should continue to be investigated as alternatives to direct delivery of growth factors.

Indications for Gene Therapy in Spinal Disorders

What kind of spinal disorders are indicated for gene therapy? A guiding principle is that one must exert caution when genetically modifying the cell. With a view toward reducing the potential risks to patients, gene transfer to the musculoskeletal system should perhaps be limited initially to conditions where patient suffering is great and there are no other safe or practical possibilities for treatment. In light of the potentially severe side effects of systemic gene therapy, local gene therapy seems a preferable first step, and, therefore, potentially applicable to the treatment of acquired, localized, chronic types of spinal disorders.

Intervertebral disc degeneration appears to be a particularly appropriate candidate for gene therapy. Many degenerative spinal disorders are strongly associated with intervertebral disc degeneration (e.g., spondylosis, development of osteophytes, and disc herniation) and are leading sources of morbidity resulting in substantial pain and increased healthcare costs (Borenstein 1992; Waddell 1996). Although disc degeneration can occur naturally with aging (Buckwalter 1995; Kraemer 1995), clinical observations (supported by basic science studies) have demonstrated early occurrence of disc degeneration secondary to annular disruptions (Lipson and Muir 1981; Osti, Vernon-Roberts, Moore, and Fraser 1992) as well as in discs adjacent to a spinal fusion (Lee 1988). Some evidence indicates that disc degeneration can originate in the nucleus pulposus (Hirsch and Schajowicz 1953; Silberberg, Aufdermaur, and Adler 1979; Adler, Schoenbaum, and Silberberg 1983), with progressive decrease in proteoglycan content leading to dehydration of the nucleus pulposus (Pearce, Grimmer, and Adams 1987; Buckwalter 1995). Because the swelling pressure resulting from a high concentration of proteoglycans in the nucleus pulposus normally helps to maintain disc height and contributes to the disc's load bearing ability, the loss of proteoglycans may directly affect the biomechanical functioning of the intervertebral disc as well as alter loading of the facet joints and other structures, leading to degenerative changes (Butler, Trafimow, Andersson, McNeill, and Huckman 1990). One potential strategy for treatment or prevention of disc degeneration might, therefore, involve genetically modifying the disc such that proteoglycan content is increased or maintained within the nucleus pulposus (Figure 4.6).

Technical Issues Relevant to Clinical Application of Gene Therapy for the Treatment of Spinal Disorders

Successful gene therapy requires judicious selection of the gene of interest and a method for delivery such that sufficient amounts and duration of gene expression may be effected within the target tissues. In the following sections, we address technical issues related to choice of vectors and gene delivery strategy, and consideration of the target tissues.

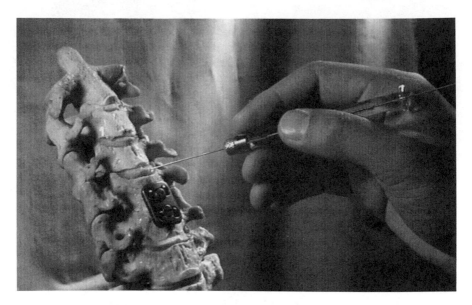

FIGURE 4.6. A potential clinical application of adenovirus mediated gene transfer, involving injection of therapeutic genes into the disc adjacent to a spinal fusion at the time of surgery with the intent of modulating the biological activity within the disc to enhance the ability of the disc to withstand elevated or abnormal mechanical loads.

Choice of Vectors and Gene Delivery Strategy (Ex Vivo vs. In Vivo Gene Transfer)

With a few exceptions (e.g., the LMP–1 gene used by Boden et al. 1998), naked DNA (i.e., DNA not associated with other molecules) is usually not well taken up and expressed by cells. This requires that *vectors* be used to facilitate the cellular uptake of genetic material in such a way that its genetic information can be expressed.

Viruses are very efficient vectors because entry into cells and high expression of virally encoded genes are parts of the normal viral life cycle. While viral vectors are typically used to deliver genes to target cells, use of non-viral vectors has also been explored. Although they are usually easier to produce than their viral counterparts and are chemically more stable, their ability to transfect cells is relatively limited, and at present, all nonviral vector systems deliver DNA episomally, thus limiting the possibility of prolonged gene expression.

The choice of vector can be a major factor in the potential success of a gene-therapy-based treatment. For direct, in vivo gene transfer to succeed, a vector that can very effectively deliver genes is required. Furthermore, the vector must diffuse appropriately within the target organ. Adenovirus has previously demonstrated excellent ability to transfer foreign genes to quiescent, nondividing, highly differentiated cells (Akli et al. 1993; Anonymous 1993; Vincent et al. 1993)—this is a prime reason why adenovirus is mainly used for direct, in vivo gene transfer.

In most gene transfer experiments, a decline in transgene expression over time is observed, and, to date, this is one of the biggest limitations to effective gene therapy. Long-term transgene expression is considered very difficult for adenovirus mediated direct gene transfer to cells within tissues where the immune system is active. Some evidence suggests that the duration of gene expression following adenoviral transfer is limited due to immune reactions to the viral proteins or to the action of foreign proteins encoded by the transgenes (Yang et al. 1994; Tripathy, Black, Goldwasser, and Leiden 1996). For example, adenoviral vectors have been shown to cause inflammation when injected into the joint space (Nita et al. 1996; Sawchuk et al. 1996) and a variety of organs (McCoy, Davidson, Roessler, Huffnagle, and Simon 1995). Furthermore, the exogenous gene delivered by adenovirus is not integrated into the chromosome, but rather is maintained in an episomal state—meaning that gene expression will decrease with cell division.

Retrovirus, widely used in clinical trials in humans, is mainly used for indirect, ex vivo gene transfer. One advantage of retrovirus is that the exogenous gene delivered by retrovirus is integrated into the chromosome of the host cell; this transgene will thus be transmitted to daughter cells. Theoretically, then, long-term transgene expression can be expected with the retrovirus. For any ex vivo technique to be successful, however, the reimplanted, genetically modified cells must survive for an appropriate length of time as well as spread throughout the target organ. Another problem in retrovirus–mediated gene transfer is that retroviruses can only transfer their genes to cells that are actively replicating at the time of infection (Miller, D.G., Adam, and Miller, A.D. 1990). Accordingly, retrovirus mediated transfer of exogenous genes to differentiated, nondividing cells is inadvisable. Other types of viruses, including adeno–associated virus and herpes simplex virus, are also under investigation.

Given these advantages and limitations associated with the different available vectors and gene delivery strategies, we turn next to a consideration of target tissues, examining the characteristics of the spinal tissues that could influence the success of gene therapy.

Consideration of Target Tissues

A consideration of the characteristics of the target tissues is one of the key points for successful gene therapy, and is a prerequisite for informed selection of vector and gene delivery strategy. For example, the intervertebral disc is anatomically and physiologically unique, being largely avascular and populated by poorly characterized cells in an extensive extracellular matrix.

Passive diffusion mechanisms result in poor nutrition as well as relatively low oxygen tension at the center of the disc (Holm, Maroudas, Urban, Selstam, and Nachemson 1981). Since the metabolism is predominantly anaerobic and significant amounts of lactate are produced at the center of the disc, the pH in the disc is low (Diamant, Karlsson, and Nachemson 1968; Nachemson 1969). The disc's low oxygen tension, poor nutrition, and low pH create a biologically severe environment, especially within the nucleus pulposus.

Based on the unique anatomical and physiological characteristics of the intervertebral disc, it would seem, intuitively, that in vivo gene transfer methods offer greater advantages than ex vivo methods. Ex vivo methods would require that disc cells be harvested, cultured, and genetically modified before reimplantation into the disc. Cultured cells (especially those that have been multiply subcultured in an environment quite different from in vivo environments) often lose part of their original character and, thus, may not survive when reimplanted back into the relatively harsh environment of the nucleus pulposus. Furthermore, retroviruses only infect actively dividing cells—nucleus pulposus cells are highly differentiated and are poorly mitotic. For these reasons, retroviral ex vivo gene therapy would be considered impractical for transferring genes directly to the cells of the nucleus pulposus. Adenovirus, on the other hand, would appear to offer several advantages for the in situ infection of nucleus pulposus cells. First, adenovirus has good ability to transfer foreign genes to quiescent, nondividing, highly differentiated cells. Second, the intervertebral disc, being relatively encapsulated and avascular, would seem to have an ideal environment for maintaining a high concentration of directly injected viral vectors, thus increasing the likelihood of transduction. Third, the relatively encapsulated and avascular environment of the nucleus pulposus may limit the access of immunocompetent cells, thereby preventing immune reactivity and permitting prolonged gene expression.

Based on these considerations, the intervertebral disc appears to be a uniquely appropriate site for adenovirus mediated transfer of exogenous genes and production of therapeutic growth factor proteins, offering a system in which the cells of the nucleus pulposus are efficiently transduced by an adenovirus, where there is little possibility of immune reaction due to the avascularity of the disc, and long term gene expression is possible.

Future Directions

For gene therapy to work in a given clinical application, the genes must be expressed at the appropriate levels and at the required times, which often requires regulated expression of the gene. Although gene expression is controlled by many different mechanisms, most attention has focused on specific gene promoters that regulate RNA synthesis—and hence regulate production of growth factors. In vivo, the production of certain growth factors is very tightly regulated. Accordingly, inducing *excessive* growth factor production via gene therapy could be harmful. In general, more basic science data are needed to understand the role of growth factors in vivo, and much of the research effort in the development of gene therapies for the treatment of spinal disorders needs to be focused not only on gene delivery systems, but also on issues related to the expression and regulation of genes.

The immune response to adenovirus vectors may be avoided if vectors are introduced into neonatal animals (which have immature immune systems) or into immuno compromised animals. However, this kind of strategy is unlikely to be of clinical relevance, and, therefore, other solutions need to be developed. Accordingly, there is also a great need for the development of vectors that can effectively transfer their DNA while minimizing side effects, such as immunoreaction or cell toxicity, and that can integrate their gene safely at the appropriate site of the genome of the host cell. Interestingly, the intervertebral disc seems to provide a protected environment for gene therapy where risk of immunoreaction is diminished.

Clearly there are numerous obstacles to overcome before gene therapy can be considered for clinical use in humans for treatment of spinal disorders (not the least of these is a lack of basic science understanding of the effect of growth factor proteins in the biological processes and mechanical functioning of the spine). Despite these obstacles, it is already apparent that gene therapy has the potential of becoming a valuable clinical treatment mode for the spine in the 21st century.

Acknowledgments. The authors gratefully acknowledge Dr. Savio L.-Y. Woo, Dr. Freddie H. Fu, and the Musculoskeletal Research Center for generous guidance and support of this research. Dr. Johnny Huard kindly assisted with the histology. Funding was provided in part by a grant from the Albert B. Ferguson, Jr., M.D., Orthopaedic Research Foundation.

References

Adler, J.H., Schoenbaum, M., and Silberberg, R. 1983. Early onset of disk degeneration and spondylosis in sand rats (Psammomys obesus). *Veterinary Pathology* 20(1):13–22.

Akli, S., Caillaud, C., Vigne, E., Stratford-Perricaudet, L.D., Poenaru, L., Perricaudet, M., Kahn, A., and Peschanski, M.R. 1993. Transfer of a foreign gene into the brain using adenovirus vectors. *Nat Genet* 3(3):224–8.

Alden, T.D., Hankins, G.R., Beres, E.J., Kallmes, D.F., and Helm, G.A. 1998. Bone morphogenetic protein gene therapy for the induction of spinal arthrodesis. *Neurosurg Focus* 4(2):1–4.

Anonymous. 1993. Adventures with adenovirus. *Nat Genet* 3(1):1–2.

Boden, S.D., Schimandle, J.H., and Hutton, W.C. 1995. 1995 Volvo award winner in basic science studies. The use of an osteoinductive growth factor for lumbar spinal fusion. Part II: Study of dose, carrier, and species. *Spine* 20(24):2633–44.

Boden, S.D., Schimandle, J.H., Hutton, W.C., and Chen, M.I. 1995. 1995 Volvo award winner in basic science studies. The use of an osteoinductive growth factor for lumbar spinal fusion. Part I: Biology of spinal fusion. *Spine* 20(24):2626–32.

Boden, S.D., Titus, L., Hair, G., Liu, Y., Viggeswarapu, M., Nanes, M.S., and Baranowski, C. 1998. 1998 Volvo award winner in basic science studies. Lumbar spine fusion by local gene therapy with a cDNA encoding a novel osteoinductive protein (LMP–1). *Spine* 23(23):2486–92.

Borenstein, D. 1992. Epidemiology, etiology, diagnostic evaluation, and treatment of low back pain. *Curr Opin Rheumatol* 4(2):226–32.

Buckwalter, J.A. 1995. Aging and degeneration of the human intervertebral disc. *Spine* 20(11):1307–14.

Butler, D., Trafimow, J.H., Andersson, G.B., McNeill, T.W., and Huckman, M.S. 1990. Discs degenerate before facets. *Spine* 15(2):111–13.

Dale, D.C. 1995. Where now for colony-stimulating factors? *Lancet* 346(8968):135–6.

Diamant, B., Karlsson, J., and Nachemson, A. 1968. Correlation between lactate levels and pH in discs of patients with lumbar rhizopathies. *Experientia* 24(12):1195–6.

Evans, C.H. and Robbins, P.D. 1995. Possible orthopaedic applications of gene Therapy. *J Bone Joint Surg* 77-A:1103–14.

Hildebrand, K.A., Woo, S.L.Y., Smith, D.W., Allen, C.R., Deie, M., Taylor, B.J., and Schmidt, C.C. 1998. The effects of platelet-derived growth factor-BB on healing of the rabbit medial collateral ligament—an in vivo study. *Am J Sports Med* 26(4):549–54.

Hirsch, C. and Schajowicz, F. 1953. Studies on structural changes in the lumbar annulus fibrosus. *Acta Orthop Scand* 22:184–231.

Holm, S., Maroudas, A., Urban, J.P., Selstam, G., and Nachemson, A. 1981. Nutrition of the intervertebral disc: solute transport and metabolism. *Connect Tissue Res* 8(2):101–19.

Kostuik, J.P. 1997. Intervertebral disc replacement. Experimental study. *Clin Orthop Rel Res* 337:27–41.

Kraemer, J. 1995. Natural course and prognosis of intervertebral disc diseases. International Society for the Study of the Lumbar Spine. Seattle, Washington: June 1994. *Spine* 20(6):635–9.

Lee, C.K. 1988. Accelerated degeneration of the segment adjacent to a lumbar fusion. *Spine* 13(3):375–7.

Lemaire, J.P., Skalli, W., Lavaste, F., Templier, A., Mendes, F., Diop, A., Sauty, V., and Laloux, E. 1997. Intervertebral disc prosthesis. Results and prospects for the year 2000. *Clin Orthop Rel Res* 337:64–76.

Lipson, S.J. and Muir, H. 1981. 1980 Volvo award in basic science. Proteoglycans in experimental intervertebral disc degeneration. *Spine* 6(3):194–210.

McCoy, R.D., Davidson, B.L., Roessler, B.J., Huffnagle, G.B., and Simon, R.H. 1995. Expression of human interleukin-1 receptor antagonist in mouse lungs using a recombinant adenovirus: effects on vector-induced inflammation. *Gene Ther* 2(7):437–42.

Miller, D.G., Adam, M.A., and Miller, A.D. 1990. Gene transfer by retrovirus vectors occurs only in cells that are actively replicating at the time of infection [published erratum appears in 1992 *Mol Cell Biol* 12(1):433]. *Mol Cell Biol* 10(8):4239–42.

Nachemson, A. 1969. Intradiscal measurements of pH in patients with lumbar rhizopathies. *Acta Orthop Scand* 40(1):23–42.

Nishida, K., Kang, J.D., Suh, J.K., Robbins, P.D., Evans, C.H., and Gilbertson, L.G. 1998. Adenovirus mediated gene transfer to nucleus pulposus cells: implications for the treatment of intervertebral disc degeneration. *Spine* 23(22):2437–43.

Nishida, K., Kang, J.D., Gilbertson, L.G., Moon, S.H., Suh, J.K., Vogt, M.T., Robbins, P.D., and Evans, C.H. 1999. Volvo award winner in basic science studies. Modulation of the biological activity of the rabbit intervertebral disc by gene therapy: an in vivo study of adenovirus mediated transfer of the human TGF–β1 encoding gene. *Spine* 24(23):2419–25.

Nita, I., Ghivizzani, S.C., Galea-Lauri, J., Bandara, G., Georgescu, H.I., Robbins, P.D., and Evans, C.H. 1996. Direct gene delivery to synovium. An evaluation of potential vectors in vitro and in vivo. *Arthritis Rheum* 39(5):820–8.

Osti, O.L., Vernon-Roberts, B., Moore, R., and Fraser, R.D. 1992. Annular tears and disc degeneration in the lumbar spine. A post–mortem study of 135 discs. *J Bone Joint Surg Br* 74(5):678–82.

Papp, T., Porter, R.W., Aspden, R.M., and Shepperd, J.A. 1997. An in vitro study of the biomechanical effects of flexible stabilization on the lumbar spine. *Spine* 22(2):151–5.

Pearce, R.H., Grimmer, B.J., and Adams, M.E. 1987. Degeneration and the chemical composition of the human lumbar intervertebral disc. *J Orthop Res* 5(2):198–205.

Sawchuk, S.J., Boivin, G.P., Duwel, L.E., Ball, W., Bove, K., Trapnell, B., and Hirsch, R. 1996. Anti-T cell receptor monoclonal antibody prolongs transgene expression following adenovirus mediated in vivo gene transfer to mouse synovium. *Hum Gene Ther* 7(4):499–506.

Silberberg, R., Aufdermaur, M., and Adler, J.H. 1979. Degeneration of the intervertebral disks and spondylosis in aging sand rats. *Arch Pathol Lab Med* 103(5):231–5.

Thompson, J.P., Oegema, Jr., T.R., and Bradford, D.S. 1991. Stimulation of mature canine intervertebral disc by growth factors. *Spine* 16(3):253–60.

Tripathy, S.K., Black, H.B., Goldwasser, E., and Leiden, J.M. 1996. Immune responses to transgene encoded proteins limit the stability of gene expression after injection of replication defective adenovirus vectors. *Nat Medicine* 2(5):545–50.

Vincent, N., Ragot, T., Gilgenkrantz, H., Couton, D., Chafey, P., Gregoire, A., Briand, P., Kaplan, J.C., Kahn, A., and Perricaudet, M. 1993. Long term correction of mouse dystrophic degeneration by adenovirus mediated transfer of a minidystrophin gene. *Nat Genet* 5(2):130–4.

Waddell, G. 1996. Low back pain: a twentieth century health care enigma. *Spine* 21(24):2820–5.

Wehling, P., Schulitz, K.P., Robbins, P.D., Evans, C.H., and Reinecke, J.A. 1997. Transfer of genes to chondrocytic cells of the lumbar spine. *Spine* 22(10):1092–7.

Yang, Y., Nunes, F.A., Berencsi, K., Furth, E.E., Gonczol, E., and Wilson, J.M. 1994. Cellular immunity to viral antigens limits E1 deleted adenoviruses for gene therapy. *Proc Nat Acad Sci USA* 91(10):4407–11.

5
The Development of Approaches Based on Gene Therapy to Improve Muscle Healing Following Injury

JACQUES MÉNÉTREY, CHANNARONG KASEMKIJWATTANA,
CHARLES S. DAY, PATRICK BOSCH, MOREY S. MORELAND,
FREDDIE H. FU, and JOHNNY HUARD

Introduction

Muscle injuries are common, with an incidence varying from 10% to 55% of all injuries sustained in sports (Lehto and Jarvinen 1991). Muscle injuries are divided into 2 types: a shearing injury, in which both the myofibers and the connective tissue framework are torn, or an in situ injury, in which only the myofibers are damaged and the basal lamina and connective tissue sheaths do not undergo significant harm. Shearing injuries, the most frequent muscle injuries related to sports, may be lacerations, contusions, or strains, depending on the mechanism of injury (Lehto and Jarvinen 1991). Contusion is sustained through a significant compressive force to the muscle, such as a direct blow, a common occurrence in contact sports. A strain occurs when a forceful eccentric contraction is applied to an overstretched muscle, especially in jumping or sprinting (Garrett, Jr. 1990; Lehto and Jarvinen 1991). Injury is common near the musculotendinous junction (MTJ) of a superficial muscle that crosses 2 joints, such as the rectus femoris, semitendinosus, and gastrocnemius muscles. Though rather rare in sports, muscle laceration is a dramatic injury that consistently incapacitates athletes for long periods of time and often jeopardizes their professional careers.

Orthopaedic surgeons face two other challenging muscle conditions: the compartment syndrome and limb lengthening. The compartment syndrome is characterized by an increased pressure within a closed compartment that is bounded by bone and fascia, resulting in damage to the muscle. Such increased pressure diminishes tissue *perfusion*, which may lead to an ischemic injury (Mubarak and Hargens 1981). Common causes of increased compartment pressure include bleeding into a compartment, edema following partial or temporary ischemia, and crushing injuries. Ischemic contracture and nerve damage constitute the major consequences of nontreated compartment syndromes. During the lengthening of a limb, despite high regenerative capabilities, the muscle is incapable of keeping pace with the distraction, resulting in contracture.

Muscle Healing Process

The healing process of an injured muscle is composed of three phases (Kalimo, Rantanen, and Jarvinen 1997). The destruction phase is characterized by hematoma formation, muscle tissue necrosis, degeneration, and an inflammatory cell response. The repair phase includes phagocytosis of the damaged tissue, regeneration of the striated muscle, production of connective scar tissue, and capillary ingrowth. In the final remodeling phase, the regenerated muscle matures and contracts, and scar tissue is reorganized; however, an incomplete restoration of the functional capacity of the muscle often occurs.

The regeneration of myofibers begins with the activation of myogenic precursor cells, or satellite cells, located between the basal lamina and the plasma membrane of each individual myofiber. When activated, these satellite cells begin to proliferate and differentiate into multinucleated myotubes and eventually into myofibers. Many of these myoblasts have the ability to fuse with existing necrosed myofibers and may prevent the muscle fibers from undergoing complete degeneration (Huard, Verreault, Roy, Tremblay, M., and Tremblay, J.P. 1994). In the meantime, fibroblasts invade the gap and begin to produce extracellular matrix in order to restore the connective tissue framework (Hurme, Kalimo, Sandberg, et al. 1991; Lehto, Duance, and Restall 1985). The physiological role of this scaffold is to transmit load across the defect, thereby enabling use of the injured limb prior to completion of the repair process (Kalimo et al. 1997). In extensive muscle trauma, proliferation of fibroblasts can quickly lead to an excessive formation of dense scar tissue that impedes muscle regeneration and results in an incomplete recovery (Jarvinen and Sorvari 1975; Hurme, Kalimo, Lehto, and Jarvinen 1991). This process has been shown to occur in several injuries, including strains, contusions, and muscle lacerations (Carlson and Faulkner 1983; Garrett, Jr., Saeber, Boswick, Urbaniak, and Goldner 1984; Nikolaou, MacDonald, Glisson, Seaber, and Garrett 1987; Garrett 1990; Crisco, Jolk, Heinen, Connell, and Panjabi 1994; Kasemkijwattana et al. 1998; Ménétrey, Kasemkijwattana, Fu, Moreland, and Huard in press).

Clinical Treatment

The treatment of muscle injuries has remained essentially unchanged for decades. Immediate care consists of Rest, Ice, Compression and Elevation (RICE) to prevent hematoma formation and interstitial oedema, and thereby decrease tissue ischemia. Nonsteroidal antiinflammatory medications should be instituted in the early phase because their long term use may be detrimental (Mishra, Friden, Schmitz, and Lieber 1995). Glucocorticoids should not be used, as they delay both muscle regeneration and the elimination of hematoma and necrotic tissue. After 3 or 4 days, physical therapy consisting of stretching, strengthening, and ultrasound should be instituted, although the use of ultrasound has yet to be proven in this setting. The final rehabilitation phase is sport-specific training.

Development of Novel Approaches in the Treatment of Muscle Injury

Autologous Myoblast Transplantation

Autologous myoblast transplantation (AMT), the implantation of myoblast precursors (satellite cells), has been extensively studied to promote muscle regeneration and to create a reservoir of normal myoblasts cabable of fusing and delivering genes to skeletal muscle. The potential use of myoblast transplantation has been investigated for the management of Duchenne muscular dystrophy (DMD) (Partridge 1991). Interestingly, this approach can deliver and restore structural protein, such as dystrophin, in DMD muscle. In this disease, AMT allows for an increase of strength.

In an attempt to improve muscle healing following injury, a muscle biopsy can be performed from noninjured muscle on the same individual, thereby establishing an autologous donor. The myoblasts are isolated by enzymatic digestion from the biopsy material, cultured in an enriched milieu, and eventually injected in the damaged muscle. The major hurdle of myoblast transplantation remains the well documented immune rejection (Huard, Guerette, et al. 1994), which may be circumvented by using AMT (see Chapter 13). In addition, studies on DMD have revealed that this approach allows for the delivery of genes, the improvement of muscle regeneration, and the enhancement of strength in dystrophic muscle. Therefore, AMT may promote muscle regeneration and improve muscle healing following a severe injury.

In a previous study, we have demonstrated an enhancement of muscle regeneration after AMT in a muscle injured with myonecrotic agents (Huard et al. 1994a). We have observed that more than 90% of the transplanted muscle was populated with myofibers formed by the fusion of the injected myoblasts (Huard, Verreault, et al. 1994). These myoblasts participated in muscle regeneration by fusing with existing necrosed myofibers and thereby preventing the muscle fibers from undergoing complete degeneration (Huard, Verreault, et al. 1994). Furthermore, the ability of these myoblasts to secrete trophic substances, which are primordial for the regeneration process to occur, might stimulate all repair processes and improve muscle healing.

Growth Factors and Muscle In Vitro

Recently, new substances, such as growth factors, have been found to play a determinant role in muscle regeneration and healing (Grounds 1991;

Lefaucheur and Sebille 1995). Growth factors are small peptides that bind to membrane receptors to influence the various stages in the growth and development of cells via several signaling pathways (Grounds 1991; Chambers and McDermott 1996). Growth factors have already been shown to be capable of stimulating the growth and protein secretion of many musculoskeletal cells (Trippel et al. 1996).

During muscle regeneration, trophic substances released by the injured muscle are presumed to activate the satellite cells (Schultz, Jaryszak, and Valliere 1985; Allamedine, Dehaupas, and Fardeau 1989; Schultz 1989; Hurme and Kalimo 1992; Bischoff 1994). During growth and development, many growth factors have been shown to be capable of eliciting variable responses from the skeletal muscle (Florini and Magri 1989; Grounds 1991; Chambers and McDermott 1996). Some preliminary data have suggested that individual growth factors play a specific role during muscle regeneration (Jennische and Hansson 1987; Jennische 1989; Anderson, Liu, and Kardami 1991; Grounds 1991; Lefaucheur and Sebille 1995; Chambers and McDermott 1996). In vitro studies have demonstrated that basic fibroblast growth factor (b–FGF) stimulates cell proliferation while inhibiting differentiation in bovine and chick myoblasts in culture (Kardami, Spector, and Strohman 1985; Gospodarowicz et al. 1987). The mechanism for b–FGF stimulated proliferation appears to be the advancing of cells from G0 to G1 in the cell cycle (Florini and Magri 1989; Grounds 1991). More recently, b–FGF has been found capable of stimulating proliferation and repressing differentiation in MM14 mouse myoblasts (Campbell et al. 1995). This inhibition of differentiation is thought to be due to repression of "cell commitment" (the irreversible fate of cells of a particular cell lineage to differentiate). Still in vitro, insulin growth factor type–1 IGF–1 has been found to be capable of highly stimulating myoblast proliferation and differentiation (Ewton and Florini 1980; Allen and Boxhorn 1989; Florini and Magri 1989; Florini, Ewton, and Roof 1991).

In our laboratory we have also investigated the effect of various growth factors on myoblast proliferation and differentiation in vitro. Myoblasts have been cultured with basic and acidic fibroblast growth factor (b–FGF, a–FGF), IGF–1, nerve growth factor (NGF), platelet derived growth factor AA (PDGF–AA), and transforming growth factor α and β (TGF–α, TGF–β) at different concentrations (1, 10, 100 ng/ml). Myoblast proliferation and differentiation have been monitored at 48 and 96 hours postincubation. We have observed that IGF–1, b–FGF, and NGF were potent stimulators of both myoblast proliferation and differentiation in vitro (Table 5.1) (Ménétrey et al. 1998). Interestingly, this stimulation was dose dependent (Ménétrey et al. 1998). The other growth factors showed no stimulating effect on myoblast proliferation and differentiation. These results suggest that b–FGF, IGF–1, and NGF enhance myoblast proliferation and differentiation, and are logical choices for delivery in a muscle injury to improve healing.

TABLE 5.1. Growth factor effect on myoblast proliferation and fusion in vitro.

Growth Factor	Proliferation	Fusion
b–FGF	stimulate*	stimulate*
IGF–1	stimulate*	stimulate*
NGF	stimulate*	stimulate*
a–FGF	inhibit	stimulate*
PDGF–AA	inhibit	inhibit
EGF	inhibit	inhibit
TGF–α	inhibit	inhibit
TGF–β	inhibit	inhibit

- * ANOVA, $p < 0.05$.
- b–FGF: basic fibroblast growth factor; IGF–1: insulin like growth factor type 1; NGF: nerve growth factor; a–FGF: acidic fibroblast growth factor; PDGF–AA: platelet derived growth factor AA form; EGF: epidermal growth factor; TGF–α: transforming growth factor α; TGF–β: transforming growth factor β.

Growth Factors and Muscle In Vivo

Despite the experimental elegance of the in vitro system, it is important to recognize that regeneration in vivo is more complex due to the involvement of circulatory and intercellular communication (Grounds 1991; Chambers and McDermott 1996). There has been, however, some preliminary characterization of the role of certain growth factors during muscle regeneration, suggesting that the individual roles of growth factors are similar to their individual effects seen in vitro (DiMario et al. 1989; Lefaucheur and Sebille 1995). Other studies have shown that during the muscle regeneration process following any injury, b–FGF is present in the extracellular space as soon as 8 hours after injury. The level of b–FGF reaches a peak at 24 hours and slowly decreases over a period of 1 week (Anderson et al. 1995). IGF–1 is present after 2 days, reaches its peak at 3 days, and decreases over a period of 1 week (Jennische and Hansson 1987; Jennische 1989).

Based on our in vitro study, we have injected human recombinant proteins (b–FGF, IGF–1, and NGF) in a mouse muscle laceration and monitored the muscle regeneration by regular and quantitative histology 1 week postinjury. This study has shown that b–FGF, IGF–1, and, to a lesser extent, NGF improved muscle regeneration in mouse muscle (Figure 5.1). We have been able to show this by documenting an increase in the number and size of the regenerating myofibers as an index of muscle regeneration. We have also shown in our laceration model that regenerating myofibers were located in the superficial area of the injured site of growth factor treated muscles only, thus demonstrating greater initial muscle healing through

Direct   Ex-vivo

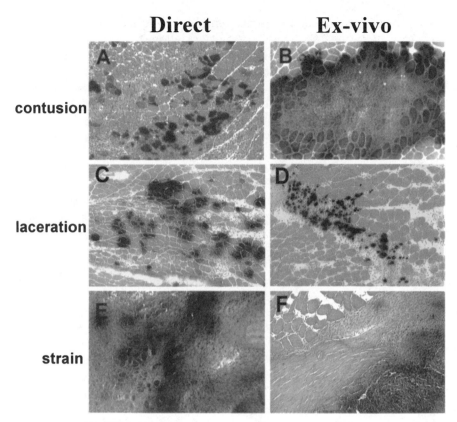

FIGURE 5.1. Hematoxylin-eosin staining of the laceration site at 7 days post-injury. A, C, E: Control with regenerating myofibers in the deep part of the muscle and infiltration of inflammatory cells in the superficial part. B: Laceration injected with NGF, D: b-FGF, F: IGF-1. Regenerating myofibers are located throughout the injured site (deep and superficial area). Magnification ×10.

treatment with specific growth factors. Exogenous growth factor administration is a promising approach to improve muscle healing; however, the successful clinical implementation of this technique is currently limited by the problem of maintaining an adequate concentration of growth factor in the lesion site or target tissue. The short half-life of growth factors and systemic lavage may lead to a rapid clearance of the substances from the desired site. To address these issues, gene therapy may provide an interesting delivery system to the muscle.

Gene Therapy in Muscle

New delivery techniques are required to achieve sustained and efficient local delivery of therapeutic proteins such as growth factors. The new delivery system should aim at circumventing the rapid clearance and non-specificity of the growth factor action, and allow for the release of an efficient concentration of protein. Gene therapy provides a promising approach to meet these requirements. Genetic information encoding for the therapeutic protein is inserted, using nonviral and viral vectors, into living cells. In turn, these genetically modified cells express the protein, encoded by the transferred DNA, in a sustained manner. By using this technology, growth factors can be delivered to the tissue locally on a long-term basis, avoiding the need for repeated injections and systemic administration.

Efficient gene transfer is the first step in addressing a successful delivery system. Extensive work, from our laboratory and others, has been performed to promote muscle growth and enhance muscle regeneration, aimed at alleviating muscle weakness associated with diseases such as Duchenne muscular dystrophy. The different gene transfer systems developed for these muscle diseases may also be used for other orthopaedic applications to muscle.

Recent effort has been directed at the development of vectors for efficient gene delivery to muscle. Actually, plasmid DNA, liposomes, and viral vectors have been used to tranfer genes into skeletal muscle. Direct transfection of naked DNA into muscle cells in vitro and in vivo has been found to be inefficient (Acsadi et al. 1991). However, transgene expression has been found to persist for up to 1 year, suggesting low immunogenicity and cytotoxicity related to the use of nonviral vectors (Wolfe et al. 1992; Katsumi et al. 1994). Recently, liposomes have been successfully used to transduce myoblasts in vitro with a promising efficiency (Wolfe et al. 1992). Although the efficiency of nonviral vectors remains inferior to viral vectors, it could be sufficient in some orthopaedic or sports medicine applications.

Retrovirus, adenovirus, and herpes simplex virus type 1 (HSV–1) have also been investigated as viral gene delivery vehicles to skeletal muscle. These viral vectors were found capable of highly transducing muscle cells; however, they were hindered by several limitations: differential viral transduction throughout muscle maturation, transient transgene expression related to cytotoxicity, and immunological problems against the viral vectors (Dunckley et al. 1993; Acsadi et al. 1994; Huard, Akkaraju, Watkins, Cavalcoli, and Glorioso 1997).

Recently, adeno–associated virus (AAV) has been investigated as another viral gene delivery vehicle to skeletal muscle. Since this viral vector bypasses some of the limitations associated with the currently used gene transfer systems to skeletal muscle, it may become a good vehicle to deliver therapeutic proteins to injured muscle (Xiao, Li, and Samulski 1996; Clark, Sferra, and Johnson 1997; Fisher et al. 1997).

Direct gene therapy, which consists of directly injecting vectors into the muscles, has been extensively used to mediate gene delivery to skeletal muscle. Direct gene therapy approaches based on naked DNA (Acsadi et al. 1991), retrovirus (Dunckley et al. 1993), adenovirus (Acsadi et al. 1994; Huard, Lochmueller, Jani, et al. 1995, Huard, Lochmueller, Acsadi, et al. 1995), HSV-1 (Huard, Goins, and Glorioso 1995; Huard et al. 1996), and AAV (Xiao et al. 1996; Reed–Clark, Sferra, and Johnson 1997) have been characterized to deliver genes to skeletal muscle. With vectors carrying reporter genes, muscle cells have been successfully transduced in vitro and in vivo using replication-defective adenovirus, retrovirus, HSV, and AAV recombinants. Although a poor level of gene transfer has been observed with many of these viral vectors in mature skeletal muscle, some of them, such as adenovirus and AAV, allow for an efficient gene transfer in adult regenerating muscle. Regarding the direct approach, adenovirus and AAV, which are capable of infecting dividing and nondividing cells, appear to be the vectors of choice.

Another gene delivery method is the ex vivo approach, which consists of establishing a primary myoblast cell culture from injured muscle; infecting it with engineered vectors; and then injecting the transduced cells into the same host. This method has already been performed with the use of recombinant adenovirus (Huard, Acsadi, Jani, Massie, and Karpati 1994; Floyd et al. 1998), retrovirus (Salvatori et al. 1993), and HSV (Booth et al. 1997) carrying reporter genes (β–galactosidase or luciferase). Several studies have shown that transduced myoblasts (isogenic myoblasts) fused and introduced reporter genes into the injected muscle (Salvatori et al. 1993; Huard, Acsadi, et al. 1994; Booth et al. 1997). Furthermore, a higher efficiency of gene transfer has been found using the ex vivo approach versus direct injection of the same amount of virus (Booth et al. 1997; Floyd et al. 1997). Since myoblast transplantation and gene therapy have primarily been hindered by immunorejection, ex vivo gene transfer offers the advantage of reducing immunological problems against the injected myoblasts. This statement is not valid for the adenoviral vector, which expresses viral proteins at the transduced cell surface and induces immunological responses (see Chapter 13). Regarding the ex vivo approach, retrovirus and AAV appear to be the vectors of choice. For orthopaedic applications, however, nonviral vectors (plasmid DNA, liposomes) may allow for a sufficient level of gene transfer to achieve an efficient delivery system.

Potential Applications in the Orthopaedic Field

The tissues forming the musculoskeletal system exhibit different healing capacities following an injury and, with the exception of bone, there is no restitutio ad integrum (Lattermann et al. 1998). The healing process in soft tissue always ends in the formation of scar tissue with lower mechanical and functional properties than the original tissue. Some tissues, such as articular cartilage, meniscus, and anterior cruciate ligament, have low healing capacities, while others, such as muscle and bone, have high healing capacities. Despite a high capacity for regeneration, muscle tissue's response to serious injury typically involves dense fibrotic scar tissue intervening between normal muscle. By enhancing muscle growth and regeneration, it may be possible to prevent the formation of dense scar tissue and improve the quality of muscle healing. This may possibly reduce the risk of reinjury and decrease the incidence of muscle pain secondary to scarring. The enhancement of muscle growth and regeneration might also prevent the occurrence of contracture secondary to limb lengthening and/or compartment syndrome.

Gene therapy is not yet an established therapeutic method in the treatment of orthopaedic or sports medicine injuries. We believe, however, that gene therapy has a great potential as a delivery system to musculoskeletal tissue, because it allows for the continuous local delivery of therapeutic proteins, such as growth factors. Recently, the feasibility of gene transfer to musculoskeletal tissues has been investigated and has shown promising results.

In our laboratory, the feasibility of gene transfer to muscle injury has been studied in recent years. After the development in mouse of muscle strain, contusion, and laceration models, direct and ex vivo approaches to gene transfer have been performed on different injured muscles. Either 10 μl of 2.5×10^6 recombinant adenovirus carrying the LacZ reporter gene or 10 μl containing 1×10^6 transduced myoblasts (MOI = 25) have been injected into injured muscles. The muscles were harvested 1 week later, cryosectioned, and assayed for β–galactosidase expression. With the use of both direct and ex vivo gene transfer, we have observed the presence of many β–galactosidase positive myofibers in the strained, contused, and lacerated muscles, which had been injected with recombinant adenovirus or transduced myoblasts (Figure 5.2 A to F). These transduced myofibers were located at the injured site (Figure 5.2 A to F).

In summary, both direct gene transfer and ex vivo gene delivery mediated by adenovirus have been found to be capable of delivering the β–galactosidase reporter gene to the injured muscle at 5 days posttrauma. The development of an approach based on the use of gene therapy to deliver growth factors for a quicker and more complete recovery may revolutionize the significant down time following a muscle injury. Indeed, the enhancement of muscle growth and regeneration, by continuous and local delivery of specific growth factors, may limit the formation of scar tissue, accelerate muscle recovery, and result in complete healing after severe muscle strain, laceration, and contusion. Furthermore, this may possibly reduce the formidable risk of reinjury at the junction between the scar tissue and regenerated muscle, thus preventing recurrences of muscle injury that may jeopardize an athlete's career.

In sports medicine, however, gene therapy must remain solely a therapeutic tool, not a new means by which to enhance performance. The entire scientific community must be concerned with this issue and act responsibly as soon as the technology is broadly available at the preclinical level.

Lacerated Muscle

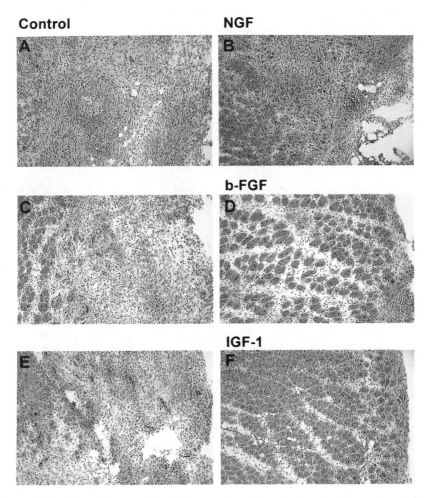

Control   **NGF**

b-FGF

IGF-1

FIGURE 5.2. Adenovirus mediated direct and ex vivo gene transfer in a contused (A, B), lacerated (C, D), and strained muscle (E, F). We observed many transduced myofibers at the injured sites following direct injection of adenoviral vector at 7 days postinjury (A, C, E). We detected also LacZ-positive myofibers at the injured sites following ex vivo gene transfer of the β-galactosidase reporter gene at 5 days postinjection (B, D, F). The presence of these transduced myofibers suggested that our injected myoblasts have fused into the injured area, mediating gene transfer in injured muscle (B, D, F). Magnification: A–F: ×10.

Scientists involved in gene therapy projects related to sports medicine should found an ethics committee, which would regulate the application of the method to the sports medicine field. There is a true danger that the misuse of gene therapy could lead to a complete and unfair manipulation of sports performance, as well as expose athletes' health to hazardous secondary effects.

The development of an approach based on the use of gene therapy to deliver growth factors may also be very helpful in controlling certain muscle conditions, such as the effects of limb lengthening, compartment syndrome and ischemia, and muscular dystrophy.

Muscle's regenerative capabilities are illustrated by the increased release of satellite cells in elongated muscle (Day, Moreland, Floyd, and Huard 1997). The regenerative response, however, is often inadequate; muscle tissue is unable to keep pace with limb lengthening and thus muscle tears occur (Day et al. 1997). By enhancing muscle growth and regeneration, the elongated muscle could follow the lengthening of the limb without structural damage, and the occurrence of contracture would be prevented.

The conditions related to compartment syndrome and ischemia are distinct from limb lengthening. Following an ischemic trauma, the muscle is partially or completely necrotic, and its regenerative capacity initially depends upon the numbers of lively myogenic precursor cells. Hence, when the regeneration and repair phase occurs, the gene transfer technique, which has the potential of continuously expressing and delivering an exogenous source of growth factors to the injured muscle, may promote healing and limit the formation of connective scar tissue.

Numerous studies completed on muscular dystrophy, especially regarding DMD, have served as basic research for the application of gene therapy to muscle in the orthopaedic and sports medicine fields. The application of this technology to muscular dystrophy diseases represents one of the most challenging topics of investigation in orthopaedic medicine. In this situation, the therapeutic protein (dystrophin) must be delivered to all the muscles of the organism and particularly to the respiratory muscles, such as the diaphragm, which is not easily accessible and represents an important muscle mass to be transduced. Furthermore, a lifelong persistence of gene transfer is necessary in order to achieve a definitive therapeutic effect, while a short-term persistence would probably be sufficient for a significant improvement in a sports medicine application.

In conclusion, we believe that gene therapy has great potential as a delivery system to the muscular tissue. Both in vitro and in vivo studies have demonstrated that specific growth factors are potent stimulators of muscle cell proliferation and differentiation. The continuous and local production, by gene delivery, of these proteins in an injured muscle may represent a promising and revolutionary way of treating severe injuries. This method may have the potential to change the treatment and outcome of severe muscle strain, extensive laceration, and large, multiple contusions. In addition these techniques may even revolutionize the surgical approach as well as impact the treatment of different muscle conditions, such as fibrosis after limb lengthening, compartment syndrome, and muscular dystrophy.

Acknowledgments. The authors wish to thank Marcelle Pellerin and Ryan Pruchic for their technical assistance, and Dana Och and Megan Mowry for their assistance in the preparation of the manuscript. This work was supported by grants to Dr. Huard from the Parent's Project, the Muscular Dystrophy Association, and the National Institutes of Health.

References

Acsadi, G., Dickson, G., Love, D., Jani, A., Walsh, F.S., Gurusing, A., Wolff, J.A., and Davies, K.E. 1991. Human dystrophin expression in mdx mice after intramuscular injection of DNA constructs. *Nature* 352:815–18.

Acsadi, G., Jani, A., Huard, J., Blaschuk, K., Massie, B., Holland, P., Lochmueller, H., and Karpati, G. 1994. Cultured human myoblasts and myotubes show markedly different transducibility by replication defective adenovirus recombinants. *Gene Ther* 1:338–40.

Allamedine, H.S., Dehaupas, M., and Fardeau, M. 1989. Regeneration of skeletal muscle fiber from autologous satellite cells multiplied in vitro. *Muscle Nerve* 12:544–55.

Allen, R.E. and Boxhorn, L.A. 1989. Regulation of skeletal muscle satellite cell proliferation and differentiation by transforming growth factor–beta, insulin–like growth factor 1, and fibroblast growth factor. *J Cell Physiol* 138:311–15.

Anderson, J.E., Liu, L., and Kardami, E. 1991. Distinctive patterns of basic fibroblast growth factor (bFGF) distribution in degenerative and regenerating areas of dystrophic (mdx) striated muscle. *Dev Biol* 147:96–109.

Anderson, J.E., Mitchell, C.M., McGeachie, J.K., and Grounds, M. 1995. The time of basic fibroblast growth factor expression in crush-injured skeletal muscles of SJL/J and BALB/c mice. *Exp Cell Res* 216:325–34.

Bischoff, R. 1994. The satellite cell and muscle regeneration. In Engel A.G. and Franzin-Amstrong C. (eds.), *Myology*. 2nd ed. 97–118. Philadelphia: McGraw–Hill.

Booth, D.K., Floyd, S.S., Day, C.S., Glorioso, J.c., Kovesdi, I., and Huard J. 1997. Myoblast mediated ex vivo gene transfer to mature muscle. *J Tissue Eng* 3: 125–33.

Campbell, J.S., Wenderoth, M.P., Hauschka, S.D., and Krebs, E.G. 1995. Differential activation of mitogen-activated protein kinase in response to basic fibroblast growth factor in skeletal muscle cells. *Proc Natl Acad Sci USA* 92:870–4.

Carlson, B.M. and Faulkner, J.A. 1983. The regeneration of skeletal muscle fibers following injury: a review. *Med Sci Sports Exerc* 15:187–96.

Chambers, R.L. and McDermott, J.C. 1996. Molecular basis of skeletal muscle regeneration. *Can J Appl Physiol* 21(3):155–84.

Clark, K.R., Sferra, T.J., and Johnson, P.R. 1997. Recombinant adeno-associated viral vectors mediate long term transgene expression in muscle. *Hum Gene Ther* 8:659–69.

Crisco, J.J., Jolk, P., Heinen, G.T., Connell, M.D., and Panjabi, M.M. 1994. A muscle contusion injury model, biomechanics, physiology and histology. *Am J Sports Med* 15:9–14.

Day, C.S., Moreland, M., Floyd, S., and Huard, J. 1997. Limb lengthening promotes muscle growth. *J Orthop Res* 15:227–34.

DiMario, J., Buffinger, N., Yamanda, S., and Strohman R.C. 1989. Fibroblast growth factor in the extracellular matrix of dystrophic (mdx) mouse muscle. *Science* 244(4905):688–90.

Dunckley, M.G., Wells, D.J., Walsh, F.S., and Dickson, G. 1993. Direct retroviral mediated transfer of a dystrophin minigene into mdx mouse muscle in vivo. *Hum Mol Gen* 2:717–23.

Ewton, D.Z. and Florini, J.R. 1980. Relative effects of somatomedins, multiplication stimulating activity, and growth hormone on myoblasts and myotubes in culture. *Endocrinology* 106:583.

Fisher, K.J., Jooss, K., Alston, J., Yang, Y., Haecker, S.E., High, K., Pathak, R., Raper, S.E., and Wilson, J.M. 1997. Recombinant adeno–associated virus for muscle directed gene therapy. *Nature Medicine* 3:306–12.

Florini, J.R., Ewton, D.Z., and Roof, S.L. 1991. Insulin–like growth factor–1 stimulates terminal myogenic differentiation by induction of myogenin gene expression. *Mol Endocrinol* 5:718.

Florini, J.R. and Magri, K. 1989. Effect of growth factors on myoblast differentiation. *Am J Physiol* 256:701–11.

Floyd Jr., S.S., Booth II, D.K., van Deutekom, J.C.T., Day, C.S., and Huard, J. 1997. Autologous myoblast transfer: A combination of myoblast transplantation and gene therapy. *Basic Appl Myol* 7(3, 4):241–50.

Floyd, S.S., Clemens, P.R., Ontell, M.R., Kochanek, S., Day, C.S., Yang, J., Hauschka, S.D., Balkir, L., Morgan, J.E., Moreland, M.S., Feero, W.G., Epperly, M., and Huard, J. 1998. Ex vivo gene transfer using adenovirus mediated full length dystrophin delivery to mature dystrophic muscles. *Gene Ther* 5:19–30.

Garrett Jr., W.E. 1990. Muscle strain injuries: clinical and basic aspects. *Med Sci Sports Exerc* 22:436–43.

Garrett Jr., W.E., Saeber, A.V., Boswick, J., Urbaniak, J.R., and Goldner, L. 1984. Recovery of a skeletal muscle after laceration and repair. *J Hand Surg* 9A: 683–92.

Gospodarowicz, D., Ferrara, N., Schweigerer, L., and Neufeld, G. 1987. Structural characterization and biological functions of fibroblast growth factor. *Endocrinol Rev* 8:95–114.

Grounds, M.D. 1991. Towards understanding skeletal muscle regeneration. *Path Res Pract* 187:1–22.

Huard, J., Acsadi, G., Jani, A., Massie, B., and Karpati, G. 1994. Gene transfer into skeletal muscles by isogenic myoblasts. *Hum Gene Ther* 5:949–58.

Huard, J., Goins, W.F., and Glorioso, S.C. 1995. Herpes simplex virus type I vector mediated gene transfer to muscle. *Gene Ther* 2:385–92.

Huard, J., Akkaraju, G., Watkins, S.C., Cavalcoli, M.P., and Glorioso, J.C. 1996. Persistent LacZ expression in skeletal muscle of immunodeficient (SCID) mice mediated by highly defective Herpes simplex virus type 1 vector. *Hum Gene Ther* 8:439–52.

Huard, J., Goins, B., and Glorioso, J.C. 1995. Herpes simplex virus type 1 vector mediated gene transfer to muscle. *Gene Ther* 2:1–9.

Huard, J., Guerette, B., Verreault, S., Tremblay, G., Roy, R., and Tremblay, J.P. 1994. Human myoblast transplantation in immunodeficient and immunosuppressed mice: evidence of rejection. *Muscle Nerve* 17:224–34.

Huard, J., Lochmueller, H., Acsadi, G., Jani, A., Massie, B., and Karpati, G. 1995. The route of administration is a major determinant of the transduction efficiency of rats tissue by adenoviral recombinants. *Gene Ther* 2:107–15.

Huard, J., Lochmueller, H., Jani, A., Holland, P., Guerin, C., Massie, B., and Karpati, G. 1995. Differential short term transduction efficiency of adult versus newborn mouse tissues by adenoviral recombinants. *Exp Mol Pathol* 62:131–43.

Huard, J., Verreault, S., Roy, R., Tremblay, M., and Tremblay, J.P. 1994. High efficiency of muscle regeneration following human myoblast clone transplantation in SCID mice. *J Clin Invest* 93:586–99.

Hurme, T. and Kalimo, H. 1992. Activation of myogenic precursor cells after muscle injury. *Med Sci Sports Exerc* 24:197–205.

Hurme, T., Kalimo, H., Lehto, M., and Jarvinen, M. 1991. Healing of skeletal muscle injury. An ultrastructural and immunohistochemical study. *Med Sci Sports Exerc* 23:801–10.

Hurme, T., Kalimo, H., Sandberg, M., Lehto, M., and Vuorio, E. 1991. Localization of type I and III collagen and fibronectin production in injured gastrocnemius muscle. *Laboratory Investigation* 64:76–84.

Jarvinen, M. and Sorvari, T. 1975. Healing of a crush injury in rat striated muscle. *Acta Path Microbiol Scand* 83A:259–65.

Jennische, E. 1989. Sequential immunohistochemical expression of IGF–1 and the transferrin receptor in regenerating rat muscle in vivo. *Acta Endocrinol* 121:733–8.

Jennische, E. and Hansson, H.A. 1987. Regenerating skeletal muscle cells express insulin-like growth factor 1. *Acta Physiol Scand* 130:327–32.

Kalimo, H., Rantanen, J., and Jarvinen, M. 1997. Muscle injuries in sports. *Balliere's Clin Orthop* 2, 1:1–24.

Kardami, E., Spector, D., and Strohman, R.C. 1985. Selected muscle and nerve extracts contain an activity which stimulates myoblast proliferation and which is distinct from transferrin. *Dev Biol* 112:353–8.

Kasemkijwattana, C., Ménétrey, J., Somogi, G., Moreland, M.S., Fu, F.H., Buranapanitkit, B., Watkins, S.C., and Huard, J. 1998. Development of approaches to improve the healing following muscle contusion. *Cell Trans* 7(6):585–98.

Katsumi, A., Emi, N., Abe, A., et al. 1994. Humoral and cellular immunity to an encoded protein induced by direct DNA injection. *Hum Gene Ther* 5:1335–9.

Lattermann, C., Baltzer, A.W.A., Whalen, J.D., et al. 1998. Gene therapy in sports medicine. *Sports Med Arthrosc Rev* 6:83–8.

Lefaucheur, J.P. and Sebille, A. 1995. Muscle regeneration following injury can be modified in vivo by immune neutralization of basic-fibroblast growth factor, transforming growth factor 1 or insulin–like growth factor 1. *J Neuroimmunol* 57:85–91.

Lehto, M. and Jarvinen, M. 1991. Muscle injuries healing and treatment. *Annales Chirurgiciae et Gynaecologiae* 80:102–9.

Lehto, M., Duance, V.J., and Restall, D. 1985. Collagen and fibronectin in a healing skeletal muscle injury. An immunohistochemical study of the effects of physical activity on the repair of the injured gastrocnemius muscle in the rat. *J Bone Joint Surg* 67:820–8.

Ménétrey, J., Kasemkijwattana, C., and Day, C.S. 1998. Characterization of trophic factors to promote muscle growth. *Trans Orthop Research Soc* 44:166.

Ménétrey, J., Kasemkijwattana, C., Fu, F.H., Moreland, M.S., and Huard, J. 1999. Suturing versus immobilization of a muscle laceration: a morphological and functional study. *Am J Sports Med* 27(2):222–9.

Mishra, D.K., Friden, J., Schmitz, M.C., and Lieber, R.L. 1995. Antiinflammatory medication after muscle injury. A treatment resulting in short-term improvement but subsequent loss of muscle function. *J Bone Joint Surg* 77A:1510–19.

Mubarak, S. and Hargens, A.R. 1981. Compartment syndromes and Volksmann's contracture. 106–18. Philadelphia: WB Saunders.

Nikolaou, P.K., MacDonald, B.L., Glisson, R.R., Seaber, A.V., and Garrett, W.E. 1987. Biomechanical and histological evaluation of muscle after controlled strain injury. *Am J Sports Med* 15:9–14.

Partridge, T.A. 1991. Myoblast transfer: a possible therapy for inherited myopathies. *Muscle Nerve* 14:197–212.

Reed–Clark, K., Sferra, T.J., and Johnson, P.R. 1997. Recombinant adeno–associated virus vectors mediate long term transgene expression in muscle. *Hum Gene Ther* 8:659–69.

Salvatori, G., Ferrari, G., Messogiorno, A., Servidel, S., Colette, M., Tonalli, P., Giarassi, R., Cosso, G., and Mavillo, F. 1993. Retroviral vector mediated gene transfer into human primary myogenic cells lead to expression in muscle fibers in vivo. *Hum Gene Ther* 4:713–23.

Schultz, E. 1989. Satellite cell behavior during skeletal muscle growth and regeneration. *Med Sci Sports Exerc* 21:181.

Schultz, E., Jaryszak, D.L., and Valliere, C.R. 1985. Response of satellite cells to focal skeletal muscle injury. *Muscle Nerve* 8:217.

Trippel, S.B., Coutts, R.D., and Einhorn, T., et al. 1996. Growth factors as therapeutic agents. *J Bone Joint Surg* 78-A:1272–86.

Wolfe, J.A., Ludkte, J.J., and Acsadi, G., et al. 1992. Long term persistence of plasmid DNA and foreign gene expression in mouse muscle. *Hum Mol Gene* 1:363–9.

Xiao, X., Li, J., and Samulski, R.J. 1996. Efficient long-term gene transfer into muscle tissue of immunocompetent mice by adeno-associated virus vector. *J Virol* 70:8098–108.

6
Nonviral Gene Therapy: Application in the Repair of Osteochondral Articular Defects

RANDAL S. GOOMER and DAVID AMIEL

Summary

Until recently, the only way to attain high efficiency gene transfer into primary mammalian cells was by using viral vectors. However, recent developments in novel receptor/liposome-based transfection systems have made nonviral gene therapy a real possibility. Described here is a novel, high-efficiency, nonviral protocol for delivery of genes into permeabilized primary cultured cells forming a first step toward ex vivo gene therapy for the repair of full thickness articular cartilage defects. To test the feasibility of the method, a plasmid carrying a marker β–galactosidase (β–gal) gene, driven by a strong mammalian promoter, was introduced into primary cells. The system consisted of a cell-receptor specific ligand attached to a polycation scaffold. The plasmid DNA attached to the polycation scaffold by ionic charge interactions. The system achieved greater than 70% efficiency by utilizing a three-step method: 1) Primary cells were permeabilized using a mild detergent (lysolecithin); 2) The β–gal plasmid was allowed to associate with a polycation (poly–L–lysine) core covalently linked to a receptor ligand (transferrin) forming the DNA/poly–L–lysine–transferrin complex (DTPLL complex); and 3) Cationic liposomes were introduced to the DTPLL complex. This system has now been used to transfect primary perichondrium cells and chondrocytes. More than 70% of the primary cells were found to be positive for β–gal activity. For in vivo assessment, D,D–L,L–polylactic acid (PLA) scaffolds (3 mm × 3.7 mm) seeded with the transfected primary perichondrial cells were implanted into experimentally created osteochondral defects in rabbit knees. The transformed cells continued to express β–gal, in vivo for the entire test period of 7 days, as determined by the β–gal assay. We have previously demonstrated that adding exogenous transforming growth factor beta 1 (TGF–β1) can enhance the chondrocytic phenotype of perichondrial cells (Amiel, Goomer, and Coutts 1997; Dounchis et al. 1997). These studies were initiated in order to assess the usefulness of transfected perichondrium cells as vehicles for the localized delivery of TGF–β1 into the repair site. To this end, we have developed a TGF–β1 expression vector and shown that cells transfected with this construct overexpress the TGF–β1 specific mRNAs. This system is now poised for the delivery of therapeutic genes into primary cultured cells to repair damaged or dysfunctional tissues.

Introduction

To date, ex vivo gene therapy has remained a challenge, primarily due to the difficulty of achieving high-efficiency gene transfer into primary cells with nonviral techniques. Ex vivo gene therapy involves removing autologous cells from the patient's body, amplifying them in primary culture, introducing the therapeutic gene, and implanting those transformed cells back into the patient's body. Although viral vectors have shown some efficacy in transforming primary cells with high efficiency (Versland, Wu, C.H., and Wu, G.Y. 1992; Dematteo et al. 1995), their use remains problematic. For instance, adenovirus vectors induce host immune response (Ginsburg et al. 1991; Trapnell and Gorziglia 1994; Yei, Mittereder, Tany, O'Sullivan, and Trapnell 1994; Yang, Li, Ertl, and Wilson 1995; Sokol and Gewirtz 1996; Yang, Su, and Wilson 1996; McCray et al. 1997; Tremblay 1997), while retroviral vectors require dividing cells for integration (Sokol and Gewirtz 1996). In addition, viral vectors may randomly integrate into host genome, posing a risk of neoplastic transformation (Colledge 1994; Fairbarn, Cross, and Arrand 1994; Sokol and Gewirtz 1996). Therefore, the quest for a high efficiency universal gene delivery system that would preclude the possibility of an immune response (i.e. nonviral) has continued with the use of either liposomes or ligand–polycations (such as transferrin–poly–L–lysine). Cationic liposomes contain a positively charged region and a hydrophobic tail. They function, presumably, by interacting with the negatively charged DNA (by charge attraction) at one end and by hydrophobic interactions with the cell membrane at the other. This enables the DNA to be taken up by the cell. The use of liposomes has not resulted in any known toxicity in vivo (Canonico, Plitman, Coanry, Mayrick, and Brigham 1994). However, liposome-based transfections suffer from low efficiencies—particularly in primary cells. Using cationic liposomes, gene transfection of mammalian cells can be achieved (Felgner et al. 1987; Felgner and Ringold 1989) with efficiencies ranging from 1% to 15% (Goomer, Holst, Jones, and Edelman 1994; Cooper 1996; Wheeler et al. 1996; Brant, Goomer, and Amiel 1997), where efficiency of transfection refers to the percentage of cells that express the added gene marker. The transfection efficiency of liposomes drops dramatically in suspension cultures, presumably due to reduced binding of the liposome–DNA complexes to cells (Labat–Moleur et al. 1996). In addition, transferrin–poly–L–lysine (containing positively charged poly–L–lysine scaffold and a ligand) mediated transfection has been shown to be 7–8%

efficient (Wagner, Zenke, Cotten, Beug, and Birnstiel et al. 1990; Taxman, Lee, and Wojchowski 1993; Lee and Huang 1996). Increased transfection efficiency could be achieved by the inclusion of poly–L–lysine polycations to the liposome transfection system (Gao and Huang 1995; Mack, Walzem, Lehmann-Bruinsma, Powell, and Zeldis 1996), while the utilization of transferrin and cationic liposomes high-efficiency transfection of a cultured cell line (NIH–3T3 cells) has been reported (Cheng 1996). To date, however, none of these techniques has been shown to be successful in attaining very high efficiency (>70%) transfection of primary mammalian cells. The development of a technique for the transfection of primary mammalian cells is important because only these cells (as autologous carriers) could be used for ex vivo gene therapy (Wakatani et al. 1994; Akentijevich et al. 1996).

Our laboratory has pioneered a protocol for articular defect repair in the rabbit model. Cells derived from the costal rib perichondrium are seeded into a biodegradable polylactic acid scaffold and implanted into an experimentally created full thickness articular cartilage defect in the rabbit femoral (Chu et al. 1997; Amiel, Chu, Sah, and Coutts 1999). However, the success of this repair is affected by the phenotype of the implanted cells. We have shown that the chondrocytic phenotype of cultured perichondrial cells can be enhanced by exogenous TGF–β1 (Amiel et al. 1997; Dounchis et al. 1997).

We initiated the present study based upon the premise that cells transfected with an expression vector carrying the TGF–β1 gene would form the most efficient vehicle for the localized delivery of this morphogen into the repair site. In this study, we report the development of a human cytomegalovirus (hCMV) promoter/enhancer driven TGF–β1 construct that up regulates TGF–β1 expression in transfected perichondrial cells. We also describe a novel method of gene delivery into primary perichondrium derived cells. Using this system, we deliver a marker gene, such as β–galactosidase (β–gal), into perichondrial cells and show that transfection efficiencies greater than 70% can be achieved. Furthermore, in an in vivo experiment these cells were implanted into an experimentally created osteochondral defect. The cells continued to express the transfected gene throughout the test period of 1 week.

Methodology

Liposomes

Cationic liposomes were prepared as follows: Equimolar amounts of L–α–phosphatidyl–ethanolamine and dimethyldioctadecyl–ammonium bromide were mixed in chloroform, dried as a film and resuspended. The aqueous solution was sonicated until the desired particle size, that is, 30 to 50 μm was achieved.

Covalent Bonding of Ligand with Poly–L–Lysine

The ligand (apo–transferrin) was covalently linked to a positively charged polymeric scaffold (poly–L–lysine (70 Kda)) using standard biochemistry protocols that have been previously published (Taxman et al. 1993).

Isolation of Rib Perichondrium

Primary perichondrium cells from rabbit costal ribs were extracted as follows: The rabbits (mature 8 to 10 mos old with closed epiphyses and aged 4 to 5 years old) were sacrificed according to the animal subject protocols at the University of California, San Diego. Costal ribs were removed using sterile procedures. Adhering tissue was cleaned away from the ribs with the use of sterile surgical instruments and breaking the rib and peeling off the perichondrium tissue isolated the rib perichondrium. The isolated perichondrium (Figure 6.1A and B) was washed three times in antibiotic-containing buffered salt solution.

Enzymatic Extraction of Perichondrial Cells

The perichondrium tissue was incubated overnight at 37°C under sterile conditions in 0.1% collagenase in cell culture media. Cells and tissue debris were isolated away from the media by passage through a sterile 0.45–μ filter. The cells and tissue debris were enzymatically digested with 0.1% hyaluronidase and in 0.25% wt/vol. trypsin for 1 to 2 hours. The cells were then isolated by passage through an 80–μ filter under sterile conditions (Figure 6.2).

Transfection of Primary Chondroprogenitor Cells With TGF–β1

A plasmid-carrying human TGF–β1 gene driven by the hCMV promoter/enhancer sequences and a neomycin resistant gene (Figure 6.3) expressing TGF–β1 gene in the active form were used for our experiments. Cationic liposomes (mixture of L–α–phosphatidyl–ethanolamine dioleoyl (DPE) and dimethyldioctadecyl–ammonium bromide (DDAB) were used for plasmid delivery. An aminoglycoside related to gentamicin (G418) was used to select for transfected cells. In order to optimize the amount of liposomes/DNA ratios that would result in high levels of transfection efficiency, titration of various ratios of liposomes to plasmid DNA was performed. Selection of transfected cells with G418 resulted in optimization of liposome: DNA ratio at 12 μg DNA per 2×10^6 cells (Figure 6.4). Primary perichondrial cells at 60 to 70% confluence were transfected with

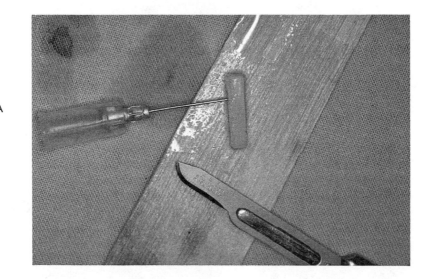

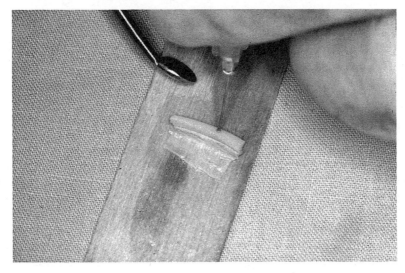

FIGURE 6.1. Rabbit rib perichondrium. (A): Illustration of anterior cartilaginous section of the rabbit rib. (B): Illustration of perichondrial tissue dissected from anterior cartilaginous section of rabbit rib.

varying ratios of liposome/DNA mixtures in serum-free medium. Eight hours post transfection, the serum-free medium was replaced with medium containing 10% fetal bovine serum. The transfected genes were allowed to express for 60 to 72 hours (Figure 6.4A), after which the cells were replated into selective medium containing G418 (0.35 µg/ml). The medium was replaced every 3 to 4 days. One week post transfection about 10 to

Enzymatic digestion of perichondrium to produce confluent cell culture in 48 hours.

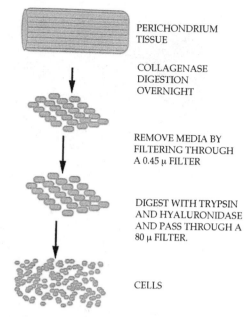

PERICHONDRIUM TISSUE

COLLAGENASE DIGESTION OVERNIGHT

REMOVE MEDIA BY FILTERING THROUGH A 0.45 μ FILTER

DIGEST WITH TRYPSIN AND HYALURONIDASE AND PASS THROUGH A 80 μ FILTER.

CELLS

FIGURE 6.2. Enzymatic extraction of primary perichondrial cells:
- Perichondrial tissue incubated at 37 C in 0.1% collagenase.
- Cells and tissue isolated through a 0.45 μ filter.
- Cells and tissue enzymatically digested with 0.1% hyaluronidase/trypsin for 1 to 2 hours.
- Cells isolated through an 80 μ filter.

20% of the cells still survived (Figure 6.4B), whereas none of the cells in control plates transfected with a plasmid lacking the neo[r] gene were still surviving. After 3 weeks, colonies of transfected cells were able to divide and amplify in selective media (Figure 6.4C). The transfected cells were harvested 60 hours post transfection, total RNA was purified, and semiquantitative reverse transcription polymerase chain reaction (RT–PCR) was performed as previously described (Amiel et al. 1997; Dounchis et al. 1997).

Transfection to Test the Efficiency of the DNA/Transferrin–Poly–L–Lysine/Liposome (DTPLL) Complexes

Primary perichondrium cells from rabbit costal ribs were cultured on tissue culture plates in media (α–MEM +10% fetal bovine serum (FBS)) under sterile conditions at 37°C. The cells were allowed to proliferate and achieve 70% to 80% confluence on the plate. These cells were permeabilized with

TGF-ß1 Gene Expression: Cell Transfection and Selection

Expression Of Introjected Growth Factor Genes In Perichondrocytes

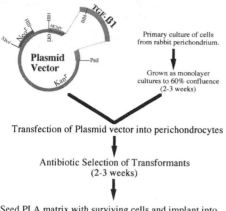

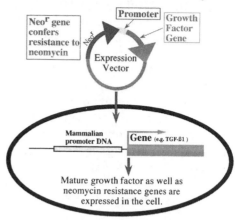

FIGURE 6.3. Transfected primary chondroprogenitor cells for localized delivery of TGF–β1.

Gene Transfection

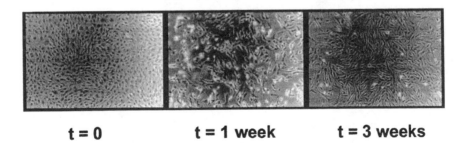

t = 0   t = 1 week   t = 3 weeks

Incubation Time in Selective Medium (G418)

FIGURE 6.4. Liposome mediated transfection of primary perichondrial cells from mature rabbit with plasmid carrying TGF–β1 and neomycin resistant gene selected with G418. (A): Transfected cells grown to confluence before selection with G418. (B): One week after G418 selection was applied. (C): Transfected cells after 3 weeks selection with G418.

lysolecithin for 2 min. The cells were subsequently washed with serum free medium. The DTPLL technique has been summarized in Figure 6.5. The purified plasmid DNA (D) was incubated with transferrin–poly–L–lysine complex (TPL). Cationic liposomes were diluted in media. The plasmid DNA/transferrin–poly–L–lysine complexes (DTPL) were added to the liposomes and allowed to form tertiary complexes for 15 to 20 min. The DNA/ligand–poly–L–lysine/liposome (DTPLL) tertiary complexes were incubated in serum free medium (α–MEM) with the previously permeabilized perichondrium derived cells and incubated for 5 hours. The medium was replaced with fresh medium containing 10% FBS. Primary rabbit chondrocytes were similarly transfected. In order to assess the transfection efficiency, the β–gal gene (the expression vector was obtained from Promega, Inc.) was amplified in culture, purified, and allowed to express for 48 hours and tested for gene expression by β–gal assay or o–nitrophenyl–β–D–galactopyranoside (ONPG) reaction. For β–gal analysis, the cells were

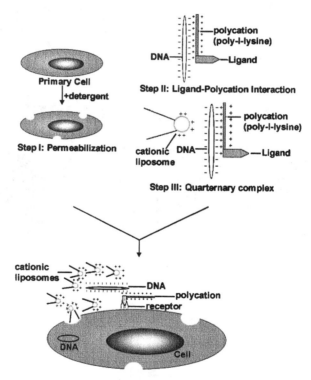

FIGURE 6.5. Transfection of the DNA/transferrin poly–L–lysine/liposome (DTPLL) complexes. *Step I*: The cell membrane is made more porous by incubation with a mild detergent (lysolecithin). *Step II*: Plasmid DNA is attached to the scaffold transferrin poly–L–lysine by ionic interactions. *Step III*: Complex from Step II added to liposome to form the quaternary DNA/transferrin poly–L–lysine/liposome (DTPLL) complex that is added to the permeabilized cells.

fixed in buffer containing 0.25% v/v glutyraldehyde in 1× PBS. The fixed cells were stained with X–gal (Sigma) solution (0.2% X–gal (from 2% stock in dimethyl formamide (DMF), 2 mM $MgCl_2$, 5 mM $K_4Fe (CN)_6 . 3H_2O$ and 5 mM $K_3Fe (CN)_6$ prepared in 1× PBS). The staining step was performed at 37°C for up to 8 hours. Efficiency of gene transfection in X-gal stained cells was determined as follows. Five fields from each plate were observed under the microscope. The number of all cells and stained cells was counted. The assessment was repeated for at least 3 plates. The assessment categories were as follows: (1) unpermeabilized cells and (a) liposomes/DNA, (b) transferrin, liposomes/DNA, (c) transferrin, poly–L–lysine (MW = 70 Kd) liposome/DNA and (d) transferrin, poly–L–lysine (MW = 150 Kd), liposome/DNA; and (2) permeabilized cells and (a) liposomes/DNA, (b) transferrin, liposomes/DNA, (c) transferrin, poly–L–lysine (MW = 70 Kd) liposome/DNA and (d) transferrin, poly–L–lysine (MW = 150 Kd), liposome/DNA.

In Vivo Study

In order to assess in vivo gene delivery into the osteochondral rabbit defect we have utilized the perichondrial cells transfected with the β–gal gene and used the polylactic acid (PLA) scaffold to deliver the cells into the defect. The gene, that is β–gal, was allowed to express for 12 hours, and the cells were seeded into PLA scaffold according to the protocol described previously (Chu, Coutts, Yoshioka, Harwood, and Amiel 1995; Chu et al. 1997). The surgical and animal care procedures published previously (Chu et al. 1995, 1997) were used to create a 3 mm × 3.7 mm full thickness articular cartilage defect in the rabbit femoral condyle (Figure 6.6). The femoral condyles were harvested 1 week postoperatively and β–gal activity was tested using X–gal reagent (Promega, Inc.). The harvested condyles were dipped into 0.2% X–gal solution, sectioned, and stained with eosin, as illustrated in Figure 6.6.

Results

Previous to this study, primary perichondrial cells were isolated and expanded in our laboratory by mincing the perichondrium and allowing the cells to migrate out of the tissue and proliferate in media until confluence was reached. This procedure required 21 days to produce a confluent plate of cultured cells (Chu et al. 1995). Here we describe a novel technique for enzymatic extraction of primary perichondrial cells. Primary cells were enzymatically extracted from the rabbit costal rib perichondrium by digesting with collagenase and hyaluronidase/trypsin using the protocol schematically represented in Figure 6.2 and described in the methods section. Utilizing this protocol, we were able to obtain $1 × 10^6$ cells from 100 mg wet weight of tissue in less than 48 hours. The gene expression of these primary

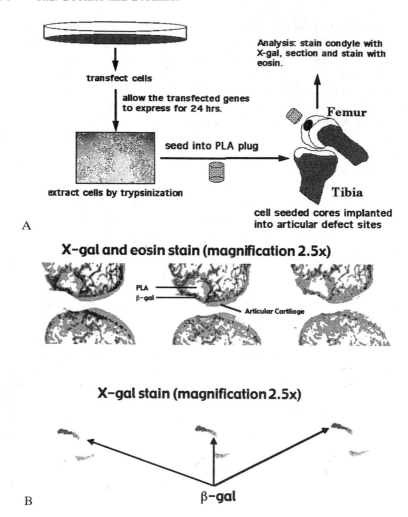

FIGURE 6.6. Protocol for ex vivo gene therapy for the repair of articular cartilage defects (A). Transfected cells were allowed to express the β–gal gene, and seeded to the PLA scaffold; cells/PLA scaffold was implanted into 3.7 mm × 3 mm defect in rabbit femoral condyle and the repair site harvested 1 week postimplantation. (B) X-gal staining for gene expression and eosin staining for cellular and collagen detail.

perichondrial cells for types I and II collagen were similarly expressed in these two techniques.

While we have previously shown that primary unpermeabilized perichondrial cells can be transfected with cationic liposomes with efficiencies as high as 15% (Brant et al. 1997), the inclusion of transferrin produced a transfection efficiency of about 25%, whereas inclusion of poly–L–lysine alone produced a transfection efficiency to 20%. Use of unbound transferrin and poly–L–lysine (70 Kd) resulted in efficiencies as high as 40%. Note that the use of poly–L–lysine of MW 150 Kd (at the concentrations tested) was toxic to perichondrial cells. However, the use of DTPLL and cationic liposomes together on cells that had been previously permeabilized with a mild detergent (lysolecithin) resulted in greater than 70% transfection efficiency (protocol summarized in Figure 6.5) (data summarized in Table 6.1). The transformation efficiency was determined by test for β–gal enzyme activity by the X–gal reagent. Blue stained cells were indicative of β–gal activity (Figure 6.7). ONPG (o–nitrophenyl–β–D–galactopyranoside) reaction (β–gal converts ONPG to o–nitrophenol and galactose) showed that the inclusion of transferrin–poly–L–lysine (TPL) enhanced transfection efficiency significantly as measured by optical density at 420 nm (Figure 6.8). For in vivo assessment, the transfected cells were seeded into a PLA scaffold and implanted into an experimentally created osteochondral defect in the rabbit femoral condyles. The femoral condyles were harvested 1 week postimplantation. The tissue was stained with X–gal, fixed, decalcified and sectioned into saggital sections. Enzyme activity was determined by X–gal assay. Since the first step of incubation in the X–gal was performed on whole femoral condyles containing tightly fitted cell-seeded PLA cores, the stain traversed only up to the first mm of the PLA scaffold. However, the cells in this region were shown to be positive for β–gal gene activity (Figure 6.6), implying that the seeded cells continued to express the transfected gene.

In Vitro Study

We have also developed a TGF–β1 expression vector driven by a CMV promoter enhancer sequence. Chondroprogenitor cells, that is perichondrial cells, were transiently transfected using liposome/DNA ratios that had been defined in selection studies (Figure 6.3). However, during transient transfection, cells were grown without selective pressure and were harvested 60 to 72 hours post transfection. Total RNA was purified as detailed above and subjected to RT–PCR with primers specific for TGF–β1 conserved region and analyzed by agarose gel electrophoresis. TGF–β1 message was significantly increased in cells transfected with the TGF–β1–containing plasmid, as compared with a mock parent plasmid in mature (Figure 6.9, compare

TABLE 6.1. Transfection efficiency of permeabilized primary perichondrial cells.

Treatment	Efficiency
• DNA + Transferrin + Liposomes	• 25.3 ± 4.0
• DNA + Transferrin + Poly–L–lysine (70 Kda) + Liposomes	• 40.8 ± 6.7
• DNA + Transferrin – Poly–L–lysine (70 Kda) + Liposomes (DTPLL)	• 71.1 ± 11.6

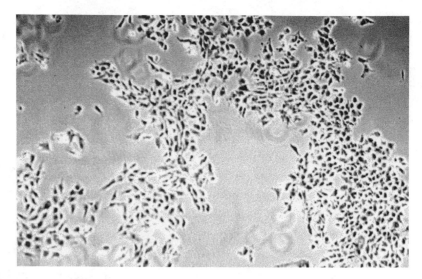

FIGURE 6.7. Transfection of primary perichondrial cells in monolayer cultures by method illustrated in Figure 6.5. β–galactosidase (β–gal) activity was monitored by in situ X–gal staining 60 to 72 hours post transfection.

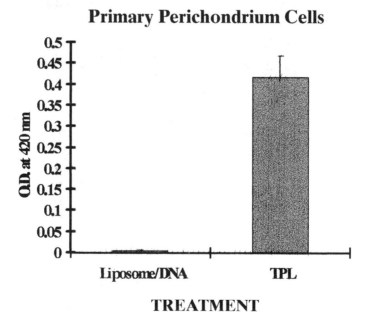

FIGURE 6.8. β–galactosidase (β–gal) activity by o–nitrophenyl–β–D–galactopyranoside (ONPG) reaction measured as optical density at 420 nm. β–gal catalyzes the breakdown of ONPG into o-nitrophenol and galactose. Note: TPL = transferrin–poly–L–lysine.

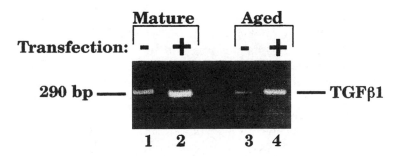

FIGURE 6.9. Expression of TGF–β1 after transfection of primary perichondrial cells from mature (Lanes 1 and 2) and aged (Lanes 3 and 4) rabbits. RT–PCR analysis was performed with exactly 1 μg of total RNA from each sample. *Lanes 1 and 3*: Untransfected controls. *Lanes 2 and 4*: Cells transfected with a plasmid carrying TGF–β1 gene driven by the human CMV promoter.

Lanes 1 and 2) and aged (Figure 6.9, compare Lanes 3 and 4) rabbits. These results show that the TGF–β1 gene was successfully transfected and was being expressed in cultured monolayer perichondrial cells from mature and aged rabbits.

Discussion

This research communication describes a method for the achievement of high-efficiency ex vivo gene transfer into primary mammalian cells while precluding the use of viral vectors, and, therefore, removing the possibility of inducing the host immune response. The efficacy of this protocol could be attributed to the sequence of events listed below and schematically outlined in Figure 6.5. 1) Permeabilization: The cell membrane was made slightly porous (etched) prior to the introduction of genes into the cells with the help of a mild detergent (lysolecithin). 2) Ligand/poly–L–lysine complex: A specific ligand (transferrin) was covalently attached to a poly-cationic core (poly–L–lysine) to create a scaffold to facilitate ionic interaction with negatively charged DNA. 3) Ligand/receptor interaction: The DNA/poly–L–lysine–ligand complex was delivered very close to the cell due to the binding of the ligand to its receptor (i.e. the receptor ligand interaction). 4) Cationic liposome mediated delivery: The DNA was putatively delivered into the cell by hydrophobic interactions between the cationic liposomes (Felgner and Ringold 1989) and the cell membrane that had been permeabilized using a mild detergent.

This gene delivery protocol has shown efficacy both in vitro and in vivo in an animal model. In vitro, more than 70% of primary perichondrium cells were determined to be positive for the transfected gene by X–gal staining (Figure 6.7) and relative reaction with ONPG (Figure 6.8). These cells were

102 R.S. Goomer and D. Amiel

implanted into an experimentally created osteochondral defect in the rabbit knee. After 1 week, the defect contained cells that continued to express the transfected gene. The staining was primarily observed on the top portion (cartilage surface) of the implanted cells (Figure 6.6). This could be explained by the inability of stain to penetrate through the implant, by the inability of cells below the upper surface to express β–gal, or by the inviability of cells in the lower part of the implant.

This system is applicable for gene therapy using chondroprogenitor cells from the perichondrium, periosteum, or chondrocytes (Kang et al. 1997). In fact, primary rabbit chondrocytes have also been transfected at high efficiency using this system. Growth factors such as TGF–β1 (Galera et al. 1992; Ballock et al. 1993; Amiel et al. 1997; Dounchis et al. 1997; Amiel et al. In press), connective tissue growth factor (CTGF) and bone morphogenetic proteins (BMP–2) (Luyten, Chen, Paralkar, and Reddi 1994; Roark and Greer 1994) and transcription factors, such as SOX–9 (Bell et al. 1997; Lefebvre, Huang, Harley, Goodfellow, and deCrombrugghe 1997), have been shown to enhance the chondrogenic phenotype of perichondrium/bone marrow/periosteum/cartilage-derived cells. Untransfected perichondrium/bone marrow/periosteum/cartilage derived cells have also been shown to partially enhance repair of osteochondral defects in animal models (Wakatani et al. 1994; Chu et al. 1995, 1997). We anticipate that, in the near future, transient ectopic expression using this system of physiologically relevant genes for growth factors such as TGF–β1, CTGF, and BMPs, or transcription factors such as SOX–9 or CART–1 into autologous perichondrium/bone marrow/periosteum/cartilage derived cells and subsequent implantation of these transfected cells in cell seeded cores will form a useful therapeutic modality for the repair of osteochondral/chondral defects and nonunion fractures.

Acknowledgments. This work was supported by NIH grants AR28467 and AG07996 and NIH Training Grant AR07484. We would like to acknowledge the technical assistance of Thira Maris, B.Sc., Karen Bowden, Myung Chul Lee, M.D., and Mike Furniss.

References

Akentijevich, I., Pastan, I., Lunardi–Iskander, Y., Gallo, R., Gottesman, M.M., and Thierry, A.R. 1996. In vitro and in vivo liposome mediated gene transfer leads to MDR1 expression in mouse bone marrow progenitor cells. *Hum Gen Ther* 7:1111–22.

Amiel, D., Chu, C.R., Sah, R.L., and Coutts, R.D. 1999. Tissue engineering of articular cartilage: perichondrial cells in osteochondral repair. *Cells and Materials* 8(1):161–74.

Amiel, D., Goomer, R.S., and Coutts, R.D. 1997. The chondrogenic phenotype of perichondrial cells used in the repair of osteochondral defects is influenced by TGF–β1 (transforming growth factor–β1). *SIROT 97 Inter—Mtng*. 34 Haifa, Israel.

Ballock R.T., Heydemann, A., Wakefield, L.M., Flanders, K.C., Roberts, A.B., and Sporn, M.B. 1993. TGF–b1 prevents hypertrophy of epiphyseal chondrocytes: regulation of gene expression for cartilage matrix protein metalloproteases. *Dev Biol* 158:414–29.

Bell, D.M., Leung, K.K.H., Wheatley, S.C., Ng, L.J., Zhou, S., Ling, K.W., Sham, M.H., Koopman, P., Tam, P.P.L., and Cheah, K.S.E. 1997. SOX9 directly regulates the type II collagen gene. *Nat Genet* 16:174–78.

Brant, W.O., Goomer, R.S., and Amiel, D. 1997. Assessment of liposome–mediated transfectional efficacy of aged human chondroprogenitor cells. *Am Fed Med Res* 45(1):159A.

Canonico, A.E., Plitman, J.D., Coanry, J.T., Mayrick, B.O., and Brigham, K.L. 1994. No lung toxicity after aerosol or intravenous delivery of plasmid–cationic liposome complexes. *J Appl Physiol* 77:415–19.

Cheng, P.-W. 1996. Receptor ligand facilitated gene transfer: Enhancement of liposome mediated gene transfer and expression by transferrin. *Hum Gene Ther* 7:275–82.

Chu, C.R., Coutts, R.D., Yoshioka, M., Harwood, F.L., and Amiel, D. 1995. Articular cartilage repair using allogenic perichondrocyte seeded biodegradable porous PLA: a tissue engineering study. *J Biomed Mat Res* 29:1147–54.

Chu, C.R., Dounchis, J.S., Yoshioka, M., Sah, R.L., Coutts, R.D., and Amiel, D. 1997. Osteochondral repair using perichondrial cells: a 1 year study. *Clin Orthop* 340:220–9.

Colledge, W.H. 1994. Cystic fibrosis gene therapy. *Curr Opin Genet Dev* 4:466–71.

Cooper, M.J. 1996. Noninfectious gene transfer and expression systems for cancer gene therapy. *Sem Oncol* 23:172–87.

Dematteo, R.P., Raper, S.E., Ahn, M., Fisher, K.J., Burke, C., Radu, A., Widera, G., Clayton, B.R., Barker, C.F., and Markmann, J.F. 1995. Gene transfer to the thymus. A means of abrogating the immune response to recombinant adenovirus. *Ann Surg* 222:229–42.

Dounchis, J.S., Goomer, R.S., Harwood, F.L., Khatod, M., Coutts, R.D., and Amiel, D. 1997. Chondrogenic phenotype of perichondrium derived chondroprogenitor cells is influenced by TGF–β1. *J Orthop Res* 15:803–7.

Fairbarn, L.J., Cross, M.A., and Arrand, J.R. 1994. Patterson Symposium 1993, *Gene Therapy Br J Cancer* 59:972–5.

Felgner, P.L., Gadek, T.R., Holm, M., Roman, R., Chan, H.W., Wenz, M., Northrop, J.P., Ringold, G.M., and Danielsen, M. 1987. Lipofection: a highly efficient, lipid mediated DNA transfection procedure. *Proc Natl Acad Sci USA* 84:7413–17.

Felgner, P.L. and Ringold, G.M. 1989. Cationic liposome–mediated transfection. *Nature* 337:387–8.

Galera, P., Redini, F., Vivien, D., Bonaventure, J., Penfornis, H., Loyau, G., and Pjuol, J.-P. 1992. Effect of transforming growth beta1 (TGF–β1) on matrix synthesis by monolayer cultures of rabbit articular chondrocytes during the differentiation process. *Exp Cell Res* 200:379–92.

Gao, X. and Huang, L. 1995. Potentiation of cationic liposome mediated gene delivery by polycations. *Biochem* 35:1027–36.

Ginsburg, H.S., Moldawer, L.L., Schgal, P.B., Redimgton, M., Kilian, D.L., Chanock, R.M., and Prince, G.A. 1991. A mouse model for investigating the molecular pathogenesis of adenovirus pneumonia. *Proc Natl Acad Sci USA* 88:1651–5.

Goomer, R.S., Holst, B.D., Jones F.S., and Edelman, G.M. 1994. The regulation of L–CAM by HOX 4.4 (d9) and HNF1α. *Proc Natl Acad Sci USA* 91:7985–9.

Kang, R., Marui, T., Ghivizzani, S.C., Nita, I.M., Georgescu, H.I., Suh, J-K., Robbins, P.D., and Evans, C.H. 1997. Ex vivo gene transfer to chondrocytes in full thickness articular cartilage defects: a feasibility study. *Osteoarthritis Cartilage* 5:139–43.

Labat–Moleur, F., Steffan, A.M., Brisson, C., Perron, H., Feugeeas, O., Furstenberger, P., Oberling, F., Brambilla, E., and Behr, J.P. 1996. An electron microscopy study into the mechanism of gene transfer with lipopolyamines. *Gene Ther* 3:1010–17.

Lee, R.J. and Huang, L. 1996. Folate targeted, anionic liposome entrapped polylysine–condensed DNA for tumor cell specific gene transfer. *J Biol Chem* 271:8481–7.

Lefebvre, V., Huang, W., Harley, V.R., Goodfellow, P.N., and deCrombrugghe, B. 1997. SOX9 is a potent activator of the chondrocyte specific enhancer of the pro I (II) collagen gene. *Mol Cell Biol* 17(4):2336–46.

Luyten, F.P., Chen, P., Paralkar, V., and Reddi, A.H. 1994. Recombinant bone morphogenetic protein–4, transforming growth factor–β1, and activin A enhance the cartilage phenotype of articular cartilage chondrocytes in vitro. *Exp Cell Res* 210:224–9.

Mack, K.D., Walzem, R.L., Lehmann-Bruinsma, K., Powell, J.S., and Zeldis, J.B. 1996. Polylysine enhances cationic liposome mediated transfection of hepatoblastoma cell line Hep G2. *Biotech Appl Biochem* 23:217–20.

McCray, P.B., Wang, G., Kline, J.N., Zabner, J., Chada, S., Jolly, D.J., Chang, S.M.W., and Davidson, B.L. 1997. Alveolar macrophages inhibit retrovirus mediated gene transfer to airway epithelia. *Hum Gene Ther* 8:1087–93.

Roark, E. and Greer, K. 1994. Transforming growth factor–β and bone morphogenetic protein–2 act by distinct mechanisms to promote chick limb cartilage differentiation in vitro. *Dev Dynamics* 200:103–16.

Sokol, D.L. and Gewirtz, A.M. 1996. Gene therapy: basic concepts and recent advances. *Crit Rev Euk Gene Expr* 6:29–57.

Taxman, D.J., Lee, E.S., and Wojchowski, D.M. 1993. Receptor targeted transfection using stable maleimido–transferrin/thio–poly–L–lysine conjugates. *Analytical Biochem* 213:97–103.

Trapnell, B.C. and Gorziglia, M. 1994. Gene therapy using adenoviral vectors. *Curr Opin Biotechnol* 5:617–25.

Tremblay, S. 1997. Immunogenecity: its role in cell transplantation and gene therapy. *SIROT 97 Mtng*. 65, Haifa, Israel.

Versland, M.R., Wu, C.H., and Wu, G.Y. 1992. Strategies for gene therapy in the liver. *Sem Liver Dis* 12:332–9.

Wagner, E., Zenke, M., Cotten, M., Beug, H., and Birnstiel, M.L. 1990. Transferrin–polycation conjugates as carriers for DNA uptake into cells. *Proc Natl Acad Sci USA* 87:3410–4.

Wakatani, S., Goto, T., Pineda, S.J., Young, R.G., Mansour, J.M., Caplan, A.I., and Goldberg, V. 1994. Mesenchymal cell based repair of large, full thickness defects of articular cartilage. *J Bone Joint Surg* 76:579–92.

Wheeler, C.J., Felgner, P.L., Tsai, Y.J., Marshall, J., Sukhu, L., Doh, S.G., Hartikka, J., Nietupski, J., Manthorpe, M., Nichols, M., Plewe, M., Liang, X., Noeman, J., Smith, A., and Cheng, S. 1996. A novel cationic lipid greatly enhances plasmid DNA delivery and expression in mouse lung. *Proc Natl Acad Sci USA* 93:11454–9.

Yang, Y., Li, Q., Ertl, H.C.J., and Wilson, J.M. 1995. Cellular and humoral immune responses to viral antigens create barriers to lung directed gene therapy with recombinant adenoviruses. *J Virol* 69:2004–15.

Yang Y., Su, Q., and Wilson, J.M. 1996. Role of viral antigens in destructive cellular immune responses to adenovirus vector transduced cells in mouse lungs. *J Virol* 70:7209–12.

Yei, S., Mittereder, N., Tany, K., O'Sullivan, C., and Trapnell, B.C. 1994. Adenovirus mediated gene transfer for cystic fibrosis: quantitative evaluation of repeated in vivo vector administration in the lung. *Gene Ther* 1:192–200.

7
The Potential of Gene Therapy for Osteogenesis Imperfecta

CHRISTOPHER NIYIBIZI, PATRICK N. SMITH, and JOEL GREENBERGER

Introduction

Osteogenesis imperfecta (OI) is a heterogeneous group of genetic disorders that affect connective tissue integrity, with bone fragility being the cardinal feature. Evidence of this debilitating disorder dates back to 1000 B.C. Skeletons of mummies from this period reveal abnormal growth patterns similar to those of OI patients. Ivar the Boneless, the Danish prince who led the Scandinavian invasion of England, was believed to have such severe OI that he had to be carried into battle on a shield because he was unable to walk. The hallmarks of OI include bone fragility, osteoporosis, dentinogenesis imperfecta, blue sclera, easy bruising, joint laxity, and scoliosis. The severity of OI symptoms ranges from prenatal death to mild osteopenia without limb deformity. Variations in the clinical manifestations of OI relate directly to the heterogeneity of genetic defects in type I collagen genes (Ibsen 1967; Kuiviniemi, Tromp, and Prockop 1991). Although bone fragility is the hallmark of OI, other tissues in which type I collagen is the primary structural protein are also affected. These include skin, tendon, ligament, and dentin. Diagnosis is based upon clinical, genetic, and radiographic findings. The incidence of OI ranges from 1 per 20,000 to 1 per 50,000 live births (Ashton et al. 1980; Byers and Steiner 1992). There is no predilection for race, gender, or ethnic origin. Treatment depends mainly on the severity of the disease, with the primary goal being to minimize fractures and maximize function. Current treatment has included systemic chemotherapy, surgical intervention with intramedullary stabilization, and use of orthotics. Recent advances in the genetic analysis of OI have opened the door for potential genetic therapy for this disease. This chapter will discuss the molecular changes seen in OI, the current treatment options, and the most recent developments in genetic therapy.

TABLE 7.1. Sillence classification of osteogenesis imperfecta (OI).

Type	Mode of Inheritance	Symptoms
I	Autosomal Dominant	Osteoporosis, distinctly blue sclera, joint hyperlaxity, mild skeletal deformities, mild short stature, easy bruising, premature arcus senilis, dentigenesis imperfecta
II	Autosomal Recessive	Lethal prenatal form, severe skeletal deformities, hypotelorism, in utero fractures, beaded ribs, cor pulmonale
III	Autosomal Recessive	Progressive skeletal deformities, severe bone fragility, normal sclera, short stature, kyphoscoliosis, triangular facies
IV	Autosomal Dominant	Variable bone deformities, normal sclera, dentinogenesis imperfecta, short stature, progressive kyphoscoliosis

Classification

The earliest classification system, based on the age at onset of clinical manifestations, was developed by Looser in 1906. Patients with fractures occurring at birth were classified as congenita and those with fractures developing later in life were classified as tarda. Seedorf further subdivided this classification as tarda gravis and tarda levis based on fractures occurring within the first year of life. The most widely used system, however, was developed by Sillence, Senn, and Danks (1979). This classification system, which is based on clinical, genetic, and radiographic findings, subdivides OI into 4 phenotypes. Although widely accepted, many patients with OI still do not readily fall into any of Sillence et al.'s four classes due to the broad spectrum of molecular abnormalities resulting in OI.

Type I OI is the most common and the mildest form (Table 7.1). It is autosomal dominantly inherited and is characterized by mild bone fragility, with fractures occurring after moderate trauma, and by blue sclera and conductive hearing loss. It is subdivided into A and B based on the presence of dentinogenesis imperfecta. The genetic mutation is such that synthesis of normal type I collagen occurs, but in reduced amounts owing to a null α I allele. Type II is the most severe form, usually resulting in prenatal fatality. Infants with type II OI experience intrauterine fractures and intracranial hemorrhage following vaginal delivery, and succumb shortly after birth to pneumonia or respiratory insufficiency secondary to decreased thoracic size. Type III is a rare autosomal recessive disorder characterized by moderate to severe bone fragility. Intrauterine fractures are characteristic along with bone deformity, short stature, and triangular facies. These patients usually require multiple orthopaedic procedures throughout life. Type IV has some similarities to type I but with more severe osseous involvement and less extraskeletal abnormalities. Patients with type IV have normal hearing and normal sclera but short stature. The genetic defect in types II, III and IV results in a structurally abnormal synthesis of one of the chains of type I collagen.

Molecular Abnormalities

Despite the wide range of clinical manifestations observed in OI, all cases of OI have biochemical abnormalities that can be traced to genetic defects in the genes encoding for type I collagen. Collagen, of which there are several types, is a fibrous, structural protein found in all multicellular organisms. Presently nineteen distinct types of collagen have been identified. Broadly, the collagen family can be divided into two groups: the fibril-forming and nonfibril-forming collagens. As the word implies, the fibril-forming collagens form fibrils whereby collagen molecules are arranged in a quarter-staggered array and cross linked. Cross-linked collagen polymers are the functional units that provide structural support and strength to connective tissues. The fibril forming collagens include type I, II, III, V, and XI collagens. Type I collagen is the most abundant and is widely distributed in almost all connective tissues with the exception of hyaline cartilage. It is the major protein in bone, skin, tendon, ligament, sclera, cornea, and blood vessels. Type I collagen comprises approximately 90% of the entire protein content of bone and 97% of the total collagen present in bone. The only other collagen identified in bone is type V. The role of this collagen in bone is not known, but it has been shown to form intermolecular cross links with type I bone collagen (Niyibizi and Eyre 1994). Type V collagen is believed to play a role in regulating type I collagen fibrils, which has been demonstrated in studies of chick cornea and in knock-out mice lacking $\alpha 2$ (V) chain (Andrikopoules, Liu, Keen, Jaenisch, and Ramirez 1995). Type I collagen is composed of two identical $\alpha 1$ polypeptide chains and one distinct $\alpha 2$ chain. The three chains are twisted around each other like strands of rope. Each polypeptide chain is composed of about 1000 amino acids arranged in uninterrupted repetitions of Gly–X–Y triplets where X can be any other amino acid, but is usually a proline, and Y is often a hydroxyproline. Glycine is an absolute requirement in every third position because it is the smallest amino acid that can occupy the limited space in the center of the triple helix. Each α chain is synthesized in a precursor form called proα chain, which consists of globular extension peptides at the amino and the carboxy terminal ends. Proα chains assemble to form a precursor form of collagen called procollagen. Extracellular processing by specific enzymes removes the carboxy and amino terminal globular domains to generate collagen molecules. The collagen molecules self assemble in a quarter staggered array, undergo intermolecular cross linking, and form fibrils. Defects in collagen processing result in other types of connective tissue disorders such as Ehlers–Danlos syndrome. OI results from mutations in COL1A1 and COL1A2 genes that affect the amino acid sequence in the helical domain, causing either the complete lack of type I collagen production or the substitution of the conserved glycine with an amino acid with a bulky side chain (Prockop 1994; Culbert and Kadler 1996).

Clinical Manifestations

The clinical expression of OI traverses a broad spectrum from mild osteoporosis to prenatal death. Some common findings within each type have been outlined by Sillence et al. (1979). Musculoskeletal abnormalities include long bone deformities with anterior bowing of the humerus, tibia, and fibula, and lateral bowing of the femur, radius, and ulna. As stated above, the hallmark of OI is bone fragility with fractures occurring with minimal to moderate trauma. The number of fractures varies according to the severity of the disease. In general, the earlier in life the fractures occur, the more severe the disease. The lower limbs are more commonly involved, as they are more susceptible to trauma. The femur bone is the most commonly fractured long bone, with the fracture usually located at the convexity of the bone appearing transverse and minimally displaced. It is not uncommon for children to complain of minimal pain, since there is usually minimal soft tissue injury and they are accustomed to frequent fractures. Multiple fractures within the same bone often occur as a result of the severe angulation in which it heals and because of disuse atrophy, both of which make the bone more susceptible to a second fracture. Bowing of the long bones results from multiple transverse fractures and the pulling of strong muscles (Figure 7.1). Cranial deformity is also common. There is flattening of the posterior cranium with a bulging calvaria and a triangular shaped face. The forehead is usually broad with prominent parietal and temporal bones. Spinal deformities include such severe kyphoscoliosis that pulmonary complications are often seen. The incidence of spinal deformities ranges from 90% for type II OI to 10% to 40% for type I OI (Benson and Newman 1981). The most common spinal deformity is a thoracic scoliosis (Figure 7.2). Ligamentous laxity results in hypermobility of the joints and frequent dislocations. Cubitus varus with flexion contractures at the elbow is another common finding. Dislocations of the radial head, the hip joint, and the patellofemoral joint are also common occurrences increasing the incidence of falls and further fractures. Muscular hypotonia is seen secondary to ligamentous abnormalities and reduced activity. Due to the deficiency of dentin, the teeth of some OI patients are extremely brittle, breaking easily and becoming susceptible to caries. The enamel is usually normal since it is of ectodermal, and not mesenchymal, origin. If the teeth are affected in OI patients, this is

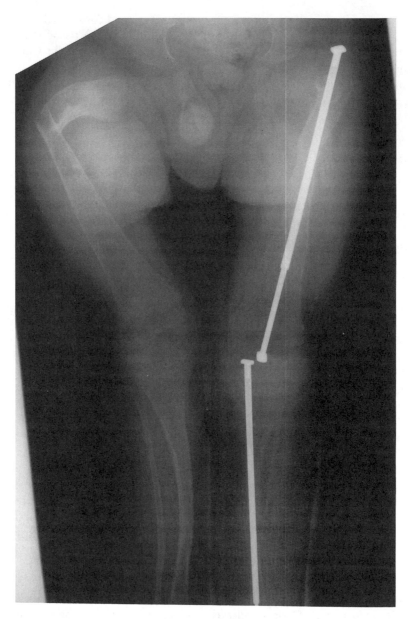

FIGURE 7.1. Radiograph of the lower extremities of an OI patient. This radiograph shows classic findings seen on X–ray, including valgus bowing of the tibia and multiple healed fractures in the left femur. Note the intermedullary fixation (Bailey rods) in the left femur and tibia. This is consistent with type III OI.

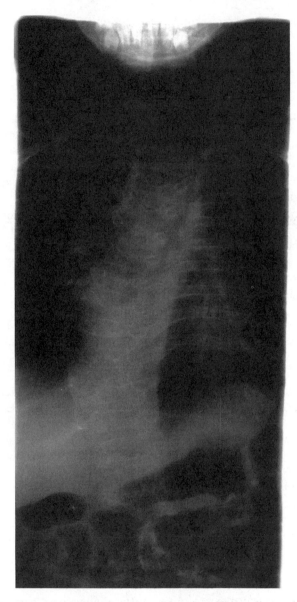

FIGURE 7.2. Radiograph of the spine of an OI patient. This radiograph shows a sco-
liotic curvature of the thoracic spine. Note the osteoporotic nature of the vertebral
bodies.

referred to as dentinogenesis imperfecta, a condition that has been used by Sillence (1981) to subclassify type I and type III OI. Extraskeletal findings include blue sclera due to abnormal corioid, hearing loss, and growth retardation. However, blue sclera, another hallmark characteristic of OI, is not found in all types. In type I, sclera are distinctly blue throughout life; in type III, they become less blue throughout life; and in type IV, the sclera are normal. Deafness occurs in approximately 40% of type I OI patients, with lower percentages in type IV. Hearing loss usually begins at adolescence and worsens with age, and results from conductive loss due to otosclerosis of the ossicles or from neurosensory loss due to compression of the auditory nerve as it exits the skull. In addition to loss of hearing, patients may also complain of tinitus and vertigo.

Radiographic findings include marked, generalized osteoporosis with thin cortices. Multiple stages of fracture healing reflect the numerous fractures sustained by these patients. Malunions of these multiple fractures result in shortened long bones with severe bowing. The cortices are characteristically thin with occasional thickened areas secondary to callus formation. Fairbank (1948) described three types of radiographic findings (1) thick bone is seen at areas of prior fractures where there is large callus formation; (2) thin bone appears with very narrow shafts, thin trabeculae, thin cortices, and marked osteopenia; and (3) cystic bone is a complication of immobilization from fracture treatment. The lack of ambulation and normal stress placed upon the bones to stimulate bone formation results in a cystic honeycomb pattern. Flaring of the metaphyses is present, indicating abnormal bone modeling. Goldman, Davidson, Pavlov, and Bullough (1980) described a classic finding in children with OI called "popcorn metaphyses" caused by the ossification of fractured physeal cartilage. Radiographs of the spine reveal flattened or biconcave vertebral bodies with resultant kyphoscoliosis.

Histological appearance of the diseased bone varies with the severity of the clinical picture. In general, there appears to be a mixture of woven and lamellar bone patterns, with a poorer clinical disease state being associated with less compact bone. Hypercellularity is also a commonly found feature, with larger than normal osteocytes (Kocher and Shapiro 1998).

Common laboratory findings include normal calcium and phosphorus, and an elevated alkaline phosphatase, although this is not always the case. Because laboratory data can vary, the diagnosis of OI is based upon clinical and radiographic findings.

Current Orthopaedic Management

The principles for management of OI are to maximize function and minimize fracture occurrence. Treatment plans are designed to correct the existing deformities and to avoid future deformities by combining state of the art fracture care with prophylactic management of brittle bones using internal supports and external orthotic devices. A multidisciplinary approach, including the orthopaedic surgeon, pediatrician, physical therapist, and social worker, can maximize the independent function of the child, provide education to the family, and improve the patient's social integration. The level of maximum function varies depending on the severity of the disease and age of the patient. Thus the extent of operative and nonoperative treatment required also varies from patient to patient. In younger children, fractures occur due to falls because of lack of coordination and adequate postural strength. As a child matures, the incidence of fractures decreases, but other complications of OI begin to manifest themselves, such as scoliosis or hearing loss. It has been reported that patients with type III and IV OI require the greatest amount of care (Bleck 1981). With adequate treatment, most patients with OI can have a normal life expectancy and lead very productive lives. Usually they can attend regular schools, enjoy a wide range of career and lifestyle choices, and experience fulfilling relationships.

Treatment starts early in life, ranging from basic concepts, such as instructions for proper handling and transfer, to complex operative procedures. Rehabilitation medicine has an important role in the treatment plan for patients with ambulation potential. It provides strength and conditioning of both the upper and lower extremities. It also assists in developing adequate head and trunk control. With the coordination of a physical therapy team, improved muscle strength allows for ambulation. This is accomplished by using such modalities as hydrotherapy, passive and active range of motion, and aerobic exercises (Gerber, Bionde, and Wentrob 1990). Increased stresses on the long bones of the lower extremities increase bone density. Repetitive fracturing and casting, particularly of the lower extremities, immobilizes the limb, causing disuse atrophy. Therefore, a strengthening program must start as soon as fracture healing occurs. Joint instability and malalignment from ligament laxity may require orthoses for ambulation. Lightweight knee–ankle–foot orthoses (KAFOs) restore proper alignment for normal ambulation. Bracing for spinal deformities has been used with limited success, although there is restricted application in patients with preexisting chest wall deformities and fragile rib cage.

Surgical management can be used for either fracture repair or limb deformity. The correct age for operative intervention is around 5 to 7 years of age. Until that age, closed treatment of fractures is the most widely accepted means of treatment. After that time, surgical correction of severe deformities that interfere with a patient's functional status may be used. Such interventions include osteoclasis, intramedullary nailing, and osteotomies, or a combination of these techniques. Elongating intramedullary rods have also been used with similar results to use of nonelongating rods (Porat et al. 1991).

Systematic Therapy

In addition to orthopaedic management, systemic therapy has been attempted, but has met with little success. Some examples of these agents include calcium, fluoride, calcitonin, anabolic steroids, and magnesium. Because the incidence of fractures decreases after puberty, sex hormones have been used, but have been of no benefit. Drugs which were initially indicated for osteoporosis have also been investigated. Calcitonin, an osteoclastic inhibitor, has been used in osteoporosis to increase total bone mass. No study, however, has shown any clinical improvement in fracture occurrence with the administration of calcitonin. Recently, however, the use of bisphosphonates in children with OI has produced some beneficial effects. Bisphosphonates, which are compounds that inhibit osteoclastic activity, have been used in the treatment of Paget's disease. Treatment is aimed at reducing bone resorption, which has been found by histomorphometric and biochemical studies to be abnormally high. Bembi et al. (1997) reported a clear clinical response with bisphosphonate over a 22- to 29-month treatment period, with a striking reduction in the frequency of new fractures. They also observed an effect on bone density; no notable adverse effects during therapy were observed. In another study, a 2-year treatment with pamidronate showed a marked reduction in bone pain in as early as 1 week in addition to a reduction in the incidence of fractures (Glorieux et al. 1998). Decreased osteoclastic activity was indicated by a reduction in serum levels of calcium, phosphate, and alkaline phosphatase. Radiographic evidence of increased bone density above normal age-related increases was noted over the 2-year treatment period as measured by x–ray absorptiometry. Other radiographic evidence of increased bone density included increased vertebral body height, formation of dense lines, and increased cortical width. Although bone density appeared to improve, no biomechanical analysis was performed to demonstrate enhanced structural integrity. These studies show promising possibilities for the use of systemic therapy in the treatment plan for OI; however, since OI is a genetic disease, systemic therapy is far from providing a cure.

Stem Cell Therapy

Although the treatment options available for OI patients have improved their lifestyle, there is still no cure for OI. Gene therapy and novel approaches using stem cells harvested from bone marrow to replace cells synthesizing defective molecules in the affected individuals are currently being investigated. Most OI mutations result from point mutations that substitute the conserved glycine with a charged amino acid or an amino acid with a bulky side chain that destabilizes the triple helix. These types of mutations lead to a dominant negative situation in which abnormal α chains are synthesized and then associate with normal chains. Consequently, either a decrease in the amount of type I collagen present in the extracellular matrix or a formation of abnormal collagen fibrils occurs, followed by abnormal mineralization. Stem cell therapy is currently being investigated for dominant negative mutations. In this approach, normal cells from a normal individual are used to replace the mutant cells of an individual with OI, the object being that enough normal cells will be supplied and that the supplied cells will synthesize sufficient normal matrix to have an effect on tissue function. Bone marrow has been shown to contain cells with the potential to differentiate into osteoblasts. The aim here is to isolate cells from the bone marrow and then transplant them into the affected individual. The stem cells of hemapoietic lineage can self renew: through transplantation of bone marrow into an osteopetrotic patient, a defect in bone resorption due to the failure of osteoclasts to resorb bone, has previously been demonstrated (Cocia et al. 1980). Since osteoclasts are believed to be derived from the cells of the hemopoietic lineage, bone marrow transplantation has been shown to reverse the defects in these patients. The treatment here for OI is, therefore, based on the same premise: there are stem cells in bone marrow that will give rise to osteoblasts in vivo once the cells are transplanted into an individual with OI. Stem cells in the marrow with the potential to give rise to a variety of cells of nonhemopoietic tissues are referred to as mesenchymal stem cells or marrow stromal cells. In the present discussion the two terms are used interchangeably.

All the organic components of bone are synthesized by *osteoblasts*, differentiated, nonproliferating cells with specific morphologic characteristics. In culture, isolated osteoblasts have been shown to synthesize several proteins localized to bone, including type I collagen, alkaline phosphatase, and osteocalcin (Beresford, Gallagher, Poser, and Russel 1984; Ashton et al. 1985). In addition, cultured osteoblasts are capable of mineralizing the extracellular matrix. Osteoblasts are believed to arise from mesenchymal stem cells that reside in the bone marrow, whereas osteoclasts, the cells involved in bone resorption, are believed to derive from hemopoietic lineage (Ash, Loutif, and Townsend 1980; Cocia 1980). The ability for bone to regenerate is attributed to quiescent stem cells in bone that undergo proliferation and differentiation. Mesenchymal stem cells have the potential to proliferate and can undergo selfrenewal. Bone marrow has long been recognized as the source of osteoprogenitor cells (Friedenstein, Chailakhjan, and Lalykina 1970; Beneyahu, Kletter, Zipori, and Weintraub 1989) as well as other cell types. Friedenstein (1976) demonstrated that bone formation occurred when bone marrow was placed in diffusion chambers and implanted intraperotineally. Histological studies demonstrated that the tissue within the chambers was calcified and morphologically resembled bone. Similar studies by Ashton et al. (1985) showed that, when placed in diffusion chambers, bone marrow isolate produced both bone and cartilage in vivo. Electron microscopy analysis revealed that some ultrastructural features of the bone formed in vivo resembled that of normal skeletal bone. In addition, implantation of adherent bone marrow derived cells from different animal species into ceramics that were implanted subcutaneously into mice demonstrated that the cells have potential to form bone and cartilage. These findings demonstrated the existence of determined precursors within the marrow stroma with the capacity for differentiation in an osteogenic direction independent of any cell–cell or cell–matrix interactions (Beresford 1989).

Marrow stromal tissue is a heterogeneous collection of several different cell types. Separation of the mesenchymal stem cells from the initial bone marrow isolate is based on the inability of hemopoietic stem cells to adhere and survive media changes. It has been shown that hemopoietic stem cells comprise less than 30% of the initial culture and significantly decrease after a few weeks in culture. As discussed above, bone marrow stromal cells can differentiate into fibroblasts, chondrocytes, adipocytes, and osteoblasts (Aubin, Turksen, and Heersch 1993). This is supported by a recent finding by Pereira et al. (1995), who showed that when bone marrow stromal cells expressing a type I collagen minigene were infused into normal mice, the cells expressing the minigene populated and persisted in several different tissues of the recipient mice. In this particular study, bone marrow cells harvested from a transgenic mouse containing the human proα 1 (I) minigene and partially enriched in mesenchymal stem cells by brief culturing were infused into irradiated mice. The fate of the cells expressing the human minigene was followed in the recipient mice up to 150 days. The cells containing the minigene were detected by polymerase chain reaction (PCR) by amplification of DNA extracted from different tissues of the recipient mice. Using this technique, the authors reported that 5% to 10% of the bone cells were derived from the donor mice in whole bone tissue. The cells were also detected in lung, spleen, and cartilage, and persisted in these tissues up to 150 days. The authors concluded that these cells could serve as long-lasting precursors for different organs (Periera et al. 1995; Prockop 1997). Based on these findings, a clinical trial was started at St. Jude Children's Research Hospital to assess the effectiveness of cells from the marrow in reversing OI defects in children with severe osteogenesis imperfecta (Horwitz et al. 1996). The effectiveness of this treatment in OI patients currently remains unknown.

Although infusion of bone marrow stromal cells into mice showed that cells engrafted in mouse bones, the mice had to be irradiated before cell infusion to achieve cell engraftment. In our own study, bone marrow stromal cells transduced with a retroviral vector expressing the Lac–Z marker gene were shown to persist in bone up to 40 days after infusion into nonirradiated mice. There are several other reports showing that infusion of whole marrow or stromal cells into dogs leads to cell engraftment without prior marrow ablation (Stewart, Crittendon, Lowry, Pearson-White, and Quesenbery 1993). These observations suggest that it may possible to infuse bone marrow stromal cells into OI patients without prior irradiation. The limitation of this approach, however, is that cell targeting would have to be performed to supply enough cells to achieve function. Although cells can be infused in mice and detected in different organs, the number of cells capable of differentiation into osteoblasts within the population of cells obtained from the marrow is very small. Presently, we are cloning such cells within the initial isolate of marrow cells. A great deal of evidence suggests that cells with the potential to differentiate toward osteoblastic lineage can be isolated from marrow by cloning of cells from the initial marrow cell isolate. Using this technique, we have shown that, by limiting dilution, the cells exhibiting higher responses to recombinant human bone morphogenetic protein-2 (rhBMP-2), based on their ability to express alkaline phosphatase activity, can be isolated from the initial isolate of cells from marrow. Other investigators have shown that cloned cells exhibiting the potential to differentiate into a variety of cells can be isolated from bone marrow by cloning (Diduch, Coe, Joyner, Owen, and Balian 1993). These observations suggest that, in order for stem cell therapy to be effective, selection of cells from the initial marrow isolate prior to infusion into the recipient, with potential to differentiate toward osteoblast lineage may lead to better cell targeting and engraftment.

Gene Therapy

As discussed above, the most common mutations in OI patients are point mutations that substitute conserved glycine with a charged amino acid or an amino acid having a bulky side chain that destabilizes the triple helix. These types of mutations lead to the abnormal production of α chains which associate with normal chains. A negative dominant mutuation occurs when the phenotype determined by the mutant allele predominates over the normal allele. The defective molecules may either be degraded intracellularly or may be secreted and assembled into defective collagen fibrils in the extracellular matrix. Supplying normal genes in this situation to substitute for the defective gene would not correct the defect. In order for the introduced gene to be functional, the expression of the mutant allele would first have to be suppressed. Several approaches are being investigated to eliminate the expression of the mutant collagen allele. One approach is antisense therapy, which is designed to suppress the expression of the mutant allele, thus reducing the severity of the disease to that of a milder phenotype. For this approach, short molecules with a sequence complementary to the mutant RNA are employed to bind to the mutant RNA. These sequences can either cleave the target RNA themselves or target the bound site for cleavage by other enzymes. Using this approach, Colige et al. (1999) demonstrated its feasibility in mouse fibroblasts transfected with an exogenous human minigene expressing truncated proα (I) chain. In this system, the mouse expressed full length proα1 (I) and internally deleted proα (I) human procollagen (Colige et al. 1994). When the oligonucleotides designed to bind to the mutant RNA were added to the cells, the expression of the human minigene for the truncated proα (I) chains were inhibited by 50 to 80% without significantly affecting the normal expression of the mouse proα1 (I) chains. In another related study, which used fibroblasts isolated from a patient affected with OI type IV, antisense oligonucleotides were shown to selectively suppress the expression of the mutant α2(I) collagen allele. In this study, the mutant protein was suppressed to 44% to 47%, and the mutant mRNA was suppressed to 37% to 43% of the level of the control fibroblasts. The authors also reported that by suppressing the mutant allele, the normal allele was also suppressed to 80% of its usual level (Wang and Marini 1996). These findings are encouraging, but in order to be of therapeutic use, specificity of binding will have to be achieved to completely eliminate the mutant allele without affecting the normal allele.

To achieve specificity, an alternative approach is the use of ribozymes (Marini and Gerber 1997), which are short RNA molecules composed of a hairpin loop with two binding arms complimentary to the mutant mRNA. The binding arms position the target mRNA so that the catalytic core on the hairpin loop is exactly opposite the cleavage site on the target mRNA. Marini and Gerber (1997) reported in vitro cleavage activity of 5 different hammerhead ribozymes directed against normal human collagen transcripts and against a mutated mouse COL1A1 gene transcript. The authors looked at the competitive effects of both total RNA transcripts and normal RNA transcripts on ribozyme cleavage activity. They found that ribozymes were able to localize and cleave specific targets, even in the presence of vast amounts of RNA. By eliminating expression of mutant genes, antisense therapy has the potential to reduce the clinical severity of OI from a severe to a mild form.

Null mutations in which a gene is absent or is inactivated would be more amenable to gene replacement than negative dominant mutations. We are currently investigating the feasibility of gene therapy for OI null mutations by studying a mouse model of human OI (*oim*) that has defective synthesis of proα 2(I) chains (Chipman et al. 1993). This naturally occurring mutation in mice, identified in the mutant mouse resource of Jackson Laboratory and called *oim*, generates phenotypic and biochemical features similar to those seen in a moderate to severe human OI. Mice homozygous for the osteogenesis imperfecta murine (*oim* +/+) mutation are deficient in proα2 (I) collagen production because of a G deletion at nucleotide 3983 of the COL1A2 gene. This mutation leads to nonincorporation of proα2 (I) chains into heterotrimers and results in accumulation of α1 (I) homotrimers in tissues. As a result, these mice exhibit cortical thinning, bowing of long bones, fractures, and callus formation characteristic of human OI type III, presenting an excellent model for investigating the feasibility of cell or gene therapy for some subsets of OI, especially those involving null mutations.

In an evaluation of the potential of gene therapy for OI null mutations, bone marrow cells were isolated from the femurs and tibias of *oim* mice. The adherent cells were characterized for potential to form bone in vitro and in vivo (Balk et al. 1997; Oyama et al. In press). The isolated cells exhibited osteoblast phenotype in vitro as indicated by the expression of alkaline phosphatase activity and synthesis of osteocalcin when treated with rhBMP-2. The cells from the OI mouse model were easily transduced with both adenoviral and retroviral vectors. The cells transduced with the retroviral vector containing the Lac–Z marker gene were assessed for potential to differentiate toward an osteoblast lineage when treated with rhBMP-2. The transduced cells exhibited alkaline phosphatase activity when treated with rhBMP-2, suggesting that they retained the ability to express osteoblast phenotype even after gene transduction. In addition, when the transduced cells were assessed for potential to form bone in vivo, cells suspended in ceramic cubes and implanted subcutaneously in mice formed cartilage at 3 weeks after implantation and bone after 6 weeks of implantation. X–gal staining of the ceramic cubes after 6 weeks of implantation indicated that the cells in ceramic cubes had differentiated into osteoblasts and osteocytes and maintained high levels of gene expression (Oyama et al. In press). These data indicate that cells transduced with exogenous genes retain the ability to form bone in vivo. When the cells transduced with the Lac—Z marker gene were infused into the femurs of the *oim* mice, the cells populated the injected femurs and contrallateral femurs of the same mice. The Lac–Z positive cells persisted in bone and were detected in the injected and contrallateral femurs up to 40 days after cell infusion. The cells that persisted in bone by 30 and 40 days after injection were located along the surface of the trabecular bone, suggesting that the cells may differentiate into osteoblasts in vivo (Figure 7.3). By PCR analysis, the cells were also detected in lung and liver of the injected mice but only at 4 and 10 days after injection, suggesting that the cells entered the circulatory system and transiently occupied the lung and liver (Figure 7.4). Besides transduction of bone marrow stromal cells with marker genes, a variety of other genes may be used as well. The cells isolated from *oim* mice were transduced with a retroviral vector expressing human growth hormone (Suzuki et al. 1998; Niyibizi et al. 1999). When cells expressing the growth hormone were injected in *oim* mouse bones, the cells expressed the human growth hormone in bone and were also detected by ELISA in mouse serum. The cells expressing human growth hormone were recovered from injected femurs as well as contrallateral femurs. These data confirmed Lac–Z gene expression and indicated not only that cells with osteogenic potential can deliver genes to bone, but also that genes can be expressed in bone with high efficiency.

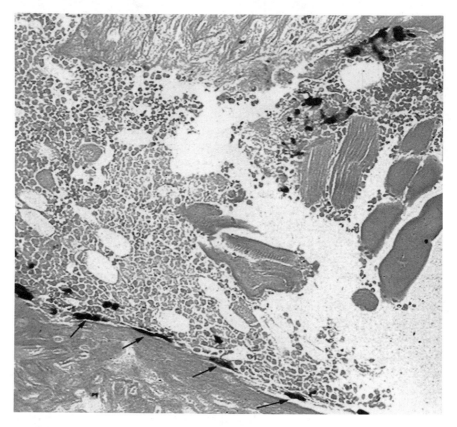

FIGURE 7.3. Histological section of femur harvested from *oim* mouse transduced with Lac–Z retrovirus. This section was taken at 40 days after injection. It shows Lac–Z positive cells along the surface of the trabecular bone. The data suggest that the cells may differentiate into osteoblasts in vivo.

Although we have shown that bone marrow stromal cells can deliver genes to bone and that the genes can be expressed in bone, the number of cells that persist in bone remains quite small. Better targeting would therefore be needed if gene therapy for OI is to be of benefit. To improve cell targeting, bone marrow stromal cells are being transduced with genes that direct the cells toward osteoblastic lineage. We are currently investigating the therapeutic potential of growth factors that induce the differentiation of bone marrow stromal cells toward osteogenic lineage, enhancing the homing and engraftment of bone marrow stromal cells into recipient mouse bones. In vitro, these cells have been shown to respond to several different factors, some of which influence their osteogenic potential. These factors include bone morphogenetic proteins (Vukicevic, Luyten, and Reddi 1989; Theis et al. 1992), glucocorticoids (Simmons et al. 1991; Kammalia, McCul-

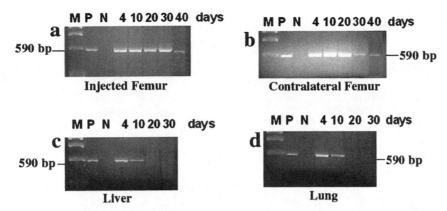

FIGURE 7.4. PCR amplification of DNA extracted from femurs, liver, and lung of *oim* mouse that received bone marrow stromal cells transduced with BAG–Lac–Z Neo[r]. The data show evidence of the persistence of the cells within the bone but cleared from other tissues. M, marker; P, positive control (Lac–Z positive cells); N, negative control (no DNA template). The expected amplified fragment is 590 bp.

louch, Tanebaum, and Lindback 1992), and basic fibroblast growth factor (Pitaru, Kotev–Emeth, Noff, and Savion 1990). To evaluate the ability of these factors to influence homing, bone marrow stromal cells isolated from normal mice were transduced with an adenovirus containing the gene for BMP–2. The cells expressing BMP–2 and Lac–Z marker genes were injected into the femurs of *oim* mice with the fate of the cells followed by X–gal staining. Seven days following injection, a large number of Lac–Z positive cells were located on the surface of trabecular bone as well as in the bone proper (Figure 7.5) (unpublished data). BMP–2 may not be an ideal growth factor to use for cell targeting because the cells treated with BMP–2 are already differentiated into osteoblasts and thus no longer regarded as stem cells. Nevertheless, this approach could be potentially applicable if factors that would maintain the cells as stem cells of osteoblast lineage without differentiating them are used to target cells to bone. One such factor, Cbfa1, an osteoblast transcription factor, may be useful in cell targeting (Ducy, Zhang, Geoffroy, and Karsenty 1997).

Currently, gene therapy for genetic diseases is hampered by lack of vectors that can deliver genes to be expressed in vivo for the life of the organism. In addition, for a structural protein like collagen, this is a formidable task since high expression of the gene would be required, especially for bone that undergoes constant remodeling. Though collagen genes are quite large, approximately 4.2 kb for COL1A1 and COL1A2 genes, retroviruses can handle genes of this size, and it has been shown previously that the proα (I) collagen gene can be inserted in a retroviral vector and expressed (Stacey, Mulligan, and Jaenisch 1987). We have constructed an adenoviral vector containing the mouse proα2 (I) gene; transduction of the

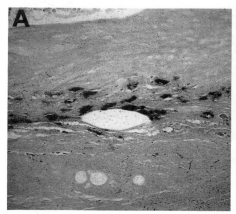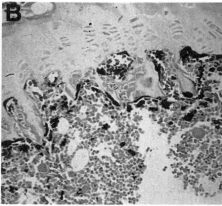

FIGURE 7.5. Histological sections of *oim* mouse femurs that received normal litter-mate bone marrow stromal cells transduced with an adenoviral vector expressing BMP–2 gene at seven days postinjection. (A) Numerous cells are located in bone proper and (B) on the surface of the trabecular bone around the growth plate.

bone marrow stromal cells from *oim* mice that have defective synthesis of proα2 (I) chains demonstrates that the gene can be expressed by the *oim* cells and the synthesized proα chain is able to associate with the proα (I) chain to form heterotrimers (unpublished data). Although these data are encouraging, more work still lies ahead before gene therapy for OI can become a reality.

Conclusion

Treatment for osteogenesis imperfecta has improved dramatically over the past 20 years. A team effort drawing researchers from several specialty fields has allowed patients with OI to lead productive lives. Currently, progress made in the fields of tissue engineering and gene therapy has brought new treatment ideas to the forefront. Much more progress is needed, however, in both of these fields before a cure by these means can be seriously discussed. Because of its current limitations, gene therapy for OI is still far from reality.

References

Andrikopoules, K., Liu, X., Keen, D.R., Jaenisch, R., and Ramirez, F. 1995. Targeted mutation in Col5a2 gene reveals a regulatory role for type V collagen during matrix assembly. *Nat Genet* 9:31–6.

Ash, P., Loutif, J.F., and Townsend, M.S. 1980. Osteoclast derived from hemopoietic cells. *Nature* 283:669–70.

Ashton, B.A., Abdullah, F., Cane, J., Williamson, M., Sykes, B.C., Couch, M., and Poser, J.W. 1985. Characterization of cells with high alkaline phosphatase activity derived from human bone and marrow: preliminary assessment of their ostegenecity. *Bone* 6:313–19.

Ashton, B.A., Allen, T.D., Howlett, C.R., Eagleson, C.C., Hattori, A., and Owen, M. 1980. Formation of bone and cartilage by marrow stromal cells in diffusioin chambers in vitro. *Clin Orthop Rel Res* 151:294–307.

Aubin, J.E., Turksen, K., and Heersch, J.N.M. 1993. Osteclastic lineage: In Noda, M. (ed). *Cellular and molecular biology of bone* 1–45. New York: Academic Press.

Balk, L., Bray, J., Day, C., Epperly, Greenberger, J.H., Evans, C.H., and Niyibizi, C. 1997. Effect of rhBMP-2 on osteogenic potential of bone marrow stromal cells from an osteogenesis imperfecta mouse. *Bone* 21:7–15.

Bembi, B., Parma, A., Bottega, M., Ceschel, S., Zanatta, M., Martini, C., and Ciana, G. 1997. Intravenous pamidronate treatment in osteogenesis imperfecta. *J Pediat* 131:622–5.

Benayahu, D., Kletter, Y., Zipori, D., and Weintraub, S. 1989. Bone marrow derived stromal cell line expressing osteoblastic phenotype in vitro and osteogenic capacity in vivo. *J Cell Physiol* 140:1–7.

Benson, D.R. and Newman, D.C. 1981. The spine and surgical treatment in OI. *Clin Orthop* 159:147–53.

Beresford, J.N., Gallagher, J.A., Poser, J.W., and Russel, R.G.G. 1984. Production of osteocalcin by human bone cells in vitro. Effects of 1,25(OH)2D3, 24,25(OH)2D3, parathyroid hormone, and glucocorticoids. *Metab Bone Dis Rel Res* 5:229–34.

Beresford, J.N. 1989. Osteogenic stem cells and the stromal system. *Clin Orthop* 240:270–9.

Bleck, E.E. 1981. Nonoperative treatment of OI: orthotic and mobility management. *Orthop Clin* 159:111–22.

Byers, P.H. and Steiner, R.D. 1992. Osteogenesis imperfecta. *Ann Rev Med* 43:269–82.

Chipman, S.D., Sweet, H.O., McBride, D.J., Davison, M.T., Marks, S.C., Shudiner, A.R., Wentstrup, R.J., Rowe, D.W., and Shapiro, J.R. 1993. Defective pro alpha 2(I) collagen synthesis in a recessive mutation in mice: a model of osteogenesis imerfecta. *Proc Natl Acad USA* 90:1701–5.

Cocia, P.F., Krivitt, W., Cervenka, J., Clawson, C., Kersey, J.H., Kim, T.H., Nesbit, N.E., Ramsey, N.K., Waskinten, P.I., Tietebaum, S.I., Kahn, A.J., and Brown, D.M. 1980. Successful bone marrow transplantation for juvenile malignant osteopetrosis. *N Engl J Med* 302:701–8.

Colige, A., Sokolov, B.P., Nugent, P., Baserga, D.J., and Prokop, D.J. 1994. Use of an antisense oligonucleotides to inhibit the expression of a mutated human procollagen gene (COL1A1(I)) in transfected mouse 3T3 cells. *Biochem* 32:7–11.

Culbert, A.A. and Kadler, K.E. 1996. Tracing the pathway between mutation and phenotype in OI: isolation of mineralization of specific gnes. *Am J Med Genet* 63:167–74.

Diduch, D.R., Coe, M.R., Joyner, C., Owen, M.E., and Balian, G. 1993. Two cell lines from bone marrow that differ in terms of collagen synthesis, osteogenic characteristics, and matrix synthesis. *J Bone Joint Surg* 75A:92–105.

Ducy, P., Zhang, R., Geoffroy, V., and Karsenty, G. 1997. Osf2/Cbfa1: a transcriptional activator of osteoblast differentation. *Cell* 89:747–54.

Fairbank, H.A.T. 1948. Osteogenesis imperfecta and osteogenesis imperfecta cystica. *J Bone Joint Surg* 30-B:164.

Friedenstein, A.J., Chailakhjan, R.K., and Lalykina, K.S. 1970. The development of fibroblast colonies in monolayer cultures in guinea pig bone marrow and spleen cells. *Cell Tissue Kinet* 3:292–403.

Friedenstein, A.J. 1976. Precursor cells of mechanocytes. *Int Rev Cytol* 47:327–59.

Gerber, C.H., Bionde, H., and Wentrob, J. 1990. Rehabilitation and infants with osteogensis imperfecta: a program for ambulation. *Clin Orthop* 251:254–62.

Glorieux, F.H., Bishop, N.J., Plotkin, H., Chabot, G., Lanoue, G., and Travers, R. 1998. Cyclic administration of pamidronate in children with severe osteogensis imperfecta. *N Engl J Med* 339:947–52.

Goldman, A.B., Davidson, D., Pavlov, H., and Bullough, P.G. 1980. "Popcorn" calcifications: a prognostic sign in OI. *Radiology* 136:351.

Horwitz, E.M., Prockop, D.J., Marini, J., Fitzpatrick, R., Peyritz, R., Sussman, M., Orchid, P., and Brenner, M.K. 1996. Treatment of severe osteogenesis imperfecta by allogenic bone marrow transplantation. *Matrix Biol* 15:188.

Ibsen, K.H. 1967. Distant varieties of OI. *Clin Orthop* 50:279.

Kamalia, N., McCullouch, C.A.G., Tanebaum, H.C., and Lindback, H. 1992. Dexamethasone recruitment of selfrenewing osteogenitor cells in chick bone marrow stromal cell cultures. *Blood* 79:320–6.

Kocher, M.D. and Shapiro, M.D. 1998. Osteogenensis Imperfecta. *J Am Acad Orthop Surg* 6:225–36.

Kuivaniemi, H., Tromp, G., and Prockop, D.J. 1991. Mutations on collagen genes: causes of rare and some common diseases in humans. *FASEB J* 5:2050–2.

Marini, J.C. and Gerber, N.L. 1997. Osteogenesis imperfecta: rehabilitation and prospects for gene therapy. *JAMA* 277:746–50.

Nibiyizi, C. and Eryre, D.R. 1994. Structural characteristics of cross linking sites in type V collagen of bone, chain specific, and heterotypic links to type I collagen. *Eur J Biochem* 224:934–50.

Niyibizi, C., Faulcon, L., Suzuki, K., Goto, H., and Mi, Z. 1999. Comparison of matrix synthesis by bone marrow stromal cells from a mouse model of human osteogenesis imperfecta (*oim*) and normal littermates. 45th Annual Meeting. *Trans Orthop Res Soc.*

Oyama, M., Tatlock, A., Johnstone, B., Nishimura, K., Evans, C.H., and Niyibizi, C. In press. Retrovirally transduced bone marrow stromal cells from a mouse model of human osteogenesis imperfecta (*oim*) retain the ability to form bone and cartilage in vivo after extended passaging. *Gene Ther.*

Pereira, R.F., Halford, K.W., O'Hara, M.D., Leeper, D.B., Sokolov, B.P., Pollard, M.D., Bagasra, O., and Prockop, D.J. 1995. Cultured adherent cells from marrow can serve as long lasting precursor cells for bone, cartilage, and lung in irradiated mice. *Proc Natl Acad Sci USA* 92:4857–61.

Pitaru, S., Kotev–Emeth, S., Noff, D., and Savion, N. 1990. Basic fibroblastic growth factor (BFGF) enhances the capacity of rat stromal bone marrow cells to form mineralized bone–like tissue in culture. *J Bone Mineral Res (Suppl 2)* 5:S78.

Porat, S., Heller, E., and Seidamn, D.S. 1991. Functional results of operations in OI: elongating and nonelongating rods with pediatric orthopaedics. *J Pediatr Orthop* 11:200–3.

Prockop, D.J. 1997. Marrow stromal cells as stem cells for non–hemopoietic tissues. *Science* 276:71–4.

Prockop, D.J. 1994. Molecular basis of osteogenesis imperfecta and related disorders of bone. *Clin Plast Surg* 21:407–13.

Sillence, D.O. 1981. OI: an expanding panorama of varients. *Clin Orthop* 159:11–12.

Sillence, D.O., Senn, A., and Danks, D.M. 1979. Genetic heterogeneity in osteogenesis imperfecta. *J Med Genet* 16:101–22.

Simmons, D.J., Seitz, P., Kidder, L., Klein, G.L., Waltz, M., Gundberg, C., Tabuchi, C., Yang, C., and Zhang, R.W. 1991. Partial characterization of rat stroam cells. *Calcif Tissue Int* 48:326–34.

Stacey, A., Mulligan, R., and Jaenisch, R. 1987. Rescue of type I collagen deficient phenotype by retroviral–vector–mediated transfer of human proα 1(I) collagen gene into Mov–13 cells. *J Virol* 61:2549–54.

Stewart, F.M., Crittendon, R.B., Lowry, P.A., Pearson–White, S., and Quesenbery, P.J. 1993. Long term engraftment of normal and post–5 flourouricil murine mouse into normal nonmyeloablated mice. *Blood* 81:2566–71.

Suzuki, K., Oyama, M., Kavalkovich, K., Faulcon, L., Robbins, P.D., Evans, C.H., and Niyibizi, C. 1998. Expression of human growth hormone in vitro and in vivo by bone marrow stromal cells from *oim* mice. 44th Annual Meeting. *Trans Orthop Res Soc* 23:15.

Theis, N., Bauduy, M., Ashton, B.A., Kurtzberg, L., Wozney, J.M., and Rosen, V. 1992. Recombinant human bone morphogenetic protein–2 induces osteoblastic differentiation in W–20–17 stromal cells. *Endocrinology* 130:1318–24.36.

Vukicevic, S., Luyten, F.P., and Reddi, A.H. 1989. Stimulation of the expression of osteogenic and chondrogenic phenotypes in vitro by osteogenin 2. *Proc Natl Acad Sci USA* 86:8793–893.

Wang, Q. and Marini, J.C. 1995. Antisense oligonucleotides selectively supress expression of the mutant α 2(I) collagen allele in type IV osteogenesis imperfecta fibroblasts. *J Clin Invest* 97:448–54.

8
The Role of Nitric Oxide as a Candidate Molecule for Gene Therapy in Sports Injuries

Scott A. Rodeo and Kazutaka Izawa

Introduction

The free radical nitric oxide (NO•) is a small molecule synthesized from the amino acid L–arginine by the enzyme nitric oxide synthase. Because of its small size and unpaired electron, this molecule is freely diffusible and highly reactive. Nitric oxide is a fundamental participant in the basic biology of sports related injury. This important molecule plays a role in the physiology of virtually all of the musculoskeletal tissues, including articular cartilage, tendon, ligament, muscle, bone, intervertebral disc, and synovium. Nitric oxide is also involved in inflammation, muscle contraction, and tendon healing. Since NO• functions as an intercellular and intracellular messenger molecule, it has been implicated in the response of these tissues to injury and repair. Because of its important role in these basic processes, NO• represents a potentially valuable candidate molecule for regional gene therapy. This chapter will examine the role of NO• in basic processes relevant to the healing of sports injuries and will explore the potential use of gene therapy to affect NO• in cartilage, tendon, ligament, and muscle (Table 8.1).

Nitric oxide synthase (NOS), the enzyme which is responsible for NO• synthesis, is a potential target for regional gene therapy. Nitric oxide synthase is present in both constitutive and inducible isoforms. Constitutive isoforms of NOS are present in the brain and vascular endothelium (Evans, Stefanovic–Racic, and Lancaster 1995). Nitric oxide produced by cardiovascular tissues is involved in the regulation of vascular tone by vasodilation, the inhibition of platelet aggregation, and the regulation of cardiac contractility. Nitric oxide produced in the central nervous system plays an important role in memory and motor function. In the peripheral nervous system, NO• is involved in mediating neurogenic vasodilatation and regulating various gastrointestinal, respiratory, and genitourinary tract functions. Inducible NOS (iNOS) is present in articular chondrocytes, macrophages, neutrophils, lymphocytes, and hepatocytes (Ambring, Benthin, Petersson, Jungersten, and Wennmalm 1994). The inducible forms of NOS release NO• in response to inflammatory mediators including

TABLE 8.1. Potential therapeutic effects of nitric oxide manipulation.

Tissue	Physiologic Process	Relevance to Orthopaedics
Articular cartilage	Proteoglycan and collagen II synthesis	Healing of chondral injury Arthritis
Ligament	Collagen synthesis	Ligament healing Differential healing of ACL vs. MCL Ligament laxity
Tendon	Collagen synthesis	Tendon healing Arthrofibrosis
Skeletal muscle	Muscle contraction and muscle blood flow	Muscle strain injury Muscle cramps
Bone	Regulation of osteoblast and osteoclast function	Response to repetitive loading Stress fracture Osteoporosis
Synovium	Synovial inflammation	Joint swelling Inflammatory arthritis
Meniscus	Matrix synthesis	Meniscal healing

interleukin-1beta (IL–1β), tumor necrosis factor-α (TNF–α), and inter-feron. Nitric oxide produced by iNOS plays a role in host defense, immuno-logical reactions, wound healing, and other physiological processes. Thus, iNOS represents a likely candidate molecule for gene therapy, as will be discussed in the sections below.

Because NO• is fundamental to many metabolic processes, alteration of systemic NO• levels may have numerous adverse side effects. Thus, manip-ulation of NO• levels should ideally be localized to the desired target tissues. Gene therapy techniques may allow such precise alteration of local NO• concentrations. The carrier with the desired gene sequence may be introduced locally (such as by injection) or, alternatively, an ex vivo approach may be used, in which specific cells are transfected outside the organism and then reintroduced into the target tissue. This technique allows manipulation of NO• levels within specific cells and tissues.

Inflammation

Inflammation is a fundamental part of the body's response to both acute traumatic injury and repetitive microtrauma, both of which occur in sports injuries. Nitric oxide is involved in virtually every phase of the inflamma-tory process (Evans et al. 1995). Inducible NOS is present in macrophages, neutrophils, lymphocytes, and peripheral blood monocytes (Hibbs, Taintor, Vavrin, and Rachoin 1988). Inducible NOS is strongly expressed in the syn-ovial lining layer, subsynovium, vascular smooth muscle and chondrocytes from patients with rheumatoid arthritis (RA), indicating that synovium and cartilage are important sources of increased NO• production in patients with inflammatory arthritis (Grabowski et al. 1997). Nitric oxide plays a complex role in inflammation, demonstrating both proinflammatory and antiinflammatory properties. For example, NO• increases vasodilatation, but decreases neutrophil adhesion. It inhibits many proinflammatory neutrophil functions, but, in some cells, promotes the synthesis of prostaglandin E–2 (Salvemini et al. 1993). The literature is contradictory on the effect of NO• on vascular permeability, chemotaxis, and release of inflammatory mediators. Nitric oxide also plays a role in the regulation of cyclooxygenase, an enzyme that is induced during inflammation (Manfield, Murrell, G., Manfield, Jang, and Murrell, G.A.C. 1996). Further studies are required to better define the complex role of NO• in the myriad processes which occur during inflammation. Such information may allow for the use of gene therapy approaches to regulate NO• at specific points in the inflammatory response.

Articular Cartilage

The role of NO• in articular cartilage synthesis and degradation has been the subject of intense research over the past several years. Sports related joint injury often results in inflammation in or around the joint, triggering the release of mediators (such as IL–1β and TNF–α) which are known to induce NO• release. Although it is clear that NO• plays a fundamental role in chondrocyte metabolism, it has become increasingly evident that the role of NO• in cartilage degradation and repair is complex. It has been established that articular chondrocytes are the principal site of NO• production in the synovial joint (Murrell et al. 1996; Stefanovic–Racic, Watkins, Kang, Turner, and Evans 1996). There is greater NO• production by chondrocytes in the superficial layer than in the deep layer (Hayashi and He 1996).

Chondrocytes produce NO• via iNOS when stimulated by inflammatory mediators. For example, increased levels of NO• are found in the synovial fluid and serum of patients with RA and osteoarthritis (Farrell, Blake, Palmer, and Moncada 1992), and inhibition of NOS results in reduced cartilage degeneration in a murine spontaneous arthritis model and a murine streptococcal cell wall induced arthritis model (McCartney–Francis et al. 1993; Weinberg et al. 1994). A recent study demonstrated that iNOS was most strongly expressed in the synovial lining layer, subsynovium, vascular smooth muscle, and chondrocytes from patients with RA (Grabowski et al. 1997). Macrophages in the synovial lining layer and, to a lesser extent, fibroblasts were the predominant source of iNOS within synovium, whereas T cells, B cells and neutrophils were negative. A similar pattern of iNOS staining was seen in osteoarthritis, but fewer cells were iNOS positive and the intensity of staining, particularly in cartilage, was much weaker than in RA. In contrast, no evidence of iNOS was detected in noninflammatory synovium or in cartilage derived from normal joints. Amin and Abramson (1998) demonstrated that osteoarthritic cartilage spontaneously produces significant amounts of NO• ex vivo, even in the absence of added stimuli such as IL–1, while normal cartilage does not produce nitric oxide unless stimulated with cytokines (Amin and Abramson 1998). This indicates the presence of nitric oxide synthase upregulating factors in osteoarthritis cartilage. These data support the hypothesis that synovium and cartilage are important sources of increased NO• production in patients with arthritis. Inhibition of iNOS using regional gene therapy approaches may present a novel treatment strategy for arthritis.

The effect of NO• on chondrocyte function has focused on three pathways: (1) inhibition of matrix synthesis; (2) matrix degradation due to upregulation of metalloprotease activity; and (3) prostaglandin synthesis via cyclooxygenase (Evans et al. 1995). There is general agreement that NO• results in inhibition of proteoglycan synthesis, as demonstrated in chondrocyte cell culture and cartilage explant culture (Hauselmann, Oppliger, Michel, Stefanovic–Racic, and Evans 1994; Stefanovic–Racic, Taskirin, Georgescu, and Evans 1995). Nitric oxide mediates the IL–1β induced inhibition of aggrecan synthesis (Lin, Hughes,Clark, and Caterson 1995). Nitric oxide has also been demonstrated to inhibit type II collagen biosynthesis by rabbit articular chondrocytes (Cao et al. 1996). This effect most likely occurs at the posttranslational level since nitric oxide does not affect type II collagen mRNA levels and NO• inhibits prolyl hydroxylase, an important enzyme in posttranslational processing of collagen (Cao, Georgescu, and Evans 1997).

The effect of NO• on cartilage degradation is less clear than its effect on cartilage synthesis. Murrell, Jang, and Williams (1995) demonstrated that NO• plays a regulatory role in the activation of two metal dependent proteolytic enzymes (collagenase and stromelysin) in articular chondrocytes and cartilage explants. These data suggest the possibility that NOS inhibition could decrease cartilage matrix degradation. However, Stefanovic–Racic, Mollers, Miller, and Evans (1997) reported that NO• prevents matrix degradation in cartilage explants. Further studies are required in order to clarify the role of NO• in cartilage degradation before it is possible to manipulate local NO• in cartilage in an effort to favorably affect cartilage matrix metabolism.

Prostaglandin E–2 (PGE–2), which is synthesized by cyclooxygenase, is an important bioactive compound involved in cartilage degradation. Nitric oxide has been found to play an important role in the regulation of cyclooxygenase (Manfield et al. 1996). This finding suggests that inhibition of NO• synthesis may represent a method of intervening in the inflammatory cascade upstream of the point of action of nonsteroidal antiinflammatory drugs (which act by cyclooxygenase inhibition), thus providing a novel therapeutic approach in arthritis treatment. Manipulation of local iNOS by gene therapy may allow regulation of cyclooxygenase production. Further studies are required to determine if NO• differentially regulates the different isoforms of cyclooxygenase (COX–1 and COX–2), since selective inhibition of COX–2 is associated with a lower incidence of such side effects as gastrointestinal irritation and platelet dysfunction.

Nitric oxide affects other aspects of chondrocyte physiology, providing further support for the critical role of this molecule in cartilage metabolism. For example, Frankel et al. (1996) demonstrated that NO• inhibits chondrocyte migration and attachment to fibronectin, and disrupts assembly of actin filaments. These effects would inhibit cartilage repair. It appears that NO• is also involved in regulation of chondrocyte response to mechanical loading. Nitric oxide mediates the increased glycosaminoglyan synthesis that occurs following fluid induced shear stress applied to chondrocytes in monolayer culture (Das, Shurman, and Smith 1997). The effect of mechanical stimuli on cartilage metabolism is relevant to joint loads which occur during sports activities, as well as in the use of modalities, such as continuous passive motion, in the rehabilitation of articular injury. However, further information is required to allow for the eventual application of these findings to the clinical arena.

The numerous actions of NO• on articular cartilage suggest that manipulation of local NO• availability by gene therapy techniques may allow for novel intervention strategies in the treatment of articular injury. Furthermore, inhibitors of NOS have been identified and also represent potential targets for gene transfer. For example, N–iminoethyl–L–lysine (L–NIL), a selective inhibitor of iNOS, has recently been reported to inhibit progression of experimental osteoarthritis in a dog model (Pelletier et al. 1998). Treatment with L–NIL also significantly decreased the production of metalloprotease and IL–1β. L–NIL is another potential candidate molecule for gene therapy.

Tendon and Ligament

Murrell et al. (1997) demonstrated that NOS is induced during tendon healing. Nitric oxide synthase was found to be expressed in healing rat Achilles tendon, and inhibition of NOS resulted in a significant inhibition of healing, as demonstrated by a reduction in tendon cross section area and failure load (Murrell et al. 1997). Additionally, superoxide free radicals (another type of free radical related to NO•) can stimulate fibroblast proliferation (Murrell, Francas, and Bromley 1990). These findings suggest the possibility that NO• donors could enhance tendon healing, while, conversely, NO• inhibition may be useful in the prevention of arthrofibrosis and scar formation. Transfection of tendon cells with cDNA for NOS may improve tendon healing, while transfection with the gene for a NOS inhibitor may be useful for the treatment of arthrofibrosis. Lou, Manske, Aoki, and Joyce (1996) have demonstrated adenovirus mediated gene transfer into tendon.

Nitric oxide may also be involved in ligament healing. Nitric oxide is produced by fibroblasts derived from the anterior cruciate, posterior cruciate, and medial collateral ligaments (Evans et al. 1995). Of possible relevance to ligament healing is the observation that anterior cruciate ligament (ACL) fibroblasts produce more NO• than fibroblasts derived from medial collateral ligament (MCL) when exposed to IL–1 (Evans et al. 1995). The response to NO• may be one factor accounting for the differential healing capacity of the MCL and the ACL. Although the data of Murrell et al. (1997) suggest that NO• may augment tendon healing, studies in healing rat MCL in vitro indicate that NO• results in decreased collagen synthesis (Case, Hurschler, Vailas, and Vanderby 1996). An important difference between these studies is that the decrease in collagen synthesis was demonstrated in healing ligaments in vitro, while Murrell et al. (1997) studied tendon healing in vivo. It is possible that the decrease in cross sectional area and failure load in healing Achilles tendons in rats that are fed a NOS inhibitor is due to a systemic effect of NO• inhibition. Alternatively, it may be that low levels of NO• increase new collagen synthesis while higher levels inhibit collagen synthesis, or possibly tendon and ligament behave differently in response to NO•. Other studies have also demonstrated decreased collagen synthesis (in rat mesangial cells and rabbit vascular smooth muscle cells) in response to increased levels of NO• (Kolpakov, Gordon, and Kulik 1995; Trachtman, Futterweit, and Singhal 1995). Induction of local NO• synthesis by gene therapy may need to be tailored to produce NO• levels that result in improved tendon healing. One further observation with relevance to ligament is the finding that phagocytosis of wear debris by synovial fibroblasts, such as may occur in the presence of a prosthetic ligament, results in increased NO• production (Evans et al. 1995). The role of NO• in tendon and ligament injury and repair is just beginning to be understood and requires further study. Manipulation of NO• may be valuable for the treatment of repetitive overuse tendon injury and tendinosis.

There has been recent interest in possible hormonal effects on ligament physiology, which may relate to the increased incidence of ligament injury in females. Ligament laxity has been reported to be altered during pregnancy and increases in NO• production have also been well documented during pregnancy. A recent study has shown that levels of mRNA for iNOS as well as some cytokines were increased in both the MCL and the ACL from multiparous rabbit during pregnancy, indicating that NO• may be involved in the biochemical alterations that result in ligament laxity (Hart, Boykiw, Sciore, and Peno 1998).

Nitric Oxide in Muscle and During Exercise

Recent studies have demonstrated that NO• is involved in the basic physiology of muscle contraction. Nakane, Schmidt, Pollock, Fostermann, and Murad (1993) have shown that an isoform of human brain NO• synthase is highly expressed in human skeletal muscle, while Balon and Nadler (1994) showed that NO• release is detectable from incubated skeletal muscle preparations, with electrical stimulation resulting in 2-fold increases in NO• release. Skeletal muscle NOS has been localized within the sarcolemma of type II (fast) fibers (Kobzik, Reid, Bredt, and Stamler 1994). The role of NO• in skeletal muscle contraction was demonstrated by Murrell, Dolan, et al. (1995) in a study in which inhibition of nitric oxide synthase in rats resulted in a significant reduction in walking speed and muscle mass, with eventual paralysis. Furthermore, exogenous NO• has been found to increase the contractile force of mouse skeletal muscle in vitro (Murrant and Barclay 1995). Further studies are needed to evaluate the role of NO• on recovery from such common sports injuries as muscle strains and muscle tears.

Nitric oxide is also involved in skeletal muscle blood flow during exercise. Vasodilation and increased blood flow in the heart and skeletal muscle during exercise is due largely to endothelial derived relaxing factor (EDRF). Nitric oxide is an integral part of EDRF. A number of studies have attempted to evaluate the possible contribution of NO• to exercise-induced hyperaemia. Results from these studies have been conflicting. Most studies report increased NO• production and vasodilation during exercise (McAllister 1995). Some studies demonstrated that hyperaemic responses during exercise were unaffected by infusion of the NOS inhibitor, while others showed marked blockade of NOS, resulting in vasoconstriction and reduced blood flow (Dyke, Proctor, Dietz, and Joyner 1995). These conflicting findings may be explained by differences in the experimental model used, differences in the timing of inhibitor administration, or differences in the intensity of the exercise. For example, NO• inhibition in exercising rats results in reduced blood flow to the hindquarter musculature (Hirai, Visneski, Kearns, Zelis, and Musch 1994). Maxwell, Schauble, Bernstein, and Cooke (1998) recently reported that endothelium-derived nitric oxide contributes significantly to limb blood flow during exercise.

By regulating blood flow to skeletal muscle during exercise, NO• participates in control of body temperature and sweating. Mills, Marlin, Scott, and Smith (1997) studied the effect of inhibition of NO• production on sweating rate (SR) and on core and peripheral temperatures in horses during exercise. Nitric oxide synthase inhibition by N^G–nitro–L–arginine methyl ester (L–NAME) diminished SR, resulting in elevated body temperatures and leading to deranged thermoregulation during exercise. L–Arginine partially reversed the inhibitory effects of L–NAME on SR. Mills et al. (1997) speculated that the inhibition of sweating by L–NAME may be related to peripheral vasoconstriction, but may also involve the neurogenic control of sweating. The important role of NO• in the regulation of skeletal muscle blood flow suggests that gene therapy techniques may be applicable to the treatment of muscle strain injury. Augmentation of local blood flow is the principal mode of action of commonly used modalities such as ultrasound and electrical stimulation, indicating the potential efficacy of using gene transfer techniques to increase local blood flow. Transfection of skeletal muscle cells with adenovirus vectors has recently been demonstrated (Floyd et al. 1998).

Only a few studies have measured NO• levels in exercising humans. Bode–Boger (1994) found that urinary nitrate excretion increased significantly during exercise, while other studies have reported increased pulmonary NO• excretion rates during exercise (Bauer, Wald, Doran, and Soda 1994). Direct measurements of serum NO• (as measured by nitrite or nitrate, the stable end products of NO•) following exercise have produced conflicting results, with both increases and decreases reported (Ambring et al. 1994; Gleim, Zeballos, Glace, Kaley, and Hintze 1994). Maddali, Rodeo, Barnes, Warren, and Murrell (1996) have recently measured changes in nitrite (the stable end product of NO•) in professional football players during preseason training. Serum nitrite was measured in players requiring intravenous rehydration for severe generalized muscle cramps that occurred during strenuous activity. There was a 300% increase in postexercise serum nitrite concentrations in comparison to baseline values. The increase in nitrite was due not simply to dehydration, but, since it was also greater than the increase in muscle enzymes, a finding not attributable to muscle breakdown alone, it appears that the increased nitrite levels are due to increased production of NO• during exercise. The limitation of this study is that postexercise nitrite measurements were not obtained from players without muscle cramps. Without this data it is not possible to state whether the postexercise increase in nitrite was due to exercise alone, muscle cramps alone, or a combination. Further studies are required to determine if NO• released during exercise is derived from skeletal muscle cells, endothelial cells, or other sources. Such information is necessary in order to be able to use gene transfer techniques to affect local NO• metabolism.

Bone

Athletic training can place extremely high demands on the skeletal system, requiring adaptation to repetitive loading. Failure of adaptation may result in "stress reaction," periostitis, or a frank stress fracture. Bone adaptation to load is controlled by a complex array of regulatory factors. Nitric oxide appears to be an important regulatory molecule in bone, with effects on cells of the osteoblast and osteoclast lineage. Both constitutive and inducible forms of NOS are expressed by bone-derived cells, although little is known about the sites of NO• synthesis in bone. Mechanical loading of bone and isolated bone cells causes a rapid and transient release of NO• (Pitsillides et al. 1995). Turner, Takano, Owan, and Murrell (1996) reported that treatment with a NO• inhibitor reduced the rate of mechanically induced bone formation by 66% in rat tibia compared to controls, while bone formation in nonloaded animals was not affected by the same treatment. These findings suggest that NO• may play a role in the transduction of a mechanical stimulus into a biological response in bone.

Nitric oxide has significant effects on bone remodeling. Significant stimulation of NO• production and inhibition of bone resorption results when bone is stimulated by a combination of interferon and other cytokines (Ralston et al. 1995). High concentrations of NO• also inhibit osteoblast proliferation, while lower concentrations of NO• stimulate bone resorption and promote the proliferation of osteoblast-like cells. Nitric oxide has been implicated in bone loss due to exercise-induced amenorrhea. Intense exercise can lead to menstrual disturbances in female athletes, and this can be accompanied by bone loss, particularly from the lumber spine (Rutherford 1993). It has been suggested that estrogen exerts its antiresorptive actions on bone via a NO• dependent mechanism (Cicinelli et al. 1996). Stacey, Korkia, Hukkanen, Polak, and Rutherford (1998) compared amenorrheic athletes and eumenorrheic athletes in terms of bone turnover and NO• level. Unlike postmenopausal women, amenorrheic athletes do not have elevated bone turnover, but do have reduced NO• metabolites and spinal osteopenia. This finding suggests the importance of NO• in estrogen mediated protection of skeletal mass and strength.

Further understanding of the role of NO• as a bone regulatory molecule and mediator of the response to strain may suggest eventual treatment strategies for osteoporosis and repetitive overuse skeletal injuries (such as stress fractures) in athletes. For example, osteoblasts may be transfected with a cDNA for NOS in order to increase local bone formation to heal a stress fracture.

Intervertebral Disc

Nitric oxide production has been reported in cells derived from the annulus fibrosis and nucleus pulposus of the intervertebral disc, as well as in the granulation tissue surrounding a herniated disc (Kawakami, Tamaki, Hayashi, Hashizumi, and Nishi 1998; Saito, Suh, Nishida, Evans, and King 1997). More NO• is produced by herniated disc compared to normal disc specimens (Kang et al. 1996). Nitric oxide mediates gelatinase (a matrix metalloprotease) production and inhibits IL–6 production in disc, suggesting its involvement in the biochemistry of disc degeneration (Kang, Sefanovic–Racic, Larkin, Georgescu, and Evans 1995). Furthermore, NOS expression has been correlated with pain related behavior that occurs in response to contact of disc material with nerve (Hashizume et al. 1997). Thus, NO• may be involved in the pathophysiology of radicular pain secondary to a herniated disc. Local inhibition of NOS using gene therapy techniques may be useful for the treatment of lumbar disc pathology.

Other Joint Tissues

Synovial cells produce NO• by both inducible and constitutive forms of NOS (Grabowski, MacPherson, and Ralston 1996; Murrell et al. 1996). Inflammatory mediators (IL–1β, TNF–α, and interferon) result in increased production of NO• (Grabowski et al. 1996). Dexamethasone inhibits NO• production by synovial cells in culture, suggesting that the antiinflammatory effects of intraarticular steroid injection may be at least partially mediated by NO• inhibition (Grabowski et al. 1996). Murrell et al. (1996) reported that, although no constitutive or inducible NOS activity was detected in human meniscal cells in culture, small amounts of nitrite were detected in the culture medium of cells derived from bovine meniscus when stimulated with endotoxin. In contrast to this report by Murrell et al. (1996), Cao et al. (1998) recently showed that rabbit meniscus can produce NO• when stimulated by cytokines. Rabbit meniscus tissue produced large amounts of NO• after stimulation with IL-1, TNF-α, or a mixture of rabbit derived cytokines known as chondrocyte-activating factors. Monolayer cultures of meniscal cells produced from the digestion of meniscal tissue also produced large amounts of NO• in response to cytokines. Two distinct cell populations were noted, one fibroblastic and the other chondrocytic, with the latter cells apparently responsible for generating most of the NO• in response to cytokines. Endogenously generated NO• suppressed the synthesis of collagen and proteoglycan by menisci, but prevented IL–1 induced proteoglycan degradation. These data suggest that manipulation of NO• levels may provide a novel method for improving meniscal healing. Hidaka et al. (1998) have recently demonstrated the transfection of meniscal fibrochondrocytes using an adenovirus vector.

Summary

Nitric oxide is a multifunctional intercellular and intracellular messenger molecule that plays a role in tissues commonly injured in sports activities. In addition to playing a fundamental role in metabolism of cartilage, tendon, ligament, muscle, and bone, NO• is also involved in basic processes such as inflammation and regulation of blood flow. Thus, NO• affects many processes active in the response to sports injury. A review of the myriad actions of nitric oxide in these important musculoskeletal tissues suggests the enormous potential for effecting tissue healing and regeneration by manipulating local NO• levels in cells and tissues. Selective inhibition and augmentation of NO• levels using gene therapy techniques represents a potentially valuable new strategy for biologic intervention in sports injuries. However, there remain several fundamental questions before this technique can be recommended for widespread application. For example, the duration of gene expression in a given tissue needs to be determined. The most ideal vector and route of transfection also need to be determined and may be different for various tissue types. The risk of adverse effects resulting from alteration of local NO• level in a given tissue or organ is largely unknown at this time. Because NO• is fundamental to many metabolic processes, alteration of systemic NO• levels may have numerous adverse side effects. Ongoing studies are beginning to provide answers to these important questions and may allow for the eventual clinical use of gene therapy techniques in the treatment of musculoskeletal injury.

References

Ambring, A., Benthin, G., Petersson, A., Jungersten, L., and Wennmalm, A. 1994. Indirect evidence of increased expression of nitric oxide synthase in marathon runners and upregulation of nitric oxide synthase activity during running. *Circulation* 90:1–137.

Amin, A.R. and Abramson, S.B. 1998. The role of nitric oxide in articular cartilage breakdown in osteoarthritis. *Curr Opin Rheumatol* 10:263–8.

Balon, T. and Nadler, J. 1994. Nitric oxide release is present from incubated skeletal muscle preparations. *J Appl Physiol* 77:2519.

Bauer, J., Wald, J., Doran, S., and Soda, D. 1994. Endogenous nitric oxide and expired air: effects of acute exercise in humans. *Life Sciences* 55:1903–9.

Bode–Boger, S. 1994. Exercise increase systemic nitric oxide production in men. *J Cardiovasc Risk* 1:173–8.

Cao, M., Georgescu, H., and Evans, C. 1997. Inhibition of prolyl hydroxylase by nitric oxide. *Trans Orthop Res Soc* 22:411.

Cao, M., Stefanovic–Racic, M., Georgescu, H.I., Miller, L.A., and Evans, C.H. 1998. Generation of nitric oxide by lapine meniscal cells and its effect on matrix metabolism: stimulation of collagen production by arginine. *J Orthop Res* 16:104–11.

Cao, M., Westerhausen–Larson, A., Niyibizi, C., Kaval–Kovich, K., Georgescu, H., Rizzo, C., Stefanovic–Racic, M., and Evans, C. 1996. Nitric oxide inhibits type II collagen biosynthesis by rabbit articular chondrocytes without altering collagen to mRNA abundance. *Trans Orthop Res Soc* 21:533.

Case, M., Hurschler, C., Vailas, A., and Vanderby, R. 1996. Increased levels of nitric oxide decrease new collagen synthesis in healing rat medial collateral ligaments in vitro. *Trans Orthop Res Soc* 21:797.

Cicinelli, E., Ignarro, L.J., Lograno, M., Galatino, P., Balzano, G., and Schonauer, L.M. 1996. Circulating levels of nitric oxide in fertile women in relation to the menstrual cycle. *Fertil Steril* 66:1036–8.

Das, P., Schurman, D., and Smith, R. 1997. Nitric oxide and G proteins mediate the response of bovine articular chondrocytes to fluid–induced shear. *J Orthop Res* 15:87–93.

Dyke, C.K., Proctor, D.N., Dietz, N.M., and Joyner, M.J. 1995. Role of nitric oxide in exercise hyperaemia during prolonged rhythmic handgripping in humans. *J Physiol (London)* 488:259–65.

Evans, C., Stefanovic–Racic, M., and Lancaster, J. 1995. Nitric oxide and its role in orthopedic disease. *Clin Orthop* 312:275–94.

Farrell, A., Blake, D., Palmer, R., and Moncada, S. 1992. Increased concentrations of nitrite in synovial fluid and serum samples suggest increased nitric oxide synthesis in rheumatic diseases. *Annal Rheum Dis* 51:1219–22.

Floyd, S.S., Booth, D.K., Clemens, P.R., Ontell, M.R., van Deutekom, J.C.T., Moreland, M.S., and Huard, J. 1998. Myoblast mediated ex vivo gene transfer to skeletal muscle. *Trans Orthop Res Soc* 23:587.

Fraenkel, S., Clancy, R., Ricci, J., DiCesare, P., Radiske, J., and Abramson, S. 1996. Effects of nitric oxide on chondrocyte migration, adhesion, and cytoskeletal assembly. *Arthritis Rheum* 39:1905–12.

Gleim, G., Zeballos, G., Glace, B., Kaley, G., and Hintze, T. 1994. Venous nitric oxide metabolites decrease in highly trained endurance athletes during maximal exercise. *Circulation* 90:I–659.

Grabowski, P., MacPherson, H., and Ralston, S. 1996. Nitric oxide production in cells derived from the human joint. *Br J Rheumatol* 35:207–12.

Grabowski, P.S., Wright, P.K., Van 't Hof, R.J., Helfrich, M.H., Ohshima, H., and Ralston, S.H. 1997. Immunolocalization of inducible nitric oxide synthase in synovium and cartilage in rheumatoid arthritis and osteoarthritis. *Br J Rheumatol* 36:651–5.

Hart, D.A., Boykiw, R., Sciore, P., and Reno, C. 1998. Complex alterations in gene expression occur in the knee ligaments of the skeletally mature multiparous rabbit during pregnancy. *Biochim Biophys Acta* 1397:331–41.

Hashizume, H., Kawakami, M., Hayashi, N., Nishi, H., Danjo, S., and Tamaki, T. 1997. Nitric oxide synthase and pain related behavior induced by autologous intervertebral disc in the rat. *Trans Orthop Res Soc* 22:689.

Hauselmann, H., Oppliger, L., Michel, B., Stefanovic–Racic, M., and Evans, C. 1994. Nitric oxide and proteoglycan biosynthesis by human articular chondrocyte in alginate culture. *FEBS Lett* 352:361–4.

Hayashi, T. and He, J. 1996. Nitric oxide production by superficial and deep articular chondrocytes. *Trans Orthop Res Soc* 21:535.

Hibbs, J., Taintor, R., Vavrin, Z., and Rachoin, E. 1988. Nitric oxide: a cytotoxic activated macrophage effector molecule. *Biochem Biophys Res Comm* 157:87–94.

Hidaka, C., Hannafin, J.A., Bhargava, M., Weiser, L., Hackett, N.R., Warren, R.F., and Crystal, R.G. 1998. Adenovirus transfection of articular chondrocytes and meniscal fibrochondrocytes in collagen gel 3-dimensional culture. *Trans Orthop Res Soc* 23:586.

Hirai, T., Visneski, M., Kearns, K., Zelis, R., and Musch, T. 1994. Effects of nitric oxide synthase inhibition on the muscular blood flow response to treadmill exercise in rats. *J Appl Physiol* 77:1288–93.

Kang, J., Georgescu, H., McIntyre–Larkin, L., Stefanovic–Racic, M., Donaldson, W., and Evans, C. 1996. Herniated lumbar intervertebral disc spontaneously produce matrix metalloproteinases, nitric oxide, interleukin–6, and prostaglandin E2. *Spine* 2:271–7.

Kang, J., Stefanovic–Racic, M., Larkin, L., Georgescu, H., and Evans, C. 1995. Nitric oxide production by human intervertebral disc in response to interleukin–1 and its effects on interleukin–6 production. *Trans Orthop Res Soc* 20:351.

Kawakami, M., Tamaki, T., Hayashi, N., Hashizume, H., and Nishi, H. 1998. Possible mechanism of painful radiculopathy in lumbar disc herniation. *Clin Orthop* 351:241–51.

Kobzik, L., Reid, M.B., Bredt, D.S., and Stamler, J.S. 1994. Nitric oxide in skeletal muscle. *Nature* 372:546–8.

Kolpakov, V., Gordon, D., and Kulik, T. 1995. Nitric oxide generating compounds inhibit total protein and collagen synthesis in cultured vascular smooth muscle cells. *Circulation Res* 76:305–9.

Lin, P., Hughes, C., Clark, J., and Caterson, B. 1995. Inhibitors of nitric oxide synthetase reverse the IL–1–induced inhibition of aggrecan biosynthesis but not its degradation. *Trans Orthop Res Soc* 20:217.

Lou, J., Manske, P.R., Aoki, M., and Joyce, M.E. 1996. Adenovirus mediated gene transfer into tendon and tendon sheath. *J Orthop Res* 14:513–7.

Maddali, S., Rodeo, S.A., Barnes, R., Warren, R.F., and Murrell, G.A.C. 1998. Post-exercise increase in nitric oxide in football players with muscle cramps. *American Journal of sports Medicine*, 26:820–4.

Manfield, L., Murrell, G., Manfield, L., Jang, D., and Murrell, G.A.C. 1996. Nitric oxide enhances cyclooxygenase activity in articular cartilage. *Inflammation Res* 45:254–8.

Maxwell, A.J., Schauble, E., Bernstein, D., and Cooke, J.P. 1998. Limb blood flow during exercise is dependent on nitric oxide. *Circulation* 98:369–74.

McAllister, R. 1995. Endothelial mediated control of coronary and skeletal muscle blood flow during exercise: introduction. *Med Sci Sports Exerc* 27:1122–4.

McCartney-Francis, N., Allen, J., Mizel, D., Albina, J., Xie, W., Nathan, C., and Wahl, S. 1993. Suppression of arthritis by an inhibitor of nitric oxide synthase. *J Exp Med* 178:749–54.

Mills, P.C., Marlin, D.J., Scott, C.M., and Smith, N.C. 1997. Nitric oxide and thermoregulation during exercise in the horse. *J Appl Physiol* 82:1035–9.

Murrant, C. and Barclay, J. 1995. Endothelial cell products alter mammalian skeletal muscle function in vitro. *Can J Physiol Pharmacol* 73:736–41.

Murrell, G.A.C., Dolan, M.M., Jang, D., Szabo, C., Warren, R.F., and Hannafin, J.A. 1996. Nitric oxide: an important articular free radical. *J Bone Joint Surg* 78A:265–74.

Murrell, G., Dolan, M., Murrell, D., Hu, A., Warren, R., and Hannafin, J. 1995. Nitric oxide and skeletal muscle. *Trans Combined Orthop Res Soc*:394.

Murrell, G., Francas, M., and Bromley, L. 1990. Modulation of fibroblast proliferation by oxygen free radicals. *Biochem J* 265:659–65.

Murrell, G.A.C., Jang, D., and Williams, R. 1995. Nitric oxide activates metalloprotease enzymes in articular cartilage. *Biochem Biophys Res Comm* 206:15–21.

Murrell, G., Szabo, C., Hannafin, J., Jang, D., Dolan, M., Deng, X., Murrell, D., and Warren, R. 1997. Modulation of tendon healing by nitric oxide. *Inflammation Res* 46:19–27.

Nakane, M., Schmidt, H.H., Pollock, J.S., Fostermann, U., and Murad, F. 1993. Cloned human brain nitric oxide synthase is highly expressed in skeletal muscle. *FEBS Lett* 316:175–80.

Pelletier, J.P., Jovanovic, D., Fernandes, J.C., Manning, P., Connor, J.R., Currie, M.G., Di Battista, J.A., and Martel–Pelletier, J. 1998. Reduced progression of experimental osteoarthritis in vivo by selective inhibition of inducible nitric oxide synthase. *Arthritis Rheum* 41:1275–86.

Pitsillides, A.A., Rawlinson, S.C.F., Suswillo, R.F.L., Bourrin, S., Zaman, G., and Lanyon, L.E. 1995. Mechanical strain induced NO production by bone cells: a possible role in adaptive bone (re)modeling? *FASEB J* 9:1614–22.

Ralston, S.H., Ho, L.P., Helfrich, M.H., Grabowski, P.S., Johnston, P.W., and Benjamin, N. 1995. Nitric oxide: a cytokine induced regulator of bone resorption. *J Bone Mineral Res* 10:1040–9.

Rutherford, O.M. 1993. Spine and total body bone mineral density in amenorrheic athletes. *J Appl Physiol* 74:2904–8.

Saito, R., Suh, J., Nishida, K., Evans, C., and King, J. 1997. Nitric oxide expression by bovine intervertebral disc cells. *Trans Orthop Res Soc* 22:417.

Salvemini, D., Misko, T., Masferrer, J,, Seibert, K., Currie, M., and Needleman, P. Nitric oxide activates cyclooxygenase enzymes. 1993. *Proc Natl Acad Sci USA* 90:7240–4.

Stacey, E., Korkia, P., Hukkanen, M.V., Polak, J.M., and Rutherford, O.M. 1998. Decreased nitric oxide levels and bone turnover in amenorrheic athletes with spinal osteopenia. *J Clin Endocrinol Metab* 83:3056–61.

Stefanovic–Racic, M., Mollers, M., Miller, L., and Evans, C. 1997 Nitric oxide and proteoglycan turnover in rabbit articular cartilage. *J Orthop Res* 15:442–9.

Stefanovic–Racic, M., Taskirin, D., Georgescu, H., and Evans, C. 1995. Modulation of chondrocyte proteoglycan synthesis by endogenously produced nitric oxide. *Inflammation Res* Suppl:S216–7.

Stefanovic–Racic, M., Watkins, S., Kang, R., Turner, D., and Evans, C.H. 1996. Identification of inducible nitric oxide synthase in humanosteoarthritic cartilage. *Trans Orthop Res Soc* 21:534.

Trachtman, H., Futterweit, S., and Singhal, P. 1995. Nitric oxide modulates the synthesis of extracellular matrix proteins in cultured rat mesangial cells. *Biochem Biophys Res Commun* 207:120–5.

Turner, C.H., Takano, Y., Owan, I., and Murrell, G.A.C. 1996. Nitric oxide inhibitior L–NAME suppresses mechanically induced bone formation in rats. *Am J Physiol* 270:E634–9.

Weinberg, J., Granger, D., Pisetsky, D., Seldin, M., Misukonis, M., Mason, S., Pippen, A., Ruiz, P., Wood, E., and Gilkeson, G. 1994. The role of nitric oxide in the pathogenesis of spontaneous murine autoimmune disease. *J Exp Med* 179:652–60.

Part III
Tissue Engineering in Orthopaedics and Sports Medicine

9
Stem Cells and Tissue Engineering

HENRY E. YOUNG

Introduction

Every year millions of people suffer tissue loss or end–stage organ failure (Langer and Vacanti 1993). Total national US health care costs for these patients exceed 400 billion dollars per year. Currently over 8 million surgical procedures requiring 40 to 90 million hospital days are performed annually in the United States to treat these disorders. Options such as tissue transplantation and surgical intervention are severely limited by critical donor shortages, long-term morbidity, and mortality. Stem cells could be used as a source of donor tissue in transplants and elective surgeries. They could also be used in the treatment of a wide variety of diseases, congenital malformations, and genetic disorders. One especially appropriate application for stem cells involves the treatment of tissue losses, which requires that large numbers of cells be available for transplantation. Similar issues arise with respect to providing sufficient numbers of cells for gene therapy.

Numerous studies have demonstrated the existence of lineage-committed "progenitor" stem cells in animals. These cells provide for the continual maintenance and repair of tissues throughout the life span of the individual. Examples of such stem cells within the mesodermal lineage include the unipotent myosatellite myoblasts of muscle (Mauro 1961; Campion 1984; Grounds, Garrett, Lai, Wright, and Beiharz 1992); the unipotent adipoblast cells of adipose tissue (Aihaud, Grimaldi, and Negrel 1992); the unipotent chondrogenic and osteogenic stem cells of the perichondrium and periosteum, respectively (Cruess 1982; Young et al. 1995); the bipotent adipofibroblast cells of adipose tissue (Vierck, McNamara, and Dodson 1996); the bipotent chondrogenic/osteogenic stem cells of marrow (Owen 1988; Beresford 1989; Caplan, Elyaderani, Mochizuki, Wakitani, and Goldberg 1997); and the multipotent hematopoietic stem cells of bone marrow and peripheral blood (Palis and Segel 1998; McGuire 1998; Ratajczak et al. 1998).

The existence of reserve stem cell populations involved in the continual maintenance and repair of mesenchymal tissues also suggests the potential for using stem cells for the repair and/or regeneration of tissues following their loss due to trauma and/or disease. Indeed, use of allogeneic and autologous progenitor stem cells for the repair and regeneration of chondrogenic and osteogenic tissue defects is currently being investigated. For example, osteochondral allografts (Mankin 1982; McDermott, Langer, Pritzker, and Gross 1985) or engraftment of cultured allogeneic chondrocytes (Chesterman and Smith 1968; Bentley and Greer 1971; Green 1977; Moskalewski 1991) have been pursued as a means to increase the amount of tissue available for grafting. However, these studies have encountered difficulties including unexpected morbidity and mortality. The rejection of allogeneic cell grafts (in an environment presumed to be protected from immunologic attack), the low degree of tissue union, and the increased risk of disease transmission (Wakitani et al. 1989; Kawabe and Yoshinato 1993; Matsusue, Yamamuro, and Hama 1993; Caplan et al. 1997) have restricted these techniques to the treatment of individuals with large chondral or osteochondral defects (Garret 1986; Matsusue et al. 1993).

In contrast, other studies have shown that grafts of autologous perichondrium (Skoog and Johansson 1976; Homminga et al. 1990) or periosteum (Rubak 1982; O'Driscoll, Keeley, and Salter 1988; Ritsila et al. 1994) can be used to repair damaged sites. These techniques depend upon the presence of mesenchymal chondrogenic progenitor cells in the perichondrium and periosteum (Caplan 1991; Young et al. 1995). Because these procedures initially resulted in the formation of a material resembling hyaline cartilage, it was suggested that the chondrogenic progenitor cells underwent metaplasia to form a chondroid tissue. However, these procedures are limited by the lack of available tissue for grafting, as well as poor graft integration and the strong possibility that the grafted tissues will eventually undergo ossification (Minas and Neher 1997).

The use of differentiated autologous chondrocytes (Grande, Pitman, Peterson, Menche, and Klein 1989; Brittberg et al. 1994; Brittberg, Nilsson, Lindahl, Ohlsson, and Peterson 1996; Frenkel, Toolan, Menche, Pitman, and Pachence 1997) or autologous osteochondral progenitor cells (Wakitani et al. 1994; Caplan et al. 1997) with subsequent expansion ex vivo prior to engraftment into the site of a chondrogenic defect has been examined in both animals and humans, with mixed results. No significant differences between experimental and controls were found in dogs (Breinen et al. 1997). However, successful regeneration has been reported in both rabbits and humans (Wakitani et al. 1994). Use of mature chondrocytes is particularly advantageous in treating young, healthy, active patients with clinically significant injuries (>2 to 3 cm^2). Unfortunately, this age group represents only a small proportion of the general population with clinically significant injuries. In addition, there are several disadvantages to using differentiated chondrocytes and chondrogenic progenitor cells. These disadvantages include limitations in the amount of autologous tissue that can be harvested for cell isolation and the limited capability for expansion ex vivo because these cells undergo programmed senescence and cell death as they approach 50 to 70 cell doublings (Hayflick's limit) (Hayflick 1965). Moreover, damage to the parts of the joint used for tissue harvest could have a negative impact on joint function (Kolettas, Buluwela, Bayliss, and Muir 1995; Minas and Nehrer 1997; Caplan et al. 1997).

The potential for using mesenchymal stem cells isolated from connective tissues to repair full-thickness cartilage defects within an articulating joint was examined in an elegant series of studies by Lucas and colleagues (Grande et al. 1995; Lucas, Calcutt, Southerland, Warejcka, and Young 1995; 1996a). Previous studies had demonstrated the potential of stem cells derived from connective tissues to form a number of different mesodermal tissues, including cartilage and bone (Pate, Southerland, Grande, Young, and Lucas 1993; Young et al. 1993; Young, Wright, et al. 1998). Lucas et al. (1995, 1996a) and Grande et al. (1995) allowed local environmental cues (presumably involving bioactive agents specific for chondrogenic and/or osteogenic tissues) to direct the differentiation of stem cells derived from connective tissue after their placement in situ. The cells were isolated from the connective tissues of skeletal muscle (including the epimysium, perimysium, and endomysium) (Pate et al. 1993; Lucas et al. (1995, 1996a) and Grande et al. (1995)) following our standard protocols for stem cell isolation (Young, Morrison, Martin, and Lucas 1991; Young, Ceballos, Smith, Lucas, and Morrison 1992), expanded ex vivo, incorporated into a polyglycolic acid (PGA) delivery vehicle, and placed into a full thickness cartilage defect in syngeneic rabbits. The defect was created by drilling a hole through the articular cartilage, the underlying subchondral bone, and the trabecular bone, and into the marrow cavity (Figure 9.1A). After 12 weeks, control animals implanted with the PGA delivery vehicle alone exhibited only fibrous tissue in the site of the defect (Figure 9.1B). After 12 weeks, experimental animals implanted with stem cells in the PGA vehicle exhibited the formation of cartilage and bone resembling that of the normal adjacent tissues (Figure 9.1C). Stem cell incorporation was assessed by the surface appearance of regenerated tissue, in combination with the histological appearance of longitudinal sections of the tissue. In addition, the glycosaminoglycan and collagen contents of the tissue were assessed by histochemistry (Figure 9.1C). The joint surface of the regenerated cartilage was irregular (Figure 9.1C). Good integration of the implant with the adjacent articular cartilage was observed (Figure 9.1C), with restoration of the underlying osseous elements complete with trabecular bone and hematopoietic tissue (Figure 9.1C, 9.1D). The regenerated surface layer displayed all five layers normally found in articular cartilage (transitional zone, tangential zone, radial zone, tidewater mark, and calcified cartilage) (compare Figure 9.1D and 9.1F). However, the inner 2 zones (tidewater mark and calcified cartilage) were much thinner than those observed in

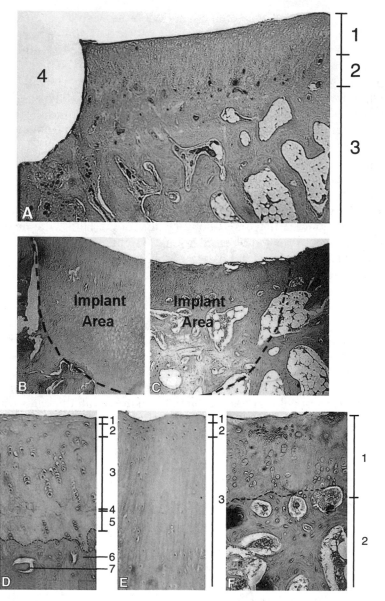

FIGURE 9.1. Use of connective tissue derived mesenchymal stem cells to repair a full-thickness cartilage defect within the articulating joint of a rabbit knee. Grande et al. 1995. *Repair of articular cartilage defect using mesenchymal stem cells. Tiss Eng 1:345–353.* Reprinted with permission from Mary Ann Liebert, Inc. Publishers. (A) Twenty-four hr after full-thickness cartilage defect created in articulating surface of knee joint. Areas: (1) articular cartilage, (2) subchondral bone, (3) trabecular bone with marrow-filled spaces, and (4) defect area. The "defect area" was created by drilling a hole through the articular cartilage, the underlying subchondral bone, and into the trabecular bone containing marrow spaces. (B) Twelve weeks after implan-

tation of polyglycolic acid (PGA) delivery vehicle without cells (control). Note fibrous tissue filling implant area and forming border along implant area. Histology of fibrous tissue does not match surrounding articular cartilage, subchondral bone, or trabecular bone with marrow-filled spaces. (C) Twelve weeks after implantation of PGA delivery vehicle containing connective tissue derived syngeneic mesenchymal stem cells. Note regeneration of articular cartilage-like tissue covering joint surface of implant area; the regeneration of underlying subchondral bone; and the regeneration of trabecular bone with marrow-filled spaces in the lower portions of the implant area adjacent to uninjured trabecular bone. Note complete annealing of regenerated articular cartilage, subchondral bone, and trabecular bone with adjacent noninjured tissues. (D) Higher magnification of control region of articular cartilage overlying subchondral bone and trabecular bone with marrow filled spaces. Note the five layers of articular cartilage: (1) tangential zone, chondrocytes within lacunae between type II collagen bundles lying parallel to surface; (2) transitional zone, chondrocytes within lacunae between interwoven type II collagen bundles lying at 45° angles to surface; (3) radial zone, chondrocytes in lacunae stacked in columns between type II collagen bundles lying perpendicular to surface; (4) tide water mark, acellular zone containing interlaced-interwoven type II collagen bundles and proteoglycans; and (5) zone of calcified cartilage, chondrocytes in lacunae stacked in columns between bundles of type II collagen ,(6) zone of cortical subchondral bone, osteocytes in lacunae within concentric rings of lamellar bone; (7) Haversian canal, within Haversian system or osteon. Carbohydrate histochemistry demonstrated chondroitin sulfate proteoglycans, keratan sulfate proteoglycans, and chondroitin sulfate/keratan sulfate proteoglycans within the extracellular matrix. The histology and histochemistry is consistent with articular cartilage. The underlying subchondral bone layer shows osteocytes within lacunae between concentric lamellae surrounding marrow-filled spaces. Histochemistry shows type I collagen fibers and chondroitin sulfate proteoglycans within the extracellular matrices, consistent with subchondral compact bone. (E) Higher magnification of site of injury after treatment with PGA delivery vehicle. Twelve weeks after transplantation, morphology shows thicker tangential, transitional, and radial zones. Histochemistry shows cells within lacunae surrounded by type I collagen bundles and chondroitin sulfate proteoglycans. Morphology and histochemistry are suggestive of an expanded articular cartilage structural framework filled with fibrocartilage. (F) Higher magnification of site of injury after treatment with mesenchymal stem cells in PGA delivery vehicle, 12 weeks after transplantation. Areas: (1) Joint surface layer, composed of 5 layers. Layers 3, 4, and 5 are not as thick as in control articular cartilage. Note residual mesenchymal stem cells within transitional layer. Type II collagen bundles, chondroitin sulfate proteoglycans, keratan sulfate proteoglycans, and chondroitin sulfate/keratan sulfate proteoglycans within surface layer consistent with articular cartilage. (2) Underlying bone with marrow-filled spaces. Cells in lacunae within concentric lamellae surrounding marrow-filled spaces. Histochemistry revealed type I collagen bundles with chondroitin sulfate proteoglycans. Morphology and histochemistry consistent with subchondral compact bone.

normal articular cartilage (compare Figure 9.1F to Figure 9.1D). The regenerated articular cartilage contained islands of apparent residual mesenchymal stem cells between the tangential zone and radial zone (Figure 9.1F). In some areas there was no histologically apparent subchondral bone beneath the regenerated articular cartilage. In these areas, trabecular bone with marrow-filled spaces was found directly adjacent to calcified articular cartilage (compare Figure 9.1D to Figure 9.1F). In other areas, subchondral bone was present (Figure 9.1C). Control animals displayed a fibrocartilaginous tissue within the area of the defect (compare Figure 9.1E to Figure 9.1F). There are at least two possible explanations for the results. First, the implanted syngeneic stem cells, directed by local environmental cues, differentiated to form articular cartilage and underlying bone. Second, the implanted stem cells released a chemotactic agent that caused a directed ingrowth of stem cells from adjacent tissues. These stem cells subsequently effected the repair process that resulted in regenerating the surface articular cartilage and underlying bone.

Lucas and colleagues (Grande et al. 1995; Lucas et al. 1995, 1996a) attempted to determine whether the regenerated tissue arose from implanted syngeneic stem cells or host stem cells. They killed syngeneic stem cells by freezing and thawing them before incorporating them into the PGA delivery vehicle. Killing the stem cells led to the development of fibrous tissue, a result comparable to that seen in the PGA controls (Figure 9.1E). In another series of experiments, smooth muscle cells were incorporated into the PGA-delivery vehicle prior to implantation. Fibrous tissue was found in the implantation site in these experiments (Figure 9.1E), resembling the results of control experiments. These studies suggest that viable stem cells are required to produce the regenerative response. To further analyze the contributions of donor stem cells versus host stem cells in the full thickness cartilage repair model, Lucas and colleagues (Grande et al. 1995; Lucas et al. 1995, 1996a) have been developing β–galactosidase transgenic animals. In future experiments, they hope to perform double labeled reciprocal syngeneic experiments utilizing β–galactosidase and the Y chromosome as permanent genomic markers. In these experiments, β–galactosidase/XY donor cells are transplanted into nontransgenic/XX hosts and normal donor/XY cells are transplanted into transgenic/XX hosts. If the regenerated tissues contain the genomic markers of the donor cells, this suggests that the donor cells were acted upon by local cues to effect the regeneration process. In contrast, if the regenerated tissues contain the genomic markers of the host cells (without β–galactosidase/XY) this suggests that donor stem cells acted in some as yet unknown fashion, possibly by releasing chemotactic agents for resident stem cells, to effect the regeneration response in the host.

The donor stem cells used in the experiments of Lucas and colleagues (Grande et al. 1995; Lucas et al. 1995, 1996a) were not manipulated experimentally (by either biochemical or molecular techniques) prior to implantation. Rather, the stem cells responded to local environmental cues to effect the regeneration process. It is hoped that these reciprocal transplantation experiments will indicate the exact origin of the regenerated articular cartilage and underlying bone after implantation of connective tissue derived mesenchymal stem cells.

Connective-Tissue-Derived Mesenchymal Stem Cells

As mentioned above, numerous studies have demonstrated the existence of lineage-committed progenitor stem cells in animals. These cells provide for the continual maintenance and repair of tissues throughout the life span of the individual. It was originally thought that lineage-specific progenitor cells were located only in specific tissues. For example, myoblasts were thought to be located only in skeletal muscle, and adipoblasts were thought be to located only in fat tissue. However, a series of landmark experiments by Young and colleagues (Young et al. 1991, 1992a, 1993, 1995) established that mesenchymal stem cells are located throughout the connective tissues of the body. Thus, all stem cell types are found in the connective tissues of muscle, bone, cartilage, and so forth. A second important finding was the existence of two categories of mesenchymal stem cells rather than just one in the connective tissues. These two categories are lineage committed progenitor cells and lineage uncommitted pluripotent cells.

Stem Cell Characteristics

Ongoing studies by Young et al. 1991; Young, Ceballos, et al. 1992; Young, Sippel, et al. 1992; Young et al. 1993, 1995; Young, Wright, et al. 1998; Young, Rogers, Adkison, Lucas, and Black, Jr., 1998; Young, Steele, et al. 1998; Young, Reeves, et al. 1998; Young, Young, et al. 1998; Young, Ceballos, et al. 1998; Young et al. 1999) have demonstrated distinct similarities and differences with respect to these 2 mesenchymal stem cell populations. Both progenitor and pluripotent stem cells prefer a type I collagen substratum for attachment followed by cryopreservation and storage at $-70°$ to $-80°C$ in medium containing 10% serum and 7.5% dimethyl sulfoxide.

Progenitor cells (e.g., precursor stem cells, immediate stem cells, and forming or –blast cells, e.g., myoblasts, adipoblasts, chondroblasts, etc.) are lineage committed. Unipotent stem cells will form tissues restricted to a single lineage (such as the myogenic, fibrogenic, adipogenic, chondrogenic, osteogenic lineages, etc.). Bipotent stem cells will form tissues belonging to two lineages (such as the chondro–osteogenic, adipo–fibroblastic lineages, etc.). Multipotent stem cells will form multiple cell types within a lineage (such as the hematopoietic lineage). Progenitor stem cells will form tissues limited to their lineage, regardless of the inductive agent that may be added to the medium. They can remain quiescent. They can also be stimulated to proliferate by various growth factors (Table 9.1). Lineage-committed prog-enitor cells are capable of self-replication but have a life span limited to approximately 50 to 70 cell doublings before programmed cell senescence occurs. They can progress down their lineage pathway and/or differentiate. If activated to differentiate, these cells require progression factors (i.e., insulin, insulin-like growth factor–1, and insulin-like growth factor–2) to stimulate phenotypic expression (Table 9.1).

In contrast, pluripotent cells are lineage uncommitted, that is, they are not committed to any particular mesodermal tissue lineage. They can remain quiescent. They can also be stimulated by growth factors (Table 9.1) to proliferate. If activated to proliferate, pluripotent cells are capable of extended self renewal as long as they remain lineage uncommitted. Pluripo-tent cells have the ability to generate various lineage-committed progeni-tor cells from a single clone at any time during their life span. For example, a prenatal pluripotent mouse clone after more than 690 doublings (Figure 9.2) (Young, Wright, et al. 1998) and a postnatal pluripotent rat clone after more than 300 doublings (Figure 9.3) (Young, Reeves, et al. 1998) were both induced to form lineage-committed progenitor cells that exhibited mor-phological and phenotypic expression markers characteristic of skeletal muscle, fat, cartilage, and bone after long-term dexamethasone exposure. This lineage-commitment process necessitates the use of either general (e.g., dexamethasone) or lineage specific (e.g., bone morphogenetic protein–2, muscle morphogenetic protein, etc.) commitment induction agents (Table 9.1). Once pluripotent cells are induced to commit to a par-ticular tissue lineage, they assume the characteristics of lineage-specific progenitor cells. They can remain quiescent or they can proliferate under the influence of specific inductive agents. Their ability to replicate is limited to approximately 50 to 70 cell doublings before programmed cell senes-cence occurs, and they require the assistance of progression factors to stim-ulate phenotypic expression (Table 9.1).

Phylogenetic Distribution of Stem Cells

Progenitor and pluripotent mesenchymal stem cells derived from various connective tissues were originally identified in prenatal chicks (Young et al. 1992a, 1995). Subsequent experiments have also identified these cells in pre-natal mice (Rogers, Adkison, Black, Jr., Lucas, and Young 1995), prenatal and postnatal rats (Lucas et al. 1995; Dixon et al. 1996; Warejcka, Harrey, Taylor, Young, and Lucas 1996; Young, Rogers, et al. 1998), postnatal rabbits (Pate et al. 1993), and prenatal, adult, and geriatric humans (Young, Steele, et al. 1998; Young et al. 1999, 2000). These stem cells can remain stellate (Figure 9.4A, 9.4B) or generate multiple mesodermal phenotypes such as skeletal muscle (Figure 9.4C, 9.4D), smooth muscle, cardiac muscle, multi-locular brown fat (Figure 9.4E, 9.4G), unilocular white fat (Figure 9.4F, 9.4G), connective tissue, scar tissue, hyaline cartilage (Figure 9.4H, 9.4I),

TABLE 9.1. Proliferation and phenotypic responses of pluripotent and progenitor cells induced by various bioactive factors.

Agent	Proliferation		Phenotypic Expression	
	Pluripotent	Progenitor	Pluripotent	Progenitor
Control	1	1	0^a	All+
PDGF–AA	16^b	16	0	All+
PDGF–BB	19	19	0	All+
PDECGF	1	1	0	All+
b–FGF	1	1	0	F++
TGF–β	$-^c$	–	0	F++
b–FGF + TGF–β	–	–	0	F++
Dex	–	–	All++	All++
MMP	2	2	M++++	M+++/All+
MMP fb^d Dex	2	2	M+++++	M+++/All++
BMP–2	2	2	C++++	C+++/All+
BMP–2 fb Dex	2	2	C+++++	C+++/All++
MMP fb^d BMP–2	2	2	M++++	M+++/C++/All+
BMP–2 fb^d MMP	2	2	C++++	M++/C+++/All+
FMP	10	10	F+++++	F++++/All+
SIF	1	1	0	All+ (F–)
FMP + SIF	10	10	0	All+ (F–)
MMP + SIF	2	2	M++++	M+++/All+ (F–)
FMP+ MMP	10	10	F+++++	F++++/All+
FMP + SIF + MMP	10	10	M++++	M+++/All+ (F–)
ADF	1	1	0	0
ADF + Dex	–	–	0	0
ADF + MMP	2	2	0	0
ADF + BMP–2	2	2	0	0
ADF + FMP	10	10	0	0
Insulin	1	2	0	All+++
IGF–I	1	1	0	All+++
IGF–II	1	1	0	All+++
Insulin + IGF–I	1	1	0	All++
Insulin + IGF–II	1	1	0	All++
IGF–I + IGF–II	1	1	0	All++
Ins + IGF–I + IGF–II	–	–	0	All++
Dex + Insulin	–	1	All+++	All+++
MMP + Insulin	2	2	M+++++	M++++/All+
BMP–2 + Insulin	2	2	C+++++	C++++/All+

[a] Presence and approximate distribution of differentiated phenotypes within the culture wells. Each individual + represents a value of up to 20% of the maximal expression for each phenotype examined: + = 0–20%, ++ = 21–40%, +++ = 41–60%, ++++ = 61–80%, and +++++ = 61–100%. 0, stellate only (no additional differentiated phenotypes noted); M, myogenic; F, fibrogenic; C, chondrogenic; All, all phenotypes (i.e., myogenic, adipogenic, fibrogenic, chondrogenic, osteogenic) expressed.

[b] 16, number of times the agent increased the DNA content per well versus its respective control.

[c] –, statistically significant decrease in DNA content per well versus its respective control.

[d] fb, followed by.

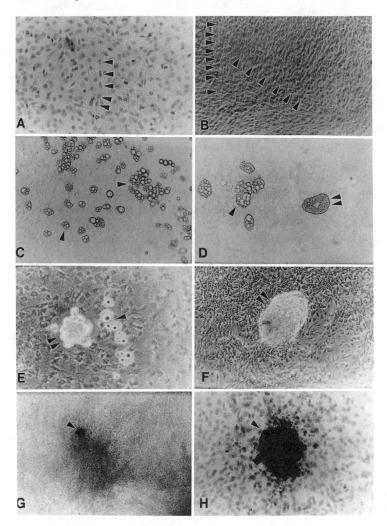

FIGURE 9.2. Swiss-XYP-7 mouse clone after 690 cell doublings. Young, Rogers, et al. 1998. *Muscle morphogenetic protein induces myogenic gene expression in Swiss-3T3 cells. Wound Rep Reg* 6(6):543–554. Reprinted with permission from Blackwell Sciences, Inc. (A) Swiss-XYP-7 clone incubated with 10^{-8}M dexamethasone. Note multinucleated linear structure (arrows showing nuclei), indicative of a myogenic structure. Orig. mag. 50×. (B) Swiss-XYP-7 clone incubated with 10^{-8}M dexamethasone. Note multinucleated linear structures (arrows showing nuclei), indicative of myogenic structures. Orig. mag. 50×. (C) Swiss-XYP-7 clone incubated with 10^{-7}M dexamethasone. Note clusters of multivesiculated cells (arrow), indicative of adipogenic cells. Orig. mag. 50×. (D) Swiss-XYP-7 clone incubated with 10^{-7}M dexamethasone. Note cluster of multivesiculated cells, indicative of adipogenic cells (single arrow) and single cell (double arrows) containing similar-sized vesicles. Orig. mag. 100×. (E) Swiss-XYP-7 clone incubated with 10^{-7}M dexamethasone. Note

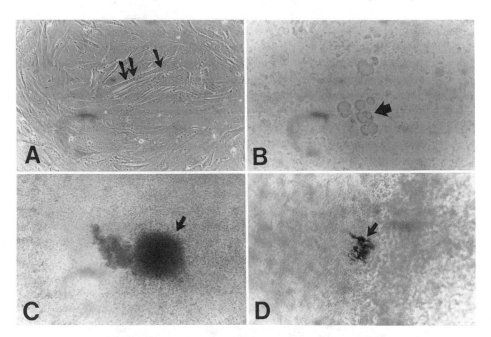

FIGURE 9.3. A2B rat clone after 300 cell doublings. (A) RAT-A2B clone incubated with 10^{-8}M dexamethasone. Note multinucleated myotube (arrows denote nuclei). Phase contrast. Orig. mag. 200×. (B) RAT-A2B clone incubated with 10^{-9}M dexamethasone. Note fluid-filled intracellular vesicles (arrowhead). Brightfield. Orig. mag. 200×. (C) RAT-A2B clone incubated with 10^{-7}M dexamethasone. Note aggregation of rounded cells with AB 1.0+ pericellular matrix halos (arrow). Brightfield. Orig. mag. 100×. (D) RAT-A2B clone incubated with 10^{-8}M dexamethasone. Note positive staining with von Kossa stain of 3–dimensional matrix overlying cell aggregation. Brightfield. Orig. mag. 100×.

clusters of multivesiculated cells, indicative of adipogenic cells (single arrow) and nodule with pericellular matrix halos (double arrows) indicative of chondrogenic structures. Orig. mag. 100×. (F) Swiss-XYP-7 clone incubated with 10^{-8}M dexamethasone. Note nodule with pericellular matrix halos (double arrows) indicative of chondrogenic structures. Orig. mag. 50×. (G) Swiss-XYP-7 clone incubated with 10^{-8}M dexamethasone. Note condensation of cells with nodule of dark stained material (arrow), indicative of mineralized matrix. Orig. mag. 10×. (H) Swiss-XYP-7 clone incubated with 10^{-6}M dexamethasone. Note nodule of dark-stained material, indicative of mineralized matrix (arrow). Orig. mag. 100×.

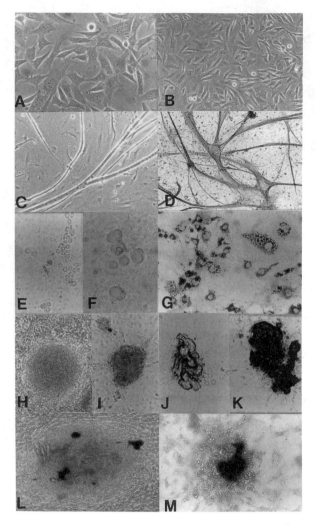

FIGURE 9.4. Mesenchymal stem cells derived from the connective tissues of chick, mouse, rat, and human. (A) RAT-A2B clone 24 hr after plating. Note the predominance of multipolar stellate cells with large ratios of nucleus to cytoplasm. Phase contrast. Orig. mag. 200×. (B) NHDF. Mesenchymal stem cells derived from the dermis of a 25-year-old human adult female after more than 80 cell doublings after harvest (greater than Hayflick's limit). 3 days after incubating in insulin-containing medium. Note increase in cell number without alteration in phenotypic expression. Phase contrast. Orig. mag. 100×. (C) Chick clone CH-Musc-64-3°-M-147 incubated for 5 days with 10^{-8} M dexamethasone. Note linear and branched multinucleated myotubes. Phase contrast. Orig. mag. 100×. (D) Chick clone CH-Musc-64-3°-M-147 incubated for 5 days with 10^{-8} M dexamethasone and then stained with MF-20 antibody to sarcomeric myosin. Note linear and branched stained myotubes. Brightfield. Orig. mag. 50×. (E) Chick clone CH-Musc-64-3°-M-147 incubated for 4 days with 10^{-9} M dexamethasone. Note cells with multiple refractile intracellular vesicles of

elastic cartilage, fibrocartilage, articular cartilage, growth plate cartilage, endochondral bone (Figure 9.4J, 9.4K), intramembranous bone (Figure 9.4L, 9.4M), tendons, ligaments, perichondrium, periosteum, endothelial cells, hematopoietic cells, and so on. Methods have been developed (Young, Dalley, and Markwald 1989a,b; Young, H.E., Young, V.E., and Caplan 1989; Young, Carrino, and Caplan 1989; Young et al. 1991; Young, Ceballos, et al. 1992; Young, Sippel, et al. 1992; Young, Wright, et al. 1998; Young, Rogers, et al. 1998; Young, Steele, et al. 2000; Young et al. 1999; Lucas, Young, and Putnam 1991; Lucas et al. 1995; Warejcka et al. 1996) for the identification of these cell types utilizing morphological, histological, histochemical, functional, immunochemical, and molecular assays (Figure 9.4, Table 9.2).

Stem Cell Location

Analysis of donor sites from the 5 animal species examined has revealed the presence of mesenchymal stem cells in the connective tissues of many organs and tissues. Organs and tissues assayed to date include whole

◄───

the same size. Brightfield. Orig. mag. 50×. (F) RAT-A2B clone incubated with 10^{-9}M dexamethasone, note multiple fluid-filled intracellular vesicles of different sizes. Brightfield. Orig. mag. 200×. (G) RAT-A2B clone incubated with 10^{-9}M dexamethasone and stained with Sudan black-B for saturated neutral lipids. Note two types of cells stained with Sudan black-B. One cell type has multiple intracellular vesicles of the same size, while the other cell type has intracellular vesicles of different sizes. Brightfield. Orig. mag. 200×. (H) RAT-A2B clone incubated with 10^{-7}M dexamethasone. Note aggregation of round cells with pericellular matrix halos. Note that the cell aggregation is surrounded by fibrous connective tissue. Phase contrast. Orig. mag. 50×. (I) Chick clone CH-Musc-64-3°-M-147 incubated with 10^{-7}M dexamethasone. Note aggregation of round cells with pericellular matrix halos stained positive with Alcian Blue at pH 1.0. Brightfield. Orig. mag. 50×. (J) Chick clone CH-Musc-64-3°-M-147 incubated with 10^{-8}M dexamethasone and pretreated with EGTA prior to staining with von Kossa for calcium phosphate. Note aggregation of cells overlain with 3–dimensional extracellular matrix. Note that the outline of the matrix is still positive for von Kossa staining. Brightfield. Orig. mag. 50×. (K) Chick clone CH-Musc-64-3°-M-147 incubated with 10^{-8}M dexamethasone and stained with von Kossa stain for calcium phosphate. Note aggregation of cells overlain with 3–dimensional extracellular matrix. Matrix stains intensely with von Kossa stain, suggesting presence of calcium phosphate within extracellular matrix. Brightfield. Orig. mag. 50×. (L) Swiss-XYP-7 clone incubated with 10^{-8}M dexamethasone and stained with von Kossa for calcium phosphate. Note aggregation of cells with discrete areas demonstrating positive staining. Phase contrast. Orig. mag. 50×. (M) RAT-A2B clone incubated with 10^{-8}M dexamethasone and stained with von Kossa stain for calcium phosphate. Note aggregation of cells overlain with 3–dimensional extracellular matrix. Matrix stains intensely with von Kossa stain, suggesting presence of calcium phosphate within extracellular matrix. Brightfield. Orig. mag. 50×.

TABLE 9.2. Histological staining, functional and/or histochemical staining, antibody staining with ELICA (enzyme-linked immunoculture assay) or flow cytometry, and northern analysis of mRNA expression.

Cell Type	Histological	Functional/ Histochemical[a]	ELICA/Flow Cytometry Antibodies[b]	Northern cDNA Probes
Skeletal Muscle	Multinucleated linear and branched structures	F: Spontaneous contractility	E: MF-20, 12/101 31–2, MF-5, C3/1, M3F7, ALD-58, CH1, 5C6, 2E8, MF-30, antimyosin, antitype IV collagen	MyoD1, myogenin, emb. myosin heavy chain, myosin light chain-3, MYD, MYF5, MYF6, MYH2, MYL1, MYF3, MYF4[c]
Smooth Muscle	Polygonal mononucleated cells with stress fibers		E: antismooth muscle α-actin	smooth muscle α-actin
Cardiac Muscle	Polygonal binucleate cells	F: Contraction rate altered with propranolol and isoproteranol	E: D76, D3, antidesmin	β–myosin heavy chain, ATP2A2
White Fat	Perinucleated cells with multiple refractile vesicles of different sizes	H: Sudan black-B		Lipoprotein lipase, adipophilin
Brown Fat	Central nucleated cells with multiple refractile vesicles of similar size	H: Sudan black-B		Lipoprotein lipase, adipophilin
Connective Tissue	spindle-shaped cells with fibrillar matrix	H: AB 1.0+, SO 2.5+, CH'ase-AC, CH'ase-ABC, MH-collagen type I	E: M-38, SP1.D8, B3/D6	CS-PG core prot.,[d] type I collagen, prepro-α 1(I) collag., collagen type I α-2, MMP-1A, MMP-1B
Scar Tissue	Spindle-shaped cells with granular matrix	H: AB 1.0+, SO 2.5+, CH'ase-AC, CH'ase-ABC, MH-collagen Type I	E: M-38, SP1.D8, B3/D6	CS-PG core prot., type I collagen, prepro-α 1(I) collag., collagen type I α-2, MMP-1A, MMP-1B
Hyaline Cartilage	Aggregates of rounded cells with pericellular matrix halos, surrounded by fibrous tissue	H: AB 1.0+, SO 2.5+, CH'ase-AC, keratanase, MH-collagen type II	E: 5-D-4, antitype II, antitype II collagen	KS-PG core protein, CS-PG core protein, CS/KS-PG core protein, type II collagen
Elastic Cartilage	Aggregates of rounded cells with pericellular matrix halos with thin interwoven fibers, with adjacent fibrous tiss.	H: AB 1.0+, SO 2.5+, CH'ase-AC, keratanase, MH-collagen type II, elastin+	E: 5-D-4, antitype II, antitype II collagen, antielastin	KS-PG core protein, CS-PG core protein, CS/KS-PG core protein, type II collagen, elastin

TABLE 9.2. *Continued*

Cell Type	Histological	Functional/ Histochemical[a]	ELICA/Flow Cytometry Antibodies[b]	Northern cDNA Probes
Fibrocartilage	Sheets of rounded cells with pericellular matrix halos intermingled with thick fibers and surrounded by fibrous tissue	H: AB 1.0+, SO 2.5+, CH'ase-AC, CH'ase-ABC, MH-collagen type I	E: B3/D6, M-38, SP1.D8	CS-PG core protein, type I collagen, prepro-α 1(I) collagen, collagen type I α-1, collagen type I α-2
Articular Cartilage	Sheets of rounded cells with pericellular matrix halos	H: AB 1.0+, SO 2.5+ CH'ase-AC, keratanase, MH-collagen type II	E: 5-D-4, antitype II, antitype II collagen	KS-PG core protein, CS-PG core protein, CS/KS-PG core protein, type II collagen
Growth Plate Cartilage	Aggregates of rounded cells with pericellular matrix halos overlain with 3-D matrix	H: AB 1.0+, SO 2.5+, CH'ase-AC, keratanase, MH-collagen types I & II	E: 5-D-4, antitype II, antitype II collagen, B3/D6, M-38, SP1.D8, antiosteocalcin	KS-PG core protein, CS-PG core protein, CS/KS-PG core protein, type II collagen, type I collagen, prepro-α 1(I) collagen, collagen type I α-1 and α-2, osteocalcin, osteonectin, osteopontine
Endo chondral Bone	Aggregates of rounded cells with pericellular matrix halos overlain with 3-D matrix	H: AB 1.0+, SO 2.5+, CH'ase-AC, keratanase, MH-collagen types I & II, von Kossa+ 3-D matrix	E: 5-D-4, antitype II, antitype II collagen, B3/D6, M-38, SP1.D8, antiosteocalcin	KS-PG core protein, CS-PG core protein, CS/KS-PG core protein, type II collagen, type I collagen, prepro-α 1(I) collagen, collagen type I α-1 and α-2, osteocalcin, osteonectin, osteopontine
Intramembran- ous Bone	Aggregations of stellate cells overlain with 3-D matrix	H: von Kossa+	E: M-38, antiosteocalcin	type I collagen, prepro-α 1(I) collagen, collagen type I α-1 and α-2, osteocalcin, osteonectin, osteopontine
Tendon/ Ligament	Cells inter-mingled with thick fibers	H: ECM: AB 1.0+, SO 2.5+, CH'ase-AC, Fibers: MH-type I collagen	E: M-38	type I collagen, prepro-α 1(I) collagen, collagen type I α-1 and α-2
Perichon- drium	Fibrous tissue surrounding cell aggregates with pericellular matrix halos	H: AB 1.0+, SO 2.5+, keratanase, CH'ase-AC, MH- type II collagen at interface with	E: 5-D-4, antitype II, antitype II collagen, SP1.D8, M-38, B3/D6	KS-PG core protein, CS-PG core protein, CS/KS-PG core protein, collagen types I & II, prepro-α 1(I) collagen, type I α-1 and α-2 collagen, MMP-1A,

TABLE 9.2. *Continued*

Cell Type	Histological	Functional/ Histochemical[a]	ELICA/Flow Cytometry Antibodies[b]	Northern cDNA Probes
		cell aggregates, collagen type I at interface with stellate cells		MMP-1B
Periosteum	Fibrous tissue surrounding aggregations of stellate cells overlain with 3-D matrix	H: AB 1.0+, SO 2.5+, CH'ase-AC CH'ase-ABC, MH-collagen type I	E: M-38, antiosteocalcin, SP1.D8, B3/D6	collagen type I, prepro-α 1(I) collagen, type I α-1 and α-2 collagen, MMP-1A, MMP-1B, osteocalcin, osteonectin, osteopontine
Endothelial Cells	Sheets of cobblestone-shaped cells	F: low density lipoprotein uptake	E: Factor-8	endothelial cell surface protein, endo-thelin-1, endothelin-3, LDL-R
Hematopoietic Cells	Floating and attached refractile cells with differing nuclear shapes		F: CD3, CD4, CD5, CD7, CD8, CD10, CD11b, CD11c, CD13, CD14, CD15, CD16, CD19, CD25, CD33, CD34, CD36, CD38, CD45, CD56, CD65, CD90, CD117, Glycophorin-A, HLA Class I, HLA-II (DR)	EPO-R, M-CSF-R, G-CSF-R, GM-CSF-R, NCAM isoform 140 kDa, monocytes, transferrin–R, neutral endopeptidase, aminopeptidase, Thy-1, HSC-GF-R, erythrocyte membrane protein band-3, spectrin α-erythrocytic-1

[a] *Histochemistry.* Sudan black-B stains saturated neutral lipids indicative of adipocytes (fat cells). Alcian Blue (AB) at pH 1.0 (AB 1.0+) stains sulfated moieties of glycosaminoglycans a blue color. Safranin–O (SO) at pH 2.5 (SO 2.5+) stains carboxylated moieties of glycosaminoglycans an orange color. Thus, AB 1.0 followed by SO 2.5 will stain the following: proteoglycans containing keratan sulfate, heparin, or heparan sulfate glycosaminoglycan chains only (containing only sulfate groups) will stain a blue color; proteoglycans containing nonsulfated chondroitin glycosaminoglycan chains or hyaluronate (containing only carboxylated groups) will stain an orange color; and proteoglycans containing chrondroitin sulfate and/or dermatan sulfate glycosaminoglycan chains (containing both sulfate groups and carboxyl groups) will stain a brown color (= blue + orange). Verification of proteoglycan content is made using selective enzyme digests. Heparinase will degrade heparin; heparanase will degrade heparan sulfate glycosaminoglycans; keratanase will degrade keratan sulfate glycosaminoglycans; Streptomyces hyaluronidase (S-HA'ase) will degrade hyaluronate and nonsulfated chondroitin; testicular hyaluronidase (t-HA'ase) will degrade hyaluronate, nonsulfated chondroitin, and chondroitin sulfate glycosaminoglycans; chondroitinase-AC (CH'ase-AC) will degrade chondroitin sulfate glycosaminoglycans; and chondroitinase-ABC (CH'ase-ABC) will degrade chondroitin sulfate and dermatan sulfate glycosaminoglycans. Von Kossa will stain divalent anions, for example Ca^{+2}, Mg^{+2}, and Zn^{+2}. Verification of the presence of calcium phosphate in mineralized tissue such as bone necessitates use of the specific calcium chelator, EGTA, in a preincubation step before staining. Use of EDTA is not recommended as a specific test for the presence of calcium because it will chelate all divalent anions. (Appendix I, Young 1983; Young et al. 1993).

[b] *Antibodies.* MF-20, sarcomeric myosin; 12/101, skeletal muscle; 31-2, laminin; MF-5, myosin light chain-2 of fast muscle; C3/1, glycoprotein of myoblast plasma membrane; M3F7, type IV collagen; ALD-58, myosin heavy chain; CH1, myosin tropomyosin; 5C6, type IV collagen; 2E8, laminin; MF-30, neonatal and adult

embryo, skeletal muscle, dermis, fat, tendon, ligament, perichondrium, periosteum, atria and ventricles (endocardium, myocardium, and epicardium), aorta, granulation tissue, nerve sheaths, dura, trachea, esophagus, marrow, stomach, small intestine, large intestine, liver, spleen, pancreas, parietal peritoneum, urinary bladder, and kidney (Young et al. 1992a, 1995; Young, Wright, et al. 1998; Lucas et al. 1993, 1995, 1996a & b; Davis et al. 1995; Warejcka et al. 1996).

The connective tissues of these organs and tissues contained fibrocytes, tissue-specific lineage-committed progenitor stem cells, and pluripotent stem cells. They also contained progenitor stem cells from other tissue lineages (Young et al. 1993, 1995; Young et al. 1995, unpublished raw data). For example, the perichondrium surrounding hyaline cartilage appeared to be segregated into three zones based on stem cell composition. The inner third (or cambial layer) contained predominantly chondrogenic progenitor cells and a few pluripotent cells. The middle third contained predominantly pluripotent stem cells, along with a few chondrogenic progenitor cells and a few nonchondrogenic progenitor cells. The outer third contained predominantly nonchondrogenic progenitor cells (e.g., myogenic, adipogenic, fibrogenic, and osteogenic progenitor cells), fibrocytes, and a few pluripotent cells. We found similar regional patterns of stem-cell distribution with respect to pluripotent cells, tissue-specific progenitor cells and nontissue specific progenitor cells in the supporting connective tissues of skeletal muscle (endomysium, perimysium, and epimysium), heart (endocardium, epicardium) and bone (periosteum).

◄───

myosin; Antimyosin, myosin; Antitype IV collagen, type IV collagen; anti-smooth muscle α-actin, smooth muscle; D76, desmin; D3, desmin; anti–desmin, desmin; M-38, type I procollagen; SP1.D8, procollagen type III; B3/D6, fibronectin; 5-D-4, keratan sulfate proteoglycan; antitype II, type II collagen; antitype II collagen, type II collagen; antielastin, elastin; antiosteocalcin, osteocalcin/bone Gla-protein; CD3, T-cells; CD4, T-cells; CD5, T-cells, B-cells; CD7, T-cells, thymocytes; CD8, T-cells; CD10, neutral endopeptidase; CD11b, T-cells, B-cells, granulocytes, monocytes, natural killer cells; CD11c, monocytes/macrophages, natural killer cells, granulocytes; CD13, aminopeptidase; CD14, granulocytes, monocytes; CD15, granulocytes; CD16, granulocytes, monocytes, natural killer cells; CD19, B-cells, follicular dendritic cells; CD25, T-cells, B-cells, monocytes; CD33, myeloid progenitor cells; CD34, hematopoietic stem cell marker; CD36, monocytes/macrophages, granulocytes; CD38, monocytes/macrophages, natural killer cells, granulocytes, myeloid progenitor cells, some neuronal cells; CD45, T-cells, B-cells, granulocytes, monocytes/macrophages, natural killer cells, mature erythrocytes, myeloid progenitor cells; CD56, neural cell adhesion molecule isoform 140kDa (Natural Killer cells); CD65, granulocytes; CD90, Thy-1; CD117, T-cells, monocytes/macrophages, natural killer cells, granulocytes, myeloid progenitor cells; glycophorin-A, erythrocytes; HLA Class I, major histocompatibility complex Class I molecules; HLA-DR-II, monocytes/macrophages, some neuronal cells.

[c] Each phenotype is probed with cDNA for PDGF-α receptor, PDGF-β receptor, β-actin (as internal control).

[d] CS-PG core protein, chondroitin sulfate proteoglycan core protein; MMP-1A, matrix metalloproteinase-1A; MMP-1B, matrix metalloproteinase-1B; KS-PG core protein, keratan sulfate proteoglycan core protein; CS/KS-PG core protein, chondroitin sulfate/keratan sulfate proteoglycan core protein; LDL-R, low density lipoprotein receptor; EPO-R, erythropoietin receptor; M-CSF-R, macrophage colony stimulating factor receptor; G-CSF-R, granulocyte colony stimulating factor receptor; GM-CSF-R, granulocyte/macrophage colony stimulating factor receptor; NCAM, neural cell adhesion molecule; NK cells, natural killer cells; transferrin-R, transferrin receptor; HSC-GF-R, hematopoietic stem cell growth factor receptor.

Cloning

Clonogenic analyses of mesenchymal stem cells isolated from prenatal avians (Figure 9.5) (Young et al. 1993), prenatal mice (Figure 9.2) (Rogers et al. 1995; Young, Rogers, et al. 1998), and postnatal rats (Figure 9.3) (Young, Reeves, et al. 1998) consistently demonstrate the presence of both lineage-committed progenitor stem cells and lineage-uncommitted pluripotent stem cells. General lineage progression agents, such as insulin (Table 9.1), have been used to accelerate phenotypic expression in lineage committed progenitor cells (Figure 9.4). General and lineage specific inductive agents, such as dexamethasone, muscle morphogenetic protein, bone morphogenetic protein–2, etc. (Table 9.1), have been used to induce 4 tissue lineages in prenatal and postnatal pluripotent stem cell clones: muscle (Figure 9.2A, 9.2B, 9.3A, 9.5A), fat (Figures 9.2C–E, 9.3B, 9.5B, 9.5D), cartilage (Figure 9.2E, 9.2F, 9.3C, 9.5C), and bone (Figure 9.2G, 9.2H, 9.3D, 9.5D).

Age of Donor

Studies are ongoing to determine the optimal age for harvesting progenitor and pluripotent stem cells for transplantation therapies (Young et al. 1991; Young, Ceballos, et al. 1992; Young et al. 1993, 1995; Young, Wright, et al. 1998; Young, Rogers, et al. 1998; Young, Steele, et al. 2000; Young, Reeves, et al. 1998; Young, H.E., Young, T.M., et al. 1998; Young et al. 1999; Young et al. 1999; unpublished raw data). In all five species examined (chicken, mouse, rat, rabbit, and human), no age related differences have been found with respect to the number of pluripotent stem cells present per species. No influence of age on the ability to proliferate or on the ability to differentiate has been found in the stem cells isolated from these five species. In addition, no influence of gender has been found in prenatal to geriatric human stem cells.

Effects of Bioactive Factors

Having access to mixed populations of progenitor stem cells, progenitor stem-cell clones, and pluripotent stem-cell clones has allowed us to address the influence of various bioactive factors on the growth characteristics and phenotypic expression of these stem cells (Young, Ceballos, et al. 1992; Young et al. 1993, 1995; Young, Wright, et al. 1998; Young, Rogers, et al. 1998; Young et al. 1998, unpublished raw data). To date 14 bioactive factors have been tested with these cells, both singly and in combination (Table 9.1). Three general categories of activity have been shown (proliferation, lineage commitment, and lineage progression). Bioactive factors could produce either stimulatory or inhibitory effects that could be either general across all the lineages or limited to 1 or more specific tissue lineages.

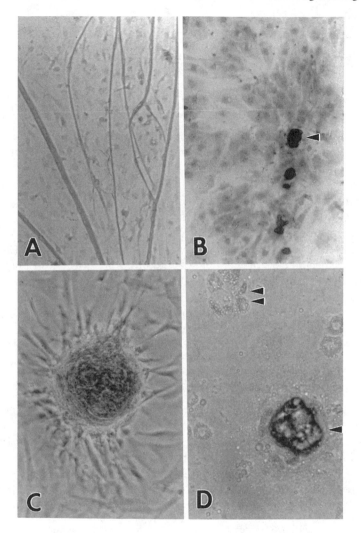

FIGURE 9.5. Chick clone CH-Musc-64-3°-M-147 generated from mesenchymal stem cells harvested from day 11 embryonic chick leg muscle connective tissue. Young et al. 1993. *Pluripotent mesenchymal stem cells reside within avian connective tissue matrices. In Vitro Animal Cell Dev Biol 29A:723–736.* Reprinted with permission from the Society for In Vitro Biology. (A) Chick clone CH-Musc-64-3°-M-147 incubated with 10^{-8} M dexamethasone for 5 days. Note linear and small branched myotubes with centrally located nuclei. Brightfield. Orig. mag. 200×. (B) Chick clone CH-Musc-64-3°-M-147 incubated with 10^{-9} M dexamethasone for 4 days. Note cells stained with Sudan black-B (arrowhead). Brightfield. Orig. mag. 400×. (C) Chick clone CH-Musc-64-3°-M-147 incubated with 10^{-7} M dexamethasone for 5 days. Note nodule of rounded cells stained with Alcian Blue pH 1.0. Brightfield. Orig. mag. 400×. (D) Chick clone CH-Musc-64-3°-M-147 incubated with 10^{-8} M dexamethasone for 14 days. Note nodule stained with von Kossa stain for calcium phosphate. Brightfield. Orig. mag. 200×.

Platelet-derived growth factor-AA (PDGF-AA) and platelet-derived growth factor-BB (PDGF–BB) stimulated proliferation in pluripotent cells and in all lineages of progenitor cells. Platelet-derived endothelial cell growth factor (PDECGF) showed no measurable effect on either progenitor or pluripotent stem cells under the assay conditions used. Basic-fibroblast growth factor (b-FGF) and transforming growth factor-β (TGF-β)-stimulated lineage progression in fibrogenic progenitor cells, inhibited lineage progression in all other progenitor cells, and had no effect on pluripotent cells. Dexamethasone depressed proliferation in pluripotent stem cells, stimulated general lineage commitment in pluripotent cells, and acted as a weak stimulator of lineage progression in all progenitor cells. Muscle morphogenetic protein (MMP) acted as a specific myogenic lineage commitment agent in pluripotent cells and as a weak stimulator of lineage progression in myogenic progenitor cells, but had no effect on progenitor cells committed to other lineages. Bone morphogenetic protein–2 (BMP–2) acted as a specific chondrogenic lineage commitment agent in pluripotent cells and as a weak stimulator of lineage progression in chondrogenic progenitor cells, but had no effect on progenitor cells committed to other lineages. Fibroblast morphogenetic protein (FMP) acted as a specific fibro-genic lineage-commitment agent in pluripotent cells and as a stimulator of lineage progression in fibrogenic progenitor cells, but had no effect on progenitor cells committed to other lineages. Scar inhibitory factor (SIF) acted as a specific inhibitor of the lineage-commitment activity of FMP on pluripotent cells and as a specific inhibitor of the lineage progression activity of FMP on progression in fibrogenic progenitor cells, but had no effect on lineage induction or lineage progression for other tissue lineages. Antidifferentiation factor (ADF) acted as a general inhibitor of lineage-commitment activity on pluripotent cells and as a general inhibitor of lineage-progression activity on progenitor cells. Insulin, insulin-like growth factor–1 (IGF–1), and insulin-like growth factor–2 (IGF–2) stimulated lineage progression in all progenitor cells, but had no measurable effect on pluripotent cells.

Cell Surface Cluster Differentiation Markers

Since cluster differentiation (CD) markers have not been reported for human mesenchymal stem cells (Kishimoto et al. 1997), we have begun to use flow cytometry to determine the CD profiles for mesenchymal stem cells derived from human male and female fetal, adult, and geriatric tissues. To date, we have analyzed 26 CD markers by flow cytometry. All human stem cell populations examined displayed phenotypic expression markers indicative of both progenitor cells and pluripotent cells (Young et al. 1998c,d) as assessed by lineage–progression (insulin) and lineage–commitment (dexamethasone) analyses (Young et al. 1998b). After 30 doublings from the time of tissue harvest, mesenchymal stem cells from all three age groups and both genders demonstrated the presence of CD markers for CD10, CD13, CD34, CD56, CD90, and HLA Class-1. Negative results were obtained for CD3, CD4, CD5, CD7, CD8, CD11b, CD11c, CD14, CD15, CD16, CD19, CD25, CD33, CD36, CD38, CD45, CD65, CD117, Glycophorin-A, and HLA-II (DR) (Young et al. 1998c,d).

Northern Analysis of Expressed mRNAs

We have used Northern blot analysis to examine the induction of myogenesis by MMP in a mouse pluripotent stem cell clone. We have also used this technique to examine CD marker transcription in human mesenchymal stem cells. MMP induced the transcription of mRNAs for myogenin and MyoD1 gene expression in Swiss–XYP–7, a prenatal mouse pluripotent stem cell clone (Rogers et al. 1995; Young, Wright, et al. 1998). Northern blot analysis also showed that the genes for aminopeptidase (CD13), neural cell adhesion molecule (CD56), and Thy–1 (CD90) were actively being transcribed at the time of cell harvest in both prenatal and postnatal human mesenchymal stem cells (Young et al. 1998; Young, Steele et al. 2000).

Gene Transfection

We have proposed the use of pluripotent stem cells as human leucocyte antigen (HLA)-matched delivery vehicles for gene therapy (Young, Reeves, et al. 1998). This proposal is predicated on the hypothesis that the insertion of an exogenous gene will not alter the developmental potential of the pluripotent stem cells. As an initial test of this proposal, RAT-A2B pluripotent stem cell clone was transfected with the insect gene β–galactosidase. The transfected clone was then examined in our standard insulin/dexamethasone assay. Incubation with insulin did not alter the phenotypic expression of the cells compared to untreated controls. This suggested that the cells had not converted to a progenitor lineage due to either the transfection procedure itself or incorporation of β–galactosidase into its genome. In contrast, dexamethasone incubation induced the expression of morphological phenotypes consistent with myotubes, fat, cartilage, and bone. Immunochemical and histochemical analyses confirmed these cell lineages. In addition, these same morphologies coexpressed the reaction product for β–galactosidase. These results suggest that gene transfected–stem cells retain pluripotent capabilities without loss of the expression of the transfected gene (Young, Dupla, et al. 1998). In vivo transplantation studies are in progress to examine the incorporation of β–galactosidase transfected cells into various mesodermal tissues in the body with subsequent coexpression of the β–galactosidase gene.

Stem Cell Isolation, Cloning, and Expression

To isolate progenitor and pluripotent stem cells, a sample containing connective tissue (see Stem Cell Location, above) is harvested aseptically and
transported in MSC-1 culture medium, containing an additional 2× antibiotic–antimycotic solution, to a sterile hood (Lucas et al. 1995). MSC-1
culture medium consists of 89% (v/v) medium [either Eagle's Minimal
Essential Medium with Earle's salts (EMEM, GIBCO, Grand Island, NY)
(Young et al. 1991) or Opti–MEM (GIBCO) containing 0.01 mM β–
mercaptoethanol (Sigma Chemical Co., St. Louis, MO) (Young, Reeves, et
al. 1998; Young et al. 1999)], supplemented with 10% serum [either prese-
lected horse serum, such as HS7 (lot #17F–0218, Sigma), HS4 (lot
#49F–0082, Sigma), HS3 (lot # 3M0338, Bio–Whittaker, Walkersville, MD)
(Young, Reeves, et al. 1998e) or any nonselected serum containing 2 U/ml
antidifferentiation factor (ADF, Morphogen Pharmaceuticals, Inc., New
York, NY) (Young, Reeves, et al. 1998; Young et al. 1999)], and 1% antibiotic–antimycotic solution [10,000 units/ml penicillin, 10,000 μg/ml strepto-
mycin, and 25 μg/ml amphotericin B as Fungizone, GIBCO] (Lucas et al.
1995), pH 7.4. Tissue samples are placed in 10 ml of MSC-1 and carefully
minced. After mincing, the tissue suspension is centrifuged at $50 \times g$ for 20
min. The supernatant is discarded and an estimate made of the volume of
the cell pellet. The cell pellet is resuspended in 7 pellet volumes of EMEM
(or Opti-MEM + 0.01 mM β–mercaptoethanol), pH 7.4, and 2 pellet
volumes of collagenase/dispase solution to release the cells by enzymatic
action (Lucas et al. 1995). The collagenase/dispase solution consists of
37,500 units of collagenase (CLS–I, Worthington Biochemical Corp., Free-
hold, NJ) in 50 ml of EMEM (or Opti-MEM + 0.01 mM β–mercaptoethanol)
added to 100 ml dispase solution (Collaborative Research, Bedford, MA).
The final concentrations are 250 units/ml collagenase and 33.3 units/ml
dispase (Young, Ceballos, et al. 1992). The resulting suspension is stirred at
37°C for 1 hr to disperse the cells and centrifuged at $300 \times g$ for 20 min. The
supernatant is discarded, and the tissue pellet resuspended in 20 ml of MSC-
1 (Lucas et al. 1995). The cells are sieved through 90 μm and 20 μm Nitex
to obtain a single cell suspension (Young et al. 1991). The cell suspension
is centrifuged at $150 \times g$ for 10 min., the supernatant discarded, and the cell
pellet resuspended in 10 ml of MSC-1 (Lucas et al. 1995). Cell viability is
determined by Trypan blue exclusion assay (Young et al. 1991). Cells are
seeded at 10^5 cells per 1% gelatinized (EM Sciences, Gibbstown, NJ) 100
mm culture dish (Falcon, Becton–Dickinson Labware, Franklin Lakes, NJ)
or T-75 culture flask (Falcon). Cell cultures are propagated to confluence at
37°C in a 95% air/5% CO_2 humidified environment. At confluence the cells
are released with trypsin and cryopreserved. Cells are slow frozen (temperature drop of 1 degree per minute) in MSC-1 containing 7.5% (v/v)
dimethyl sulfoxide (DMSO, Morton Thiokol, Danvers, MA) until a final
temperature of −70° to −80°C is reached (Young et al. 1991).

Clonogenic Analysis

Clonogenic analysis by limiting serial dilutions is undertaken to determine the composition of cells within the identified populations of mesenchymal stem cells.

Preconditioned Medium

Cloning studies with prenatal chicks (Young et al. 1993) and prenatal mice (Rogers et al. 1995; Young, Wright, et al. 1998) revealed that a higher efficacy of cloning could be achieved if individual cells were grown in medium "preconditioned" by highly proliferating cells of the same parental line. Therefore, each time the stem cells are harvested at confluence, during log-phase growth, the culture medium is pooled, filtered twice through 0.2 μm filters, divided into aliquots, and stored at 4°C.

Propagation Past 50 Cell Doublings

If only pluripotent stem cells are desired, prepropagation for more than 50 cell doublings is performed. If progenitor cells are also wanted, this step is eliminated. Previous cloning studies in prenatal mice (Rogers et al. 1995; Young, Wright, et al. 1998) and postnatal rats (Young, Rogers, et al. 1998) revealed that a higher efficacy of cloning could be achieved if cells were propagated for more than 50 cell doublings prior to cloning. When such stem cells were incubated with insulin, less than 1% of the cells displayed phenotypic markers for differentiated cells of the various mesodermal lineages. These observations suggested that a majority of the progenitor stem cells were removed from the population by prepropagating the cells for more than 50 cell doublings. Presumably propagating the cells past Hayflick's limit (50 to 70 doublings) caused the lineage-committed stem cells to undergo programmed cell senescence and death (Hayflick, 1965; Young, Rogers, et al. 1998; Young, Young, Floyd, Reeves, Davis, and Mancini 2000; Young, Young, Floyd, Reeves, Davis, Eaton, et al. 2000).

In the studies by Young and colleagues (Young, Rogers, et al. 1998; Young, Reeves, et al.), larger-sized cells (with high ratios of cytoplasm to nuclei) were observed to undergo apoptosis between 40 and 50 cell doublings. The majority of the cells remaining after 50 cell doublings were of smaller size, with smaller ratios of cytoplasm to nuclei (Figure 9.4A). Thus, the standard protocol of thawing cryopreserved cells, culturing them to confluence, collecting them in preconditioned medium, releasing them with trypsin, and subjecting them to cryopreservation is repeated until the stem cell population has undergone a minimum of 50 cell doublings. Aliquots of cells propagated for more than 50 doublings are cryopreserved for cloning.

Cloning

Frozen aliquots of cells are thawed, grown to confluence, released with trypsin, and centrifuged. The supernatants are discarded, cell pellets resuspended, and the viability of the cells determined. Cells are diluted to clonal density (1 cell per 5 µl) with cloning medium. Cloning medium is prepared by mixing equal volumes of MSC-1 medium and preconditioned medium. Five µl of cell suspension is placed into the center of each well of gelatinized 96-well plates (Costar, Curtain–Matheson Scientific, Atlanta, GA) and incubated at 37°C. After 6 hr an additional 200 µl of cloning medium is added to each well. Eighteen hr after initial seeding, the number of cells per well is determined. Only those wells having a single cell are allowed to propagate further. The medium is removed from all other wells. These wells are incubated with 70% (v/v) ethanol for 5 min., and dried in room air. Two hundred µl of sterile Dulbecco's Phosphate Buffered Saline (DPBS, GIBCO), pH 7.4, containing 0.03% (w/v) sodium azide is added to retard contaminant growth.

For those wells allowed to propagate further, the initial cloning medium is replaced with fresh cloning medium after 10 or more cells appear within the wells. Cloning-medium replacement thereafter is dependent upon the percentage of confluence of the cultures, with a maximum lapse of 5 days between feedings. Cultures are allowed to grow past confluence. Only those cultures displaying retention of a stellate morphology and loss of contact inhibition are chosen for continued cloning. Each culture meeting these criteria is released with trypsin, plated in toto into a well of gelatinized 6-well plates (Falcon), fed MSC-1 medium every other day, and allowed to grow past confluence. Only those cultures continuing to display a stellate morphology and loss of contact inhibition are chosen for subcloning. These cultures are released with trypsin and cryopreserved for a minimum of 24 hours. The process of seeding at clonal density in 96-well plates in cloning medium, propagation through confluence, culture selection (using criteria of stellate morphology and loss of contact inhibition), trypsin release, propagation through confluence in 6-well plates in MSC-1 medium, culture selection, trypsin release, and cryopreservation is repeated a minimum of three times after the initial cloning to ensure that each isolated clone is derived from a single cell. The resultant clones are propagated, released with trypsin, aliquoted, and cryopreserved.

Insulin–Dexamethasone Analysis for Phenotypic Expression

Cryopreserved clones are thawed and plated in MSC-1 at 5, 10, or 20×10^3 cells per well of gelatinized 24-well plates following the standard protocol. Twenty-four hours after initial plating, the medium is changed to testing medium (TM) 1 to 6 (TM-1, TM-2, TM-3, TM-4, TM-5, or TM-6). TM-1 to

TM-4 consisted of Ultraculture (cat. no. 12–725B, lot. nos. OMO455 [TM-1], 1M1724 [TM-2], 2M0420 [TM-3], or 2M0274 [TM-4], Bio–Whittaker, Walkersville, MD)[1], medium (EMEM or Opti-MEM + 0.01 mM β–mercaptoethanol)[1], and 1% (v/v) antibiotic-antimycotic, pH 7.4. TM-5 consists of 98% (v/v) medium, 1% (v/v) HS, and 1% (v/v) antibiotic-antimycotic, pH 7.4. TM-6 consists of 98.5% (v/v) medium, 0.5% (v/v) HS, and 1% (v/v) antibiotic-antimycotic, pH 7.4. Preincubation for 24 hr in testing medium alone is used to wash out any potential synergistic components in the MSC-1 medium. Twenty-four hours later, the testing medium is changed to one of the following. For controls, TM-1 to TM-6 alone is used. To identify clones of progenitor cells, the medium is replaced with TM-1 to TM-6 containing 2 µg/ml insulin (Sigma), an agent that accelerates the appearance of phenotypic expression markers in progenitor cells (Table 9.1). To identify clones of pluripotent cells, the medium is replaced with TM-1 to TM-6 containing 10^{-10} to 10^{-6} M dexamethasone (Sigma), a general nonspecific lineage-inductive agent (Table 9.1). Control and treated cultures are propagated for an additional 30 to 45 days with medium changes every other day. Four culture wells are used per concentration per experiment. During the 30 to 45 day time period the cultures are examined subjectively for changes in morphological characteristics on a daily basis. Alterations in phenotypic expression are correlated with the days of treatment and associated insulin or dexamethasone concentrations. The experiment is then repeated utilizing these parameters to confirm objectively the phenotypic expression markers using established histological, functional/histochemical, ELICA (enzyme-linked immuno-culture assay)/flow cytometry, and molecular assays (Table 9.2).

Summation

We have shown that mesenchymal stem cells are present in both prenatal and postnatal animals. Mesenchymal stem cells can be found in any tissue or organ possessing a connective tissue component. No detectable influence

---

[1] Testing medium containing ratios of Ultraculture: medium (EMEM or Opti-MEM + 0.01 mM β–mercaptoethanol): antibiotics (+ antimycotics) maintained both progenitor and pluripotent cells in "steady–state" conditions for a minimum of 30 days in culture, and as long as 120 days in culture. Four testing media (TM#'s 1 to 4), each containing various concentrations of Ultraculture, were used. The ratios of Ultraculture to medium to antibiotics present in each testing medium were determined empirically for each lot of Ultraculture, based on its ability to maintain steady-state culture conditions in both populations of avian progenitor and pluripotent cells. The 4 Ultraculture based testing media were: TM-1 = 15% (v/v) Ultraculture (Lot no. OMO455): 84% (v/v) medium: 1% (v/v) antibiotics; TM-2 = 15% (v/v) Ultraculture (Lot no. 1M1724): 84% (v/v) medium: 1% (v/v) antibiotics; TM-3 = 50% (v/v) Ultraculture (Lot no. 2M0420): 49% (v/v) medium: 1% (v/v) antibiotics; and TM-4 = 75% (v/v) Ultraculture (Lot no. 2M0274): 24% (v/v) medium: 1% (v/v) antibiotics.

of age or gender has been noted. Mesenchymal stem cells are composed of both lineage-committed progenitor stem cells and lineage-uncommitted pluripotent stem cells. Once pluripotent stem cells are committed to a particular tissue lineage, the resulting progenitor stem cells will not revert back to form pluripotent stem cells. Thus, the commitment of pluripotent stem cells to progenitor stem cells is final. Progenitor stem cells are limited to a life span of 50 to 70 cell doublings before programmed cell senescence occurs. Appropriate bioactive factors can regulate the processes of proliferation, lineage commitment, and lineage progression.

These studies lead us to propose that autologous pluripotent (and progenitor) stem cells can be used as HLA-matched donor tissue for mesodermal tissue transplantation, regeneration, and gene therapies. These stem cells could be especially useful where large numbers of cells are needed and transplant tissues are in short supply.

Current and Future Directions

The following are current and future directions pertaining to our research with pluripotent stem cells.

(1) Identifying cell surface antigenic epitopes for stem cell isolation.
(2) Serum-free defined media.
(3) Progenitor and pluripotent stem cells as a comparison/contrast assay system for bioactive factors.
(4) Cartilage repair paradigm for tissue repair using naive postnatal pluripotent stem cells with endogenous environmental cues.
(5) Postnatal pluripotent stem cells as autologous delivery vehicles for gene therapy.
(6) Hematopoietic reconstitution using autologous pluripotent stem cells.

Acknowledgments. The author would like to thank P.A. Lucas, A.C. Black, Jr., R.P. Wright, M.L. Mancini, J.C. Smith, T.M. Young, L.R. Adkison, K. Detmer, C. Duplaa, T.A. Steele, R.A. Bray, J.J. Rogers, C.R. Reagan, D. Warejcka, J. Floyd, D.C. Morrison, S. Southerland, L.W. Blake, E.M. Ceballos, G.J. Mancini, M.E. Eaton, J.D. Hill, J. Sippel, J.D. Martin, M.L. Reeves, K.H. Davis, K. Hawkins, B.L. Ragan, I. Bushell, B.J. Taylor, L.S. Putnam, A.F. Calcutt, D. Grande, D.W. Pate, M.L. Dalton, K. Dixon, R.W. Murphy, E. Davis, R. Harvey, and W. Newman as collaborators; D. Odom, C. Klausmeyer, and S. Pedersen for technical assistance; M. Cates, C. Chappell, and K. Lee for technical assistance with respect to flow cytometry; J. Hamby, A. Dawson, A. Parker, J. Elgin, R.E. Breving, E. Wolfsberger, M. Allvine, D. Groover, B. Horten, K. Ballard, B. Brown, G. Golden, L. Farmer, M. Kang, D. Estes, J.A. Wilson, P. Ossi, D.J. Mulvaney, S.G. Bowerman, J. Padrta, S. Troum, R. Estes, J.A. Stevik, C.C. Kavali, E. Jones, S.A. Tincher, R.P. Walker, J.C. Lee,

J.C. Wynn, J.T. Williams IV, and J. Souza for running assays; F. Frazer, J. Reichert, and J. Knight for photographic assistance; P. Green (Proctor & Gamble), I. Moutsatsos (Genetics Institute), G. Pierce (Amgen), and L. Rifkin (MorphoGen Pharmaceuticals, Inc.) for bioactive factors; and D.M. Ciombor, T.A. Partridge, J.E. Morgan, M.D. Grounds, M. Maley, I. Moutsatsos, R. Garner, H. Horst, R. Sawyer, B. Kacsoh, S. Leeper–Woodford, and D. Innes for critical suggestions throughout this research. These studies were supported by grants from Rubye Ryle Smith Charitable Trust, L.M. and H.O. Young Research Trust, the MedCen Foundation of the Medical Center of Central Georgia, Proctor & Gamble, Advanced Tissue Sciences, Genetics Institute, and MorphoGen Pharmaceuticals, Inc.

References

Ailhaud, G., Grimaldi, P., and Negrel, R. 1992. Cellular and molecular aspects of adipose tissue development. *Annu Rev Nutr* 12:207–34.

Bentley, G. and Greer, G.B. 1971. Homotransplantation of isolated epiphyseal and articular chondrocytes into joint surfaces. *Nature* 230:385–8.

Beresford, J.N. 1989. Osteogenic stem cells and the stromal system of bone and marrow. *Clin Orthop Rel Res* 240:270–80.

Breinan, H.A., Minas, T., Hsu, H.-P., Nehrer, S., Sledge, C.B., and Spector, M. 1997. Effect of cultured autologous chondrocytes on repair of chondral defects in a canine model. *J Bone Joint Surg Am* 79:1439–51.

Brittberg, M., Lindahl, A., Nilsson, A., Ohlsson, C., Isaksson, O., and Peterson, L. 1994. Treatment of deep cartilage defects in the knee with autologous chondrocyte implantation. *N Eng J Med* 331(4):889–95.

Brittberg, M., Nilsson, A., Lindahl, A., Ohlsson, C., and Peterson, L. 1996. Rabbit articular cartilage defects treated with autologous cultured chondrocytes. *Clin Orthop Rel Res* 326:270–83.

Campion, D.R. 1984. The muscle satellite cell: a review. *Int Rev Cytol* 87:225–51.

Caplan, A.I. 1991. Mesenchymal stem cells. *J Orthop Res* 9:641–50.

Caplan, A.I., Elyaderani, M., Mochizuki, Y., Wakitani, S., and Goldberg, V. 1997. Principles of cartilage repair and regeneration. *Clin Orthop Rel Res* 342:254–69.

Chesterman, P.J. and Smith, A.U. 1968. Homotransplantation of articular cartilage and isolated chondrocytes. *J Bone Joint Surg Br* 50:184–97.

Cruess, R.L. 1982. *The musculoskeletal system embryology, biochemistry, and physiology*. New York: Churchill Livingston.

Davis, E., Williams, J.T., IV, Souza, J., Southerland, S.S., Warejka, D., Young, H.E., and Lucas, P.A. 1995. Cells isolated from adult rat marrow are capable of differentiating into several mesenchymal phenotypes in culture. *FASEB J* 9:A590.

Dixon, K., Murphy, R.W., Southerland, S.S., Young, H.E., Dalton, M.L., and Lucas, P.A. 1996. Recombinant human bone morphogenetic proteins–2 and–4 (rhBMP–2 and rhBMP–4) induce several mesenchymal phenotypes in culture. *Wound Rep Reg* 4:374–80.

Frenkel, S.R., Toolan, B., Menche, D., Pitman, M.I., and Pachence, J.M. 1997. Chondrocyte transplantation using collagen bilayer matrix for cartilage repair. *J Bone Joint Surg Br* 79:831–6.

Garret, J.C. 1986. Treatment of osteochondral defects of the distal femur with fresh osteochondral allografts: a preliminary report. *Arthroscopy* 2:222–6.

Grande, D.A., Pitman, M.I., Peterson, L., Menche, D., and Klein, M. 1989. The repair of experimentally produced defects in rabbit articular cartilage by autologous chondrocyte implantation. *J Orthop Res* 7:208–18.

Grande, D.A., Southerland, S.S., Manji, R., Pate, D.W., Schwartz, R.E., and Lucas, P.A. 1995. Repair of articular cartilage defect using mesenchymal stem cells. *J Tiss Eng* 1:345–53.

Green, W.T. 1977. Articular cartilage repair: behavior of rabbit chondrocytes during tissue culture and subsequent allografting. *Clin Orthop* 124:237–50.

Grounds, M.D., Garrett, K.L., Lai, M.C., Wright, W.E., and Beilharz, M.W. 1992. Identification of muscle precursor cells in vivo by use of MyoD1 and myogenin probes. *Cell Tiss Res* 267:99–104.

Hayflick, L. 1965. The limited in vitro lifetime of human diploid cell strains. *Exper Cell Res* 37:614–36.

Homminga, G.N., Bulstra, S.K., Bouwmeester, P.S.M., and Van Der Linden, A.J. 1990. Perichondrial grafting for cartilage lesions of the knee. *J Bone Joint Surg Br* 72:1003–7.

Kawabe, N. and Yoshinato, M. 1991. The repair of full thickness articular cartilage defects. Immune responses to reparative tissue formed by allogeneic growth plate chondrocytes. *Clin Orthop* 268:279–93.

Kishimoto, T., Kikutani, H., Borne, AEGKrvd, Goyert, S.M., Mason, D., Miyasaka, M., Moretta, L., Okumura, K., Shaw, S., Springer, T., Sugamura, K., and Zola, H. 1997. *Leucocyte typing VI, white cell differentiation antigens.* Hamden: Garland Publishing.

Kolettas, E., Buluwela, L., Bayliss, M., and Muir, H. 1995. Expression of cartilage–specific molecules is retained on long–term culture of human articular chondrocytes. *J Cell Sci* 108:1991–9.

Langer, R. and Vacanti, J.P. 1993. Tissue engineering. *Science* 260:920–6.

Lucas, P.A., Calcutt, A.F., Ossi, P., Young, H.E., and Southerland, S.S. 1993. Mesenchymal stem cells from granulation tissue. *J Cell Biochem* 17E:122.

Lucas, P.A., Calcutt, A.F., Southerland, S.S., Warejcka, D., and Young, H.E. 1995. A population of cells resident within embryonic and newborn rat skeletal muscle is capable of differentiating into multiple mesodermal phenotypes. *Wound Rep Reg* 3:457–68.

Lucas, P.A., Grande, D.A., and Young, H.E. 1996a. Use of pluripotent mesenchymal stem cells for tissue repair. *Program of the Keystone Symposia on Tissue Engineering and Wound Repair in Context.* 1:15.

Lucas, P.A., Warejcka, D.J., Zhang, L.–M., Newman, W.H., and Young, H.E. 1996b. Effect of rat mesenchymal stem cells on the development of abdominal adhesions after surgery. *J Surg Res* 62:229–32.

Lucas, P.A., Young, H.E., and Putnam, L.S. 1991. Quantitation of chondrogenesis in culture using Alcec blue staining. *FASEB* J. 5(4).

Mankin, H.J. 1982. The response of articular cartilage to mechanical injury. *J Bone Joint Surg Am* 64:460–6.

Matsusue, Y., Yamamuro, T., and Hama, H. 1993. Arthroscopic multiple osteochondral transplantation to the chondral defect in the knee associated with anterior cruciate ligament disruption. *Arthroscopy* 9:318–21.

Mauro, A. 1961. Satellite cell of skeletal muscle fibers. *J Biophys Biochem Cytol* 9:493–8.

McDermott, A.G.P., Langer, F., Pritzker, K.P.H., and Gross, A.E. 1985. Fresh small fragment osteochondral allografts. Long term follow–up study on first 100 cases. *Clin Orthop* 197:96–102.

McGuire, W.P. 1998. High dose chemotherapy and autologous bone marrow or stem cell reconstitution for solid tumors. *Curr Probl Cancer* 22:135–77.

Minas, T. and Nehrer, S. 1997. Current concepts in the treatment of articular cartilage defects. *Orthopedics* 20(6):525–38.

Moskalewski, S. 1991. Transplantation of isolated chondrocytes. *Clin Orthop* 272:16–20.

O'Driscoll, S.W., Keeley, F.W., and Salter, R.B. 1988. Durability of regenerated articular cartilage produced by free autologous periosteal grafts in major full thickness defects in joint surfaces under the influence of continuous passive motion. *J Bone Joint Surg Am* 70:595–606.

Owen, M. 1988. Marrow stromal cells. *J Cell Sci Suppl* 10:63–76.

Palis, J. and Segel, G.B. 1998. Developmental biology of erythropoiesis. *Blood Rev* 12:106–14.

Pate, D.W., Southerland, S.S., Grande, D.A., Young, H.E., and Lucas, P.A. 1993. Isolation and differentiation of mesenchymal stem cells from rabbit muscle. *Surgical Forum*, XLIV:587–9.

Ratajczak, M.Z., Pletcher, C.H., Marlicz, W., Machlinski, B., Moore. J., Wasik, M., Ratajczak, J., and Gewirtz, A.M. 1998. CD34+, kit+, rhodamine 123(low) phenotype identifies a marrow cell population highly enriched for human hematopoietic stem cells. *Leukemia* 12:942–50.

Ritsila, V.A., Santavira, S., Alhopuro, S., Poussa, M., Jaroma, H., Rubak, J.M., Eskola, A., Hoikka, V., Snellman, O., and Osterman, K. 1994. Periosteal and perichondrial grafting in reconstructive surgery. *Clin Orthop* 302:259–65.

Rogers, J.J., Adkison, L.R., Black, A.C., Jr, Lucas, P.A., and Young, H.E. 1995. Differentiation factors induce expression of muscle, fat, cartilage, and bone in a clone of mouse pluripotent mesenchymal stem cells. *Amer Surg* 61(3):1–6.

Rubak, J.M. 1982. Reconstruction of articular cartilage defects with free periosteal grafts. *Acta Orthop Scand* 53:175–9.

Skoog, T. and Johansson, S.H. 1976. The formation of articular catilage from free perichondrial grafts. *Plast Reconstr Surg* 57:1–6.

Vierck, J.L., McNamara, J.P., and Dodson, M.V. 1996. Proliferation and differentiation of progeny of ovine unilocular fat cells (adipofibroblasts). *In Vitro Cell Dev Biol—Animal* 32:564–72.

Wakitani S., Goto, T., Pineda, S.J., Young, R.G., Mansour, J.M., Caplan, A.I., and Goldberg, V.M. 1994. Mesenchymal cell based repair of large, full thickness defects of articular cartilage. *J Bone Joint Surg Am* 76:579–92.

Wakitani, S., Kimura, T., Hirooka, A., Ochi, T., Yoneda, M., Yasui, N., Owaki, H., and Ono, K. 1989. Repair of rabbit articular surfaces with allograft chondrocytes embedded in collagen gel. *J Bone Joint Surg Br* 71:74–80.

Warejcka, D.J., Harvey, R., Taylor, B.J., Young, H.E., and Lucas, P.A. 1996. A population of cells isolated from rat heart capable of differentiating into several mesodermal phenotypes. *J Surg Res* 62:233–42.

Young, H.E. 1983. A Temporal Examination of *Glycoconjugates During the Initiation Phase of Limb Regeneration in Adult Ambystoma*. Lubbock: Texas Tech University Library.

Young, H.E., Blake, L.W., Floyd, J.A., and Black, A.C., Jr. 1998. Progenitor stem cells and pluripotent stem cells as a comparison/contrast bioassay for identifying proliferative factors, progression factors, inhibitory factors, and inductive factors for tissue restoration. Unpublished raw data.

Young, H.E., Carrino, D.A., and Caplan, A.I. 1989. Histochemical analysis of newly synthesized and resident sulfated glycosaminoglycans during musculogenesis in the embryonic chick leg. J *Morph* 201:85–103.

Young, H.E., Ceballos, E.M., Smith, J.C., Lucas, P.A., and Morrison, D.C. 1992. Isolation of embryonic chick myosatellite and pluripotent stem cells. *J Tiss Cult Meth* 14:85–92.

Young, H.E., Ceballos, E.M., Smith, J.C., Mancini, M.L., Wright, R.P., Ragan, B.L., Bushell, I., and Lucas, P.A. 1993. Pluripotent mesenchymal stem cells reside within avian connective tissue matrices. *In Vitro Cell Dev Biol* 29A: 723–36.

Young, H.E., Dalley, B.K., and Markwald, R.R. 1989a. Effect of selected denervations on glycoconjugate composition and tissue morphology during the initiation phase of limb regeneration in adult Ambystoma. *Anat Rec* 223:231–41.

Young, H.E., Dalley, B.K., and Markwald, R.R. 1989b. Glycoconjugates in normal wound tissue matrices during the initiation phase of limb regeneration in adult Ambystoma. *Anat Rec* 223:223–30.

Young, H.E., Duplaa, C., Floyd, J.A., Hawkins, K., Thomas, K., Austin, T., Edwards, C., Couzzart, J., Lucas, P.A., Hudson, J., and Black, A.C., Jr. 2000. Postnatal epiblastic-like stem cells retain pluripotency after gene transfection. (submitted)

Young, H.E., Duplaa, C., Hawkins, K., Floyd, J.A., Thomas, K., Austin, T., Edwards, C., Couzzart, J., Lucas, P.A., Hudson, J., and Black, A.C., Jr. 2000. Postnatal pluripotent mesenchymal stem cells retain pluripotency after gene transfection. (submitted)

Young, H.E., Floyd, J.A., and Black, A.C., Jr. 1996b. Progenitor stem cell numbers decrease with increasing age of the individual, in contrast, pluripotent stem cell numbers remain constant regardless of age. Unpublished raw data.

Young, H.E., Mancini, M.L., Wright, R.P., Smith, J.C., Black, A.C., Jr, Reagan, C.R., and Lucas, P.A. 1995. Mesenchymal stem cells reside within the connective tissues of many organs. *Dev Dynamics* 202:137–44.

Young, H.E., Mancini, M.L., Wright, R.P., Smith, J.C., Black, A.C., Jr., Reagan, C.R., and Lucas, P.A. 1995b. Tissue-specific progenitor cells, non-tissue specific progenitor cells, and pluripotent stem cells reside within the connective tissue matrices of many organs. Unpublished raw data.

Young, H.E., Morrison, D.C., Martin, J.D., and Lucas, P.A. 1991. Cryopreservation of embryonic chick myogenic lineage-committed stem cells. *J Tiss Cult Meth* 13:275–84.

Young, H.E., Rogers, J.J., Adkison, L.R., Lucas, P.A., and Black, A.C., Jr. 1998b. Muscle morphogenetic protein induces myogenic gene expression in Swiss-3T3 cells. *Wound Rep Reg* 6(6):534–54.

Young, H.E., Sippel, J., Putnam, L.S., Lucas, P.A., and Morrison, D.C. 1992. Enzyme linked immuno–culture assay. *J Tiss Cult Meth* 14:31–6.

Young, H.E., Steele, T., Bray, R.A., Detmer, K., Blake, L.W., Lucas, P.A., and Black, A.C., Jr. 1999. Human progenitor and pluripotent cells display cell surface cluster differentiation markers CD10, CD13, CD56, and MHC Class-I. *Proc Soc Exp Biol Med* 221:63–71.

Young, H.E., Steele, T., Bray, R.A., Hudson, J., Floyd, J.A., Hawkins, K., Thomas, K., Austin, T., Edwards, C., Couzzart, J., Duenzl, M., Lucas, P.A., and Black, A.C., Jr. 2000. Mesenchymal stem cells derived from the connective tissues of postnatal humans display cluster differentiation markers CD34 and CD90. (submitted)

Young, H.E., Wright, R.P., Mancini, M.L., Lucas, P.A., Reagan, C.R., and Black, A.C., Jr. 1998a. Bioactive factors affect proliferation and phenotypic expression in pluripotent and progenitor mesenchymal stem cells. *Wound Repair and Regeneration* 6(1):65–75.

Young, H.E., Young, T.M., Floyd, J.A., Reeves, M.L., Davis, K.H., Eaton, M.E., Hill, J.D., Mancini, G.J., Thomas, K., Austin, T., Edwards, C., Couzzart, J., Blake, L.W., Detmer, K., Lucas, P.A., Hudson, J., and Black, A.C., Jr. 2000. Clonogenic analysis reveals reserve stem cells in postnatal mammals. II. Pluripotent epiblastic-like stem cells. (submitted)

Young, H.E., Young, T.M., Floyd, J.A., Reeves. M.L., Davis, K.H., Mancini, G.J., Eaton, M.E., Hill, J.D., Thomas, K., Austin, T., Edwards, C., Couzzart, J., Blake, L.W., Detmer, K., Lucas, P.A., Hudson, J., and Black, A.C., Jr. 2000. Clonogenic analysis reveals reserve stem cells in postnatal mammals. I. Pluripotent mesenchymal stem cells. (submitted)

Young, H.E., Young, V.E., and Caplan, A.I. 1989. Comparison of fixatives for maximal retention of radiolabeled glycoconjugates for autoradiography, including use of sodium sulfate to release unincorporated [35S]sulfate. *J Histochem Cytochem* 37:223–8.

10
Tissue Engineering of Ligament Healing

Savio L-Y. Woo, Nobuyoshi Watanabe, and Kevin A. Hildebrand

Introduction

Ligaments are parallel fibered, dense connective tissue structures that play an important role in mediating normal joint mechanics. These tissues also share in the transmission of forces with other articular tissues in order to provide joint stability. Thus, the rupture of a ligament will upset normal mobility and stability, causing abnormal joint kinematics and clinical symptoms. In the knee, ligament injuries frequently result from a direct blow or an indirect twisting motion during strenuous or sports activities. Ruptures of the medial collateral ligament (MCL), anterior cruciate ligament (ACL), and posterior cruciate ligament (PCL) are common and often result in chronic knee instability and eventual joint degeneration. The MCL is one of the most frequently injured ligaments in the knee. Miyasaka, Daniel, Stone, and Hirshman (1991) studied the incidence of the acute knee ligament injury in the general population. Of these injuries, 48% were isolated ACL injuries, 29% were isolated MCL injuries, and 13% were combined ACL/MCL injuries. If grade I injuries were included, 55% involved MCL injuries.

Because of the high incidence of injury as well as the abnormal biomechanical properties of the healed MCL even after 1 year (Woo, Inoue, McGurk-Burleson, and Gomez 1987; Ohland, Woo, Weiss, Takai, and Shelley 1991), special attention has been paid to tissue engineering of the healing MCL. Its spontaneous healing capacity as well as its relatively simple geometric structure have made the MCL a popular model for a variety of studies. Recent scientific investigations have begun to define the important contribution of growth factors to enhance the healing response of this ligament (Schmidt et al. 1995; Deie et al. 1997; Marui et al. 1997; Scherping, Jr, et al. 1997). In turn, these efforts have spurred an interest in potential biological therapies aimed at improving ligament healing through the stimulation of fibroblast proliferation and collagen and matrix synthesis (Letson and Dahners 1994; Kobayashi et al. 1995; Hildebrand et al. 1998; Woo, Suh, Parsons, Wang, and Watanabe 1998).

174

Delivery of these growth factors may be possible using gene therapy to induce or enhance new or endogenous protein synthesis. This chapter will review the role of various growth factors and cells in promoting ligament healing, and will also discuss the potential of gene therapy for engineering ligament healing.

Healing of the MCL

Although the clinical management of combined ligament injuries (MCL + ACL) in the knee is still under debate, there is general agreement that isolated injuries of the MCL can heal spontaneously with conservative treatment (Indelicato 1995; Shelbourne and Patel 1995a; Hillard–Sembell, Daniel, Stone, Dobson and Fithian 1996). Because of its healing potential in vivo, the MCL is frequently used as a model for both in vitro and in vivo studies of ligament healing. Such studies have provided a more comprehensive understanding of the complex interaction of biology, biochemistry, and biomechanics that regulates and modulates the MCL healing process.

Phases of Ligament Healing

The biological response to ligament injury is an integrated process with a complicated interaction involving cells and the biochemical or morphological environment. Ligament healing can be roughly divided into four overlapping phases—hemorrhagic, inflammatory, proliferative, and remodeling (Frank et al. 1994).

The hemorrhagic phase starts with breakage of a capillary, which leads to clot formation. Cytokines within this clot activate the inflammatory cascade, involving the interaction between cells and cytokines. Immediate influx of polymorphonuclear leukocytes (PMNs) occurs. PMNs respond to local autocrine and paracrine signals to expand the inflammatory response and recruit other inflammatory cell types to the wound. Growth factors are released by a number of different inflammatory cells and regulate the ligament healing response. Platelets have been shown to release platelet-derived growth factor (PDGF), transforming growth factor-beta (TGF-β) and epidermal growth factor (EGF) (Wilder, Lafyatis, and Remmers 1990; Murphy, Frank, and Hart 1993).

In the inflammatory phase, macrophages become the principal inflammatory cell in the wound. They engulf necrotic tissue and wound debris. In addition, macrophages secrete cytokines which induce neovascularization and initiate the formation of granulation tissue. Macrophages, once activated by the inflammatory response, produce basic fibroblast growth factor (bFGF), TGF-β and PDGF (Murphy et al. 1993). Platelets, macrophages, PMNs, and lymphocytes are present in the wound bed.

In the proliferative phase, undifferentiated mesenchymal cells or mesenchymal stem cells (MSCs) begin migration into the wound sites. Whether they arise from adjacent tissue or the blood stream is still under debate (Prockop 1997). Reorganization of the initial blood clot results in the formation of a friable, vascular granulation tissue. Fibroblastic cells become the major constituent cells in this phase. They proliferate and are responsible for the production of matrix proteins (Ross 1968). Growth factors stimulate fibroblast differentiation, proliferation and synthesis of collagen types I, III, and V as well as noncollagen proteins (Roberts et al. 1986; Pierce et al. 1989).

During the final, remodeling phase, tissue vascularity decreases and the number of fibroblasts and macrophages within the gap of the torn ligament also diminishes. This phase extends over a period of 2 or more years, when matrix maturation is responsible for improving ligament stiffness and tensile strength. As with the inflammatory and proliferative phases, an integrated sequence of biochemical and biomechanical signals are critical to ligament remodeling. These signals regulate the expression of structural and enzymatic proteins, including degradation enzymes involved in tissue remodeling, such as collagenase, stromelysin, and plasminogen activator (Murphy and Hart 1994).

Biomechanical Properties of the Healed Ligament

Laboratory studies have demonstrated that the properties of the healing MCL are biomechanically inferior to the intact structure (Woo et al. 1987; Chimich, Frank, Shrive, Dougall, and Bray 1991; Weiss, Woo, Ohland, Horibe, and Newton 1991). The healing outcome of a ruptured MCL in combined MCL and ACL injuries is worse than in isolated MCL rupture (Anderson, Weiss, Takai, Ohland, and Woo 1992; Engle, Noguchi, Ohland, Shelley, and Woo 1994; Woo et al. 1997). In addition, suture repair with immobilization has been shown to adversely affect the biomechanical properties of the healing MCL when compared to no MCL repair without immobilization (Gomez et al. 1989; Inoue, Woo, Amiel, Ohland and Kitabayashi 1990).

A review of the mechanical properties of MCL and the structural properties of the femur–MCL–tibia complex (FMTC) reveals how these structures behave under loading. The mechanical properties of the MCL substance are represented by the stress–strain relationship. Stress is defined as the load per unit cross-sectional area of the ligament substance. Strain is measured by the change in length in the central region of the ligament and is defined as the change in length divided by the original length. The stress–strain curve for a ligament is nonlinear and begins with a toe region, in which stress increases slowly with respect to increasing strain. This portion of the stress–strain curve is followed by a linear region, in which the slope represents the modulus of elasticity. Tensile strength (ultimate tensile strength), ultimate strain, and strain energy density are parameters calculated from the stress–strain curve and are used to fully describe the mechanical properties of the ligament.

The structural properties of the FMTC are represented by a load-elongation curve obtained from the same uniaxial tensile test. The load elongation curve is also nonlinear and can be divided into a toe region, with low stiffness, in which load increases slowly with elongation, followed by a linear region. Stiffness, defined as the slope of the curve, increases slowly in the toe region and reaches a relatively constant value in the linear region. Stiffness, ultimate load, and energy absorbed to failure are parameters calculated from the load-elongation curve.

Many studies have sought to correlate these mechanical properties with biochemical and histologic properties in the healing ligament. Biochemical factors have been demonstrated to be associated with the inferior mechanical properties of the healing MCL. Using a rabbit model of MCL healing, Ohland et al. (1991) found no improvement in the mechanical properties of the healing MCL substance with time. Tensile strength of the healing MCL was 24 MPa compared to the sham level of 90 MPa and elastic modulus was 600 MPa less than sham at 52 weeks after injury. Frank et al. (1995) showed that the tensile strength of the healing MCL was 29% of sham control values and elastic modulus was only 22% of controls at 40 weeks after injury. On the other hand, the cross-link densities remained low and were only 45% of the control values after 40 weeks of healing. Ng, Oakes, Deacon, Mclean, and Eyre (1996) also demonstrated a high correlation between the levels of hydroxylysyl pyridinoline cross-links and the elastic modulus of normal and autograft goat patellar tendons. Average fibril diameter size was demonstrated to gradually increase during the initial remodeling phase, but remains uniformly smaller relative to normal values 2 years following injury (Frank, McDonald, and Shrive 1997; Kavalkovich, Yamaji, Woo, and Niyibizi 1997). Because collagen concentration returns to normal within 14 weeks of injury, the inferior biomechanical properties of the healing ligament have been linked to both change of collagen fibril diameter and cross-linking. In fact, tensile strength of ligament is demonstrated to depend on large-diameter collagen fibrils as well as normal concentrations of hydroxylysyl pyridinoline fibril cross-links (Parry, Craig, and Barnes 1978; Frank et al. 1992, 1997).

Further, the distribution of collagen types does not return to normal (Frank et al. 1983). Type III collagen production increases in the initial phases of matrix proliferation, but levels decrease as gradual remodeling favors production of the predominant type I collagen (Frank et al. 1994). Type V collagen also increases in the early phases of MCL healing and may remain elevated by as much as 18% over normal values for 1 year following injury (Kavalkovich et al. 1997). Prior studies on chick corneas have suggested that collagen type V plays an important role in regulating fibril diameters (Linsenmayer et al. 1993). Persistently high levels of collagen type V at 1 year after injury are a potential reason that small diameter fibrils predominate over larger firbrils. The small leucine-rich proteoglycans, such as decorin, are also elevated following injury (Sciore, Boykiw, and Hart 1998). Decorin has also been found to affect in fibril diameters of type I collagen structures and may contribute to predominant fibril diameter distribution.

Differential Healing Capacities of the MCL and ACL

The different healing potential and cellular characteristics of MCL and ACL have been demonstrated in numerous studies. While the MCL heals spontaneously, the suture of a ruptured ACL fails to mount a healing response. Although the femoral insertion of the ACL has some potential to heal (Kurosaka, Yoshiya, Mizuno, T., and Mizuno, K. 1998), surgical reconstruction using autogenous and allogeneic replacement grafts is still standard treatment (Shelbourne and Patel 1995b).

Anatomically and histologically, the MCL is a thin, flat extraarticular structure surrounded by loose areolar connective tissue. The ACL, in contrast, is an intraarticular structure composed of fiber bundles surrounded by a synovial sheath and synovial fluid (Lyon, Akeson, Amiel, Kitabayashi, and Woo 1991). Ultrastructurally, these ligaments also differ in their matrix composition and the characteristics of their resident fibroblasts. Transmission electron microscopy studies have demonstrated differences in the intrinsic characteristics of MCL and ACL fibroblasts and in their pericellular regions as well (Lyon et al. 1991; Hart, Woo, and Newton 1992). MCL fibroblasts are more spindle shaped and possess long cytoplasmic processes that intimately contact neighboring collagen fibrils. In contrast, some ACL fibroblasts are oval and devoid of cytoplasmic extensions into the pericellular matrix (Lyon et al. 1991).

In vitro, different characteristics of MCL and ACL fibroblasts have been demonstrated. Geiger, Green, Monosov, Akeson, and Amiel (1994) showed that MCL fibroblasts migrate out of explants more rapidly than ACL fibroblasts. Furthermore, Nagineni, Amiel, Green, Berchuck, and Akeson (1992) performed an in vitro wound closure assay simulating ligament healing, and demonstrated that MCL fibroblasts more rapidly regenerate the cell-free area in comparison to ACL fibroblasts.

In our research center, the different responses of MCL and ACL fibroblasts to the growth factors, measuring cell proliferation and matrix synthesis were studied (Schmidt et al. 1995; Marui et al. 1997, Scherping, Jr., et al. 1997). The results demonstrated that the amount of matrix synthesis by ACL fibroblasts in response to TGF-β1 was approximately half that by MCL fibroblasts. There was a significant difference in the magnitude of proliferation rate between the MCL and ACL fibroblasts from skeletally mature rabbits when they were exposed to bFGF or PDGF-BB. These results indicated that responses to growth factors are likely tissue dependent and that biological manipulation of ligament healing as a potential therapeutic measure must account for differences in response to injury by different ligament tissues.

The Effect of Growth Factors on Ligament Healing

Growth factors are small polypeptides synthesized by a variety of cells. They function by binding to specific cell surface receptors that activate complex intracellular signal transduction pathways. Growth factors and their receptors are expressed at different periods and intervals during wound healing (Burgess 1989; Gold, Sung, Siebert, and Longaker 1997; Green, Usui, Hart, Ammons, and Narayanan 1997; Panossian, Liu, Lane, and Finerman 1997). They regulate cell migration, proliferation, differentiation, and matrix synthesis, and aid in repairing a damaged ligament by regulating cellular behavior and modulating the wound environment (Murphy et al. 1993; Murphy and Hart 1994). Many studies examining the effects of growth factors have utilized both in vitro and in vivo methods in order to determine strategies of tissue engineering in ligament healing.

In Vitro Studies

In vitro experiments have been performed to address the biological effects of growth factors on proliferation, matrix synthesis, and migration of ligament fibroblasts. The implication of these studies is that growth factors which have an impact on the function of fibroblasts could lead to promising treatment for a damaged ligament.

Studies at our research center examined the effect of growth factors on proliferation of MCL and ACL fibroblasts from skeletally immature and mature rabbits in vitro. PDGF–BB, bFGF, and EGF were found to have a stimulatory effect. However, the effect of these growth factors on proliferation of ligament fibroblasts from mature rabbit were demonstrated to be smaller compared to that of immature rabbit (Schmidt et al. 1995; Scherping, Jr., et al. 1997) (Figure 10.1A and B). Comparison of results suggests that age has a significant effect on fibroblast proliferation.

Similar studies were performed by DesRosiers, Yahia, and Rivard (1996) in which both PDGF-AB and EGF demonstrated a comparable stimulatory effect on canine ACL fibroblast proliferation in vitro. In addition, the effect of TGF-β1 on cell proliferation was found to be dose dependent, although the effect of was shown to be lower than with PDGF-AB, EGF and insulin-like growth factor–(IGF-1). Smaller doses of TGF-β1 could act synergistically with PDGF-AB to promote fibroblast proliferation while higher concentrations inhibited the stimulatory effect of PDGF. On the other hand, Spindler, Dawson, Stahlman, and Davidson (1996) showed that explants from sheep ACL specimens had no proliferative

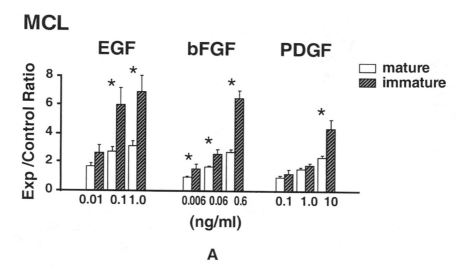

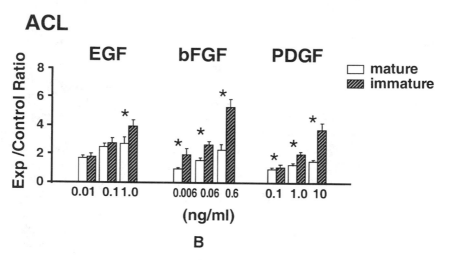

FIGURE 10.1. Effect of growth factors on cell proliferation by MCL (A) and ACL fibroblasts (B). *Significant difference between immature and mature fibroblasts. ($p < 0.05$).

response to PDGF-AB while TGF-β2 could induce a significant proliferative response.

We have also examined the effect of several growth factors on matrix synthesis of MCL and ACL fibroblasts to determine the optimal growth factor on collagen and noncollagenous protein production (Marui et al. 1997). Our findings revealed that TGF-β1 was the only growth factor that increased the collagen synthesis of both MCL and ACL fibroblasts. The level of response to TGF-β1 at 1 ng/ml dosage level was 1.6 and 1.5 times versus controls for MCL and ACL fibroblasts; the majority of this increase was for collagen type I. Increased collagen synthesis by EGF was limited to ACL fibroblasts. Both bFGF and aFGF had no effect. Although the relative increase in protein production was similar for both types of cells, the absolute increase in protein synthesis was twice as high for MCL fibroblasts as it was for ACL fibroblasts. Des Rosiers et al. (1996) found similar results. These data suggest that TGF-β1 may help improve ligament healing by increasing matrix synthesis during the healing process (Marui et al. 1997).

The effect of aging on collagen and total protein synthesis was also investigated in our research center using MCL fibroblasts from skeletally immature, mature and senescent rabbits (Deie et al. 1997). Fibroblasts from senescent rabbits produced significantly less collagen in response to TGF-β1 and EGF in all experimental doses when compared to those from immature rabbits (Figure 10.2A). However, the experimental versus control increase in both collagen and total protein production due to stimulation with TGF–β1 and EGF was greater for senescent cells than for skeletally immature fibroblasts (Figure 10.2B).

The influence of various growth factors on cell migration has also recently been addressed in an in vitro cell outgrowth model using MCL and ACL explants. Lee, Green, and Amiel (1995) showed that the presence of growth factors significantly increased the cell outgrowth from explants of both ligaments compared to controls. In addition, MCL explants exhibited greater cell outgrowth than those of the ACL. Hannafin, Attia, Warren, and Bhargava (1997) demonstrated the effect of several growth factors on migration of the fibroblasts from both intra and extraarticular knee ligaments. Previous in vitro models have been used to examine the cellular response of isolated fibroblasts in a defined environment. However, various cells appeared at the healing site in vivo, which implicated the complex biochemical, physiological, and biomechanical environment. In vivo studies have thus been critical to gaining a better understanding of the potential of growth factors in mediating and modulating ligament repair.

In Vivo Studies

Exogenous growth factors have been expected to promote ligament healing based on the common finding that growth factors cause cell signaling by their interaction with receptors that regulate adequate cellular interactions and the environment for healing. In vitro models are thus limited in that the effects of growth factors on cell proliferation and protein synthesis are studied in the absence of the extrinsic and intrinsic mechanical influences which define the true functional environment of ligament tissues. For this reason, in vivo models are necessary to examine the effects of

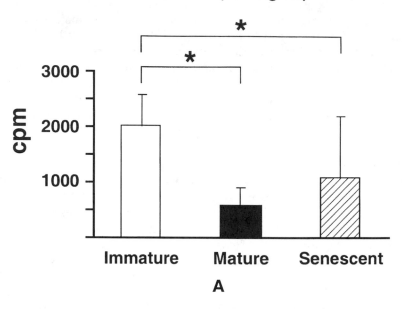

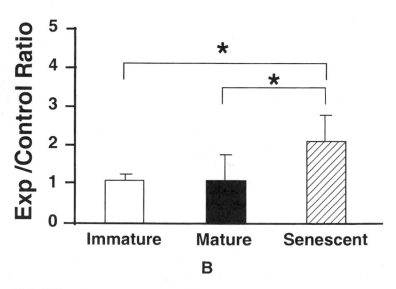

FIGURE 10.2. Effect of 0.01 ng/ml of TGF-β1 on total collagen synthesis (A) by MCL fibroblasts and the ratio between experimental over control (B) *Significant difference by age. (p < 0.05).

growth factors where the physiological and functional environments are involved.

Several investigators have used animal models to study the in vivo effect of exogenous growth factors on the healing process and biomechanical properties of the healing MCL. Letson and Dahners (1994) found that 1.2 μg of PDGF delivered by a collagen emulsion vehicle caused a 73% increase in the ultimate load, a 94% increase in stiffness, and a 101% increase in energy absorbed at failure in FMTC of rats at 12 days post-injury. Furthermore, additions of bFGF and IGF-1 were shown to effect no additional improvement in the structural properties of FMTC compared to application of PDGF-BB alone. Weiss, Beck, Levine, and Greenwald (1995) measured a significant increase in both stiffness and ultimate load of the FMTC 6 weeks after injury in response to 100 μg of PDGF-BB using a bioresorbable patch delivery vehicle in a rabbit model. In contrast, Stahlman, Spindler, Dawson, and Davidson (1994) showed no statistical difference in the structural properties of healing rabbit FMTC treated with 21 μg of PDGF-AB delivered by a time-release pellet with an elution rate of 1 μg/day and suggested that prolonged exposure to growth factors may be detrimental to ligament healing. Batten, Hansen, and Dahners (1996) examined the effect of dose dependency and timing of growth factor application on the biomechanical properties of healing MCL. Exogenous PDGF-BB at concentrations of 1.0 μg and 5.0 μg were demonstrated to be more effective than a concentration of 0.5 μg in restoring structural properties of the FMTC. The 5.0 μg dosage level did not demonstrate any improvement in structural properties compared to the dosage of 1.0 μg. Furthermore, administration of 1.2 μg of PDGF-BB more than 24 hours after the injury markedly decreased the effectiveness of treatment on healing.

Regarding the effect of TGF-β on MCL healing, Spindler, Imro, Mayes, and Davidson (1996b) compared the structural properties of FMTC after exposure to TGF-β2 at a dosage level of 1.0 μg. Structural properties of the treated group were similar to the no growth factor group at 3 weeks. However, at 6 weeks postinjury, the ultimate load and energy absorbed to failure of the treated group were decreased compared to the control. Recently, Spindler et al. (1998) evaluated the effect of TGF-β2 at 0.1 μg, 1.0 μg and 5.0 μg dosage levels on the structural properties of the FMTC. They found that only the 0.1 μg dosage increased stiffness of the FMTC, but all other treatments failed to show a difference in structural properties. Stahlman et al. (1994) also showed that treatment of rabbit MCL specimens with 21 μg of TGF-β2 did not increase the structural properties of the FMTC at 6 and 12 weeks postinjury. Conti and Dahners (1993) studied the effects of exogenous growth factors on MCL healing in an in vivo rat model. Administration of 1.0 μg of TGF-β1 to the damaged ligament produced a 66.4% increase in stiffness and a 91% increase in the load to failure of the healed tissues FMTC relative to controls. These results contrast with those of Spindler et al. (1998) and Stahlman et al. (1994) and suggest that the effects of growth factors may depend on the animal models.

Based on our finding that, in vitro, PDGF-BB provides modest stimulation of fibroblast proliferation while TGF-β1 promotes collagen synthesis (Schmidt et al. 1995; Deie et al. 1997; Marui et al. 1997; Scherping, Jr., et al. 1997), we sought to determine the effects of PDGF-BB alone and in combination with TGF-β on rabbit MCL healing in vivo. Fibrin sealant was used as a vehicle for the growth factor (Hildebrand et al. 1998). At 6 weeks, the varus–valgus laxity of each experimental group was significantly different from those of the sham control groups, except for the fibrin sealant, low-dose PDGF-BB plus TGF-β1, and low- and high-dose EGF plus TGF-β1 groups. Tensile testing of the FMTC of the higher dose PDGF-BB treated group demonstrated ultimate load, energy absorbed to failure, and ultimate elongation values that were 1.6, 2.4 and 1.6 times greater than controls, respectively. Stiffness values were also higher. Addition of TGF-β1 did not lead to any further increases in the structural properties of the FMTC. Furthermore, in groups treated with low-dose PDGF alone and in combination with TGF-β1, stiffness values were inferior to those of the control group at 6 weeks postinjury (Figure 10.3). Cross-sectional areas of all experimental groups were significantly greater than in control groups, and high-dose PDGF plus TGF-β1 increased cross-sectional areas at a rate of 2.5 times greater than sham. A related study in our laboratory addressed the effects of EGF in combination with TGF-β1 on the structural properties of healed

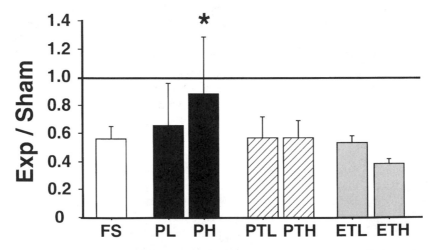

FIGURE 10.3. Ultimate load of the femur–MCL–tibia complexes. FS, fibrin sealant; PL, low-dose PDGF-BB; PH high-dose PDGF-BB; PTL low-dose PDGF-BB plus TGF-β1; PTH high-dose PDGF-BB plus TGF-β1; ETL low-dose EGF plus TGF-β1; ETH high-dose EGF plus TGF-β1; *Significance greater than other groups. ($p < 0.05$).

MCLFMTC tissue 6 weeks after injury (Woo, Yamaji, Taylor, Hildebrand, and Foy 1996). The low dose combination of these growth factors resulted in a significantly higher elastic modulus and smaller cross-sectional area than the high dose combination. Because of the biochemical network of signaling events with stimulatory and inhibitory feedback loops, further studies are necessary to determine the optimal dosage, timing, and combination of growth factors that provide an adequate environment for cells to enhance the strength of the healing ligament.

At present, we know that growth factors have been shown to affect the response of ligament fibroblasts in vitro. Based on the in vitro results, application of exogenous growth factors in vivo to enhance ligament healing have not yielded consistent results toward improving biomechanical properties. Therefore, much work remains to be done in order to better understand the mechanisms of enhancement by growth factors.

Delivery of Growth Factors: The Potential Use of Gene Transfer Technology

Although delivery of cytokines, such as growth factors, into damaged ligaments has shown some promise, the fact that these proteins are not stable at the injured site due to clearance by circulation or metabolism is a problem. It has been demonstrated in vivo that the half life of the growth factors in the wound is limited to only a few days. Hence, repeated application, or application of different growth factors at different phases of healing, may be warranted. To address these issues and maintain the potency of these exogenous molecules in the enhancement of ligament healing, novel delivery vehicles other than collagen, cellulose sponges, and fibrin sealant are necessary (Buckley, Davidson, Kameroth, Wolt, and Woodward 1985; Mustoe et al. 1987; Hildebrand et al. 1998).

One novel approach is gene transfer, a concept based on the premise that incorporation of whole genes or regulatory sequences into the host genome could stimulate or regulate protein expression over an extended period. Thus, along with the development of gene transfer research, interest has arisen for developing a new vehicle that could maintain adequate concentrations of growth factors over a longer term.

Gene Transfer Into the Ligament

In the 1960s and early 1970s, gene therapy was considered solely as a strategy to compensate for heritable genetic disorders. Thus, the potential for using gene transfer as a delivery vehicle to promote ligament healing is a record development. Using mammalian viruses, naked DNA, and cationic liposomes, direct gene transfer involves one-step delivery of the genes into the host cells in vivo while indirect gene transfer is accomplished through two steps, that is in vitro transfection of the desired genes into the cells followed by the transplantation of the genetically modified cells into the host tissue. Ligament and tendon fibroblasts were successfully transfected in vitro with the LacZ marker gene and by direct and indirect gene transfer (Gerich, Kang, Fu, Robbins, and Evans 1996). Indirect gene transfer using a retrovirus demonstrated the persistent expression of the LacZ transferred genes for 6 weeks after administration in the rabbit patellar tendon (Gerich et al. 1996). Other investigators have implanted LacZ gene and PDGF-BB gene into patellar tendon with hemagglutinating virus of Japan (HVJ) conjugated liposomes (Nakamura et al. 1996, Nakamura, Shino, et al. 1998).

In our research center, direct and indirect techniques of gene transfer were performed using adenovirus and BAG retrovirus into the MCL and ACL of rabbit knees (Hildebrand et al. In press). Both techniques resulted in expression of the LacZ marker gene by fibroblasts from intact and injured ligaments. Fibroblasts from injured ligaments showed transduction both in the wound site and in the ligament substance. Gene expression lasted for up to 10 days with the indirect technique and up to 6 weeks with the direct method (Figure 10.4A and B). Thus, direct gene transfer using adenovirus appeared to be a more effective delivery vehicle for long term protein expression compared to indirect gene transfer.

Therapeutic methods using gene transfer in ligament healing were investigated by Nakamura, Timmerman, et al. (1998) using HVJ-liposome complex. The immune response to foreign proteins could be reduced by encapsulating the plasmid DNA and HVJ spikes with liposome. Nakamura, Timmerman et al. (1998) injected HVJ-liposome complex containing a labeled antisense nucleotide for the protein decorin into the healing site of injured MCLs. Significant suppression of decorin mRNA expression was seen at both 2 days (42.7%) and 3 weeks (60.3%). In addition, protein levels of decorin were significantly decreased. Follow-up studies by Nakamura, Timmerman, et al. (1998) have found that collagen fibril diameters have a more normal biomodal distribution with an appearance of larger fiber diameters. Mechanical properties showed a significant improvement in ultimate stress compared to controls in association with these fibril diameter findings.

Despite such promising results, several biological obstacles currently impede the practical implementation of gene transfer as a biological intervention in ligament healing. Although direct gene transfer to ligament using adenovirus is possible, immune response to the viral proteins and transgene-encoded proteins could decrease gene expression (Yang et al. 1994). Immune responses to β–galactosidase have also been reported (Gong et al. 1997; Michou et al. 1997). Indirect gene transfer using retrovirus could reduce the immune response and enable the selection of transfected cells in vitro. However, retrovirus gene transfer requires that the transfected cells remain viable in vivo after retransplantation, bringing the source of

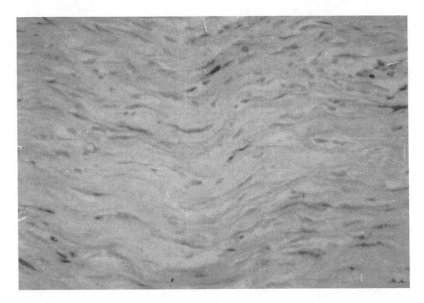

A

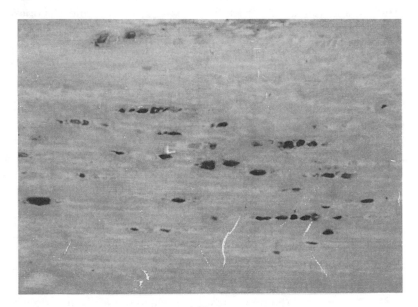

B

FIGURE 10.4. Photomicrographs of rabbit ligament sections showing intracellular staining with X–gal. (A) adjacent to injured site of MCL 10 days after retroviral ex vivo gene transfer. (B) ACL 3 weeks after adenoviral in vivo gene transfer.

cells directly into tissues. Additionally, loss of gene expression from retrovirus by methylation of promoter has been demonstrated (Challita and Kohn 1994). Regarding liposomal vectors, limited integration of genes into the genome is a problem which may lead to subsidence of the gene expression through cell division (Nakamura et al. 1996). However, for ligament healing, expression on the order of weeks may be all that is necessary. Thus, the best strategy for gene transfer into ligament is still under consideration.

Mesenchymal Stem Cells: Therapy for Ligament Healing

Because healing of ligament injuries involves a complex interaction of many cell types and local cellular environments, a strategy to enhance ligament healing could involve cell therapy, which can provide a source of cells for ligaments or provide a mechanism to introduce genes for gene transfer. Cell therapy to accelerate the repair of the ligament relies on the viability of the transplanted cells, affinity of the cells to the host tissue, and delivery technique. A candidate for cell therapy for musculoskeletal tissues is mesenchymal stem cells (MSCs). MSCs have been demonstrated to differentiate into various cells that form mesenchymal tissues under different culture conditions in vitro (Caplan 1991; Lazarus, Haynesworth, Gerson, Rosenthal, and Caplan 1995; Kadiyala, Young, Thiede, and Bruder 1997; Ferrari et al. 1998). Multipotent differentiation of these cells into mesenchymal tissues has also been demonstrated in vivo (Wakitani et al. 1994; Ferrari et al. 1998; Young et al. 1998). These cells can be obtained from bone, periosteum, and muscle (Nakahara et al. 1990; Caplan 1994; Huard et al. 1995).

In our research center, we employed nucleated cells obtained by the centrifugation of bone marrow as transplantation donor cells. As a transplantation model to eliminate the immune response to the donor cells, transplantation between female inbred rats was performed (Watanabe, Takai, Morita, Kawata, and Hirasawa 1998). The donor female transgenic rats had transgenes introduced in the chromosome 4, but transgenes were constructed so that foreign proteins were not expressed. Homozygous incorporation of transgenes into the chromosome of donor cells enabled us to track and recognize them in recipient tissues as long as they survived.

The nucleated cells of transgenic rats were injected with phosphate buffered saline (PBS) into a pocket made around the transected MCL of recipient inbred rats. Preliminary results revealed the survival of donor cells at the healing site of the MCL 3 and 7 days postoperatively (Figure 10.5A and B). Transplanted donor cells could also be identified in the midsubstance of ligaments at 7 days. This finding may demonstrate the migration potential of transplanted cells since they were not injected in these areas. Migration would be an attractive attribute for these cells when considering

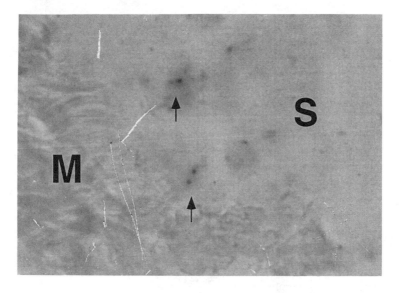

A

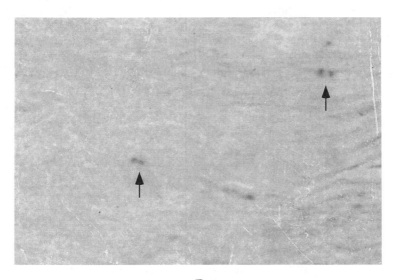

B

FIGURE 10.5. Photomicrographs of rat ligament sections showing intracellular 2 dot-positive signal of transgenes of transplanted nucleated cells. (A) injured site of healing MCL 3 days after injury and injection (M: midsubstance of the MCL, S: scar tissue). (B) midsubstance of the healing MCL 7 days after injury and injection.

ligament healing. While further work is necessary, these early results indicate the potential of cell therapy for ligament healing using nucleated cells from bone marrow.

Discussion and Conclusion

Ligament healing fails to regenerate a normal tissue. Although evidence has been gathered indicating that cells from ligaments are responsive to growth factors, and that growth factors applied in vivo can improve ligament healing, normal ligament properties have not been restored to date.

Strategies to deliver growth factors over sustained periods or certain intervals have turned to gene transfer as a potential candidate. Viral vectors have successfully implanted markers and therapeutic genes, but development of modern molecular biology has led to the discovery of newer and less immunogenic liposomal vectors. However, further studies that examine the regulation of the interval of gene expression and transfection efficiency are needed. Other possible strategies to alter the ligament healing environment include modulation of gene expression for certain target matrix molecules. Cell therapy using mesenchymal stem cells to enhance ligament healing represents another potential approach. Use of optimal cells is necessary if this therapy is to become a source of ligament healing. Furthermore, such cells can act as delivery vehicles to introduce molecules to the healing ligament.

Engineering the healing ligament has indeed generated much excitement. Potential delivery of optimal cells represents a method to improve the quality of the healing process which involves both a complex cellular interaction and specific local environmental conditions. To make this strategy more effective, scaffolding should be a potential method of improving the quality of ligament by maintaining applied cells at the site during healing. Regulation of the mechanical environment could be critical to the function of the applied cells because mechanical loading affects ligament healing.

Ultimately, the goal of tissue engineering is to promote normal function and mechanical properties of the ligament after injury. Interdisciniplinary studies involving biology and bioengineering should provide advancements in the understanding of ligament biology that foster newer ideas and enhance the utility of these alternative strategies. This philosophy will help in the successful search for the optimal treatment of ligament injuries.

Acknowledgments. We wish to acknowledge Ted Clineff, B.S., for his technical assistance, and NIH grant #AR41820, and ASIAM Institute for Research and Education for financial support.

References

Anderson, D.R., Weiss, J.A., Takai, S., Ohland, K.J., and Woo, S.L-Y. 1992. Healing of the medial collateral ligament following a triad injury: a biomechanical and histological study of the knee in rabbits. *J Orthop Res* 10:485–95.

Batten, M.L., Hansen, J.C., and Dahners, L.E. 1996. Influence of dosage and timing of application of platelet derived growth factor on early healing of the rat medial collateral ligament. *J Orthop Res* 14:736–41.

Buckley, A., Davidson, J.M., Kameroth, C.D., Wolt, T.B., and Woodward, S.C. 1985. Sustained release of epidermal growth factor accelerates wound healing. *Proc Natl Acad Sci USA* 82:7340–4.

Burgess, A.W. 1989. Epidermal growth factor and transforming growth factor alpha. *Br Med Bull* 45:401–24.

Caplan, A.I. 1991. Mesenchymal stem cells. *J Orthop Res* 9(5):641–50.

Caplan, A.I. 1994. The mesengenic process. *Clinics in Plastic Surgery* 21(3):429–35.

Challita, P.M. and Kohn, D.B. 1994. Lack of expression from a retroviral vector after transduction of murine hematopoietic stem cells is associated with methylation in vivo. *Proc Natl Acad Sci* 91(7):2567–71.

Chimich, D., Frank, C., Shrive, N., Dougall, H., and Bray, R. 1991. The effects of initial end contact on medial collateral ligament healing: a morphological and biomechanical study in a rabbit model. *J Orthop Res* 9:37–47.

Conti, N.A. and Dahners, L.E. 1993. The effect of exogenous growth factors on the healing of ligaments. *Trans Orthop Res Soc* 18:60.

Deie, M., Marui, T., Allen, C.R., Hildebrand, K.A., Georgescu, H.I., Niyibizi, C., and Woo, S.L-Y. 1997. The effects of age on rabbit MCL fibroblast matrix synthesis in response to TGF–B1 or EGF. *Mech Age Dev* 97:121–30.

DesRosiers, E.A., Yahia, L., and Rivard, C.–H. 1996. Proliferative and matrix synthesis response of canine anterior cruciate ligament fibroblasts submitted to combined growth factors. *J Orthop Res* 14:200–8.

Engle, C.P., Noguchi, M., Ohland, K.J., Shelley, F.J., and Woo, S.L-Y. 1994. Healing of the rabbit medial collateral ligament following an O'Donoghue triad injury: effects of anterior cruciate ligament reconstruction. *J Orthop Res* 12: 357–64.

Ferrari, G., Cusella–De Angelis, G., Coletta, M., Paolucci, E., Stornaiuolo, A., Cossu, G., and Mavilio, F. 1998. Muscle regeneration by bone marrow–derived myogenic progenitors. *Science* 279(5356):1528–30.

Frank, C.B., Bray, R.C., Hart, D.A., Shrive, N.G., Loitz, B.J., Matyas, J.R., and Wilson, J.E. 1994. "Soft tissue healing." In F.H. Fu, C.D. Harner, and K.G. Vince (eds.), *Knee Surgery.* 1st Ed. Vol. 1. 189–229. Baltimore, MD: Williams and Wilkins.

Frank, C., McDonald, D., Bray, D., Bray, R., Rangayyan, R., Chimich, D., and Shrive, N. 1992. Collagen fibril diameters in the healing adult rabbit medial collateral ligament. *Connect Tissue Res* 27:251–63.

Frank, C., McDonald, D., and Shrive, N. 1997. Collagen fibril diameters in the rabbit medial collateral ligament scar: a longer term assessment. *Connect Tissue Res* 36(3):261–9.

Frank, C., McDonald, D., Wilson, J., Eyre, D., and Shrive, N. 1995. Rabbit medial collateral ligament scar weakness is associated with decreased collagen pyridinoline crosslink density. *J Orthop Res* 13:157–65.

Frank, C., Woo, S.L-Y., Amiel, D., Harwood, F., Gomez, M., and Akeson, W. 1983. Medial collateral ligament healing: A multidisciplinary assessment in rabbits. *Am J Sports Med* 11:379–89.

Geiger, M.H., Green, M.H., Monosov, A., Akeson, W.H., and Amiel, D. 1994. An in vitro assay of anterior cruciate and medial collateral ligament cell migration. *Connect Tissue Res* 30(3):215–24.

Gerich, T.G., Kang, R., Fu, F.H., Robbins, P.D., and Evans, C.H. 1996. Gene transfer to ligaments and menisci. *Gene Ther* 3:1089–93.

Gold, L.I., Sung, J.J., Siebert, J.W., and Longaker, M.T. 1997. Type I (RI) and type II (RII) receptors for transforming growth factor–beta isoforms are expressed subsequent to transforming growth factor-beta ligands during excisional wound repair. *Am J Pathol* 150:209–22.

Gomez, M.A., Woo, S.L-Y., Inoue, M., Amiel, D., Harwood, F.L., and Kitabayashi, L. 1989. Medial collateral ligament healing subsequent to different treatment regimens. *J Appl Physiol* 66(1):245–52.

Gong, J., Chen, L., Chen, D., Kashiwaba, M., Manome, Y., Tanaka, T., and Kufe, D. 1997. Induction of antigen specific antitumor immunity with adenovirus transduced dendritic cells. *Gene Ther* 4:1023–8.

Green, R.J., Usui, M.L., Hart, C.E., Ammons, W.F., and Narayanan, A.S. 1997. Immunolocalization of platelet derived growth factor A and B chains and PDGF–alpha and beta receptors in human gingival wounds. *J Periodont Res* 32:209–14.

Hannafin, J.A., Attia, E., Warren, R.F., and Bhargava, M.M. 1997. The effect of cytokines on the chemotactic migration of canine knee ligament fibroblasts. *Trans Orthop Res Soc* 22:50.

Hart, R.A., Woo, S.L-Y., and Newton, P.O. 1992. Ultrastructural morphometry of anterior cruciate and medial collateral ligaments: an experimental study in rabbits. *J Orthop Res* 10:96–103.

Hildebrand, K.A., Deie, M., Allen, C.R., Smith, D.W., Georgescu, H.I., Evans, C.H., Robbins, P.D., and Woo, S.L-Y. 1999. The early expression of marker genes in the rabbit medial collateral and anterior cruciate ligaments: the use of different viral vectors and the effects of injury. *J Orthop Res* 17(1):37–42.

Hildebrand, K.A., Woo, S.L., Smith, D.W., Allen, C.R., Deie, M., Taylor, B.J., and Schmidt, C.C. 1998. The effects of platelet derived growth factor–BB on healing of the rabbit medial collateral ligament. An in vivo study. *Am J Sports Med* 26:549–54.

Hillard–Sembell, D., Daniel, D.M., Stone, M.L., Dobson, B.E., and Fithian, D.C. 1996. Combined injuries of the anterior cruciate and medial collateral ligaments of the knee. *J Bone Joint Surg* 78-A:169–76.

Huard, J., Goins, W.F., and Glorioso, J.C. 1995. Herpes symplex virus mediated gene transfer to muscle. *Gene Ther* 2:1–9.

Indelicato, P.A. 1995. Isolated medial collateral ligament injuries in the knee. *J Am Acad Orthop Surg* 3:9–14.

Inoue, M., Woo, S.L-Y., Amiel, D., Ohland, K.J., and Kitabayashi, L.R. 1990. Effects of surgical treatment and immobilization on the healing of the medial collateral ligament: a long term multidisciplinary study. *Connect Tissue Res* 25:13–26.

Kadiyala, S., Young, R.G., Thiede, M.A., and Bruder, S.P. 1997. Culture expanded canine mesenchymal stem cells possess osteochondrogenic potential in vivo and in vitro. *Cell Trans* 6:125–34.

Kavalkovich, K.W., Yamaji, T., Woo, S.L-Y., and Niyibizi, C. 1997. Type V collagen levels are elevated following MCL injury and in long term healing. *Trans Orthop Res Soc* 22:485.

Kobayashi, D., Kurosaka, M., Yoshiya, S., Hashimoto, J., Saura, R., Akamatu, T., and Mizuno, K. 1995. The effect of basic fibroblast growth factor on primary healing of the defect in canine anterior cruciate ligament. *Trans Orthop Res Soc* 20:630.

Kurosaka, M., Yoshiya, S., Mizuno, T., and Mizuno, K. 1998. Spontaneous healing of a tear of the anterior cruciate ligament—a report of two cases. *J Bone Joint Surg* 80A:1200–3.

Lazarus, H.M., Haynesworth, S.E., Gerson, S.L., Rosenthal, N.S., and Caplan, A.I. 1995. Ex vivo expansion and subsequent infusion of human bone marrow derived stromal progenitor cells (mesenchymal progenitor cells): implications for therapeutic use. *Bone Marrow Transplant* 16:557–64.

Lee, J., Green, M.H., and Amiel, D. 1995. Synergistic effect of growth factors on cell outgrowth from explants of rabbit anterior cruciate and medial collateral ligaments. *J Orthop Res* 13:435–41.

Letson, A.K. and Dahners, L.E. 1994. The effect of combinations of growth factors on ligament healing. *Clin Orthop* 308:207–12.

Linsenmayer, T.F., Gibney, E., Igoe, F., Gordon, M.K., Fitch, J.M., Fessler, L.I., and Birk, D.E. 1993. Type V collagen: molecular structure and fibrillar organization of the chicken a1(V) NH2–terminal domain, a putative regulator of corneal fibrillogenesis. *J Cell Biol* 121:1181–9.

Lyon, R.M., Akeson, W.H., Amiel, D., Kitabayashi, L., and Woo, S.L-Y. 1991. Ultrastructural differences between the cells of the medial collateral and the anterior cruciate ligaments. *Clin Orthop* 272:279–86.

Marui, T., Niyibizi, C., Georgescu, H.I., Cao, M., Kavalkovich, K.W., Levine, R.E., and Woo, S.L-Y. 1997. The effect of growth factors on matrix synthesis by ligament fibroblasts. *J Orthop Res* 15:18–23.

Michou, A.I., Santoro, L., Christ, M., Julliard, V., Pavirani, A., and Mehtali, M. 1997. Adenovirus mediated gene transfer: influence of transgene, mouse strain, and type of immune response on persistence of transgene expression. *Gene Ther* 4:473–82.

Miyasaka, K.C., Daniel, D.M., Stone, M.L., and Hirshman, P. 1991. The incidence of knee ligament injuries in the general population. *Am J Knee Surg* 4:3–8.

Murphy, P.G., Frank, C.B., and Hart, D.A. 1993. The cell biology of ligaments and ligament healing. In: Jackson, D.W. *The Anterior Cruciate Ligament: Current and Future Concepts* chapter 14: New York: Raven Press, 165–77.

Murphy, P.G. and Hart, D.A. 1994. Influence of exogenous growth factors on the expression of plasminogen activators and plasminogen activator inhibitors by cell isolated from normal and healing rabbit ligaments. *J Orthop Res* 12:564–75.

Mustoe, T.A., Pierce, G.F., Thomson, A., Gramates, P., Sporn, M.B., and Deuel, T.F. 1987. Accelerated healing of incisional wounds in rats induced by transforming growth factor–b. *Science* 237:1333–6.

Nagineni, C.N., Amiel, D., Green, M.H., Berchuck, M., and Akeson, W.H. 1992. Characterization of the intrinsic properties of the anterior cruciate and medial collateral ligament cells: an in vitro cell culture study. *J Orthop Res* 10:465–75.

Nakahara, H., Bruder, S.P., Haynesworth, S.E., Holecek, J.J., Baber, M.A., Goldberg, V.M., and Caplan, A.I. 1990. Bone and cartilage formation in diffusion chambers by subcultured cells derived from the periosteum. *Bone* 11:181–8.

Nakamura, N., Horibe, S., Matsumoto, N., Tomita, T., Natsuume, T., Kaneda, Y., Shino, K., and Ochi, T. 1996. Transient introduction of a foreign gene into healing rat patellar ligament. *J Clin Invest* 97:226–31.

Nakamura, N., Shino, K., Natsuume, T., Horibe, S., Matsumoto, N., Kaneda, Y., and Ochi, T. 1998. Early biological effect of in vivo gene transfer of platelet derived growth factor (PDGF)–B into healing patellar ligament. *Gene Ther* 5:1165–70.

Nakamura, N., Timmermann, S.A., Hart, D.A., Kaneda, Y., Shrive, N.G., Shino, K., Ochi, T., and Frank, C.B. 1998. A comparison of in vivo gene delivery methods for antisense therapy in ligament healing. *Gene Ther* 5:1455–61.

Ng, G.Y.F., Oakes, B.W., Deacon, O.W., McLean, I.D., and Eyre, D.R. 1996. Long–term study of the biochemistry and biomechanics of anterior cruciate ligament–patellar tendon autografts in goats. *J Orthop Res* 14:851–6.

Ohland, K.J.M., Woo, S.L-Y.P., Weiss, J.M., Takai, S.M., and Shelley, F.J.M. 1991. Healing of combined injuries of the rabbit medial collateral ligament and its insertions: a long term study on the effects of conservative vs. surgical treatment. *ASME* 20 (Advances in Bioengineering):447–8.

Panossian, V., Liu, S.H., Lane, J.M., and Finerman, G.A. 1997. Fibroblast growth factor and epidermal growth factor receptors in ligament healing. *Clin Orthop* 342:173–80.

Parry, D.A., Craig, A.S., and Barnes, G.R. 1978. Tendon and ligament from the horse: an ultrastructural study of collagen fibrils and elastic fibres as a function of age. *Proc Royal Soc London—Series B: Biological Sciences* 203:293–303.

Pierce, G.F., Mustoe, T.A., Lingelbach, J., Masakowski, V.R., Griffin, G.L., Senior, R.M., and Deuel, T.F. 1989. Platelet derived growth factor and transforming growth factor–b enhance tissue repair activities by unique mechanisms. *J Cell Biol* 109:429–40.

Prockop, D.J. 1997. Marrow stromal cells as stem cells for nonhematopoietic tissues. *Science* 276:71–4.

Roberts, A.B., Sporn, M.B., Assoian, R.K., Smith, J.M., Roche, N.S., Wakefield, L.M., Heine, U.I., Liotta, L.A., Falanga, V., Kehrl, J.H., and Fauci, A.S. 1986. Transforming growth factor type b: rapid induction of fibroblasts and angiogenesis in vivo and stimulation of collagen formation in vitro. *Proc Natl Acad Sci* 83:4167–71.

Ross, R. 1968. The fibroblast and wound repair *Biological Reviews of the Cambridge Philosophical Society* 43:51–96.

Scherping, Jr, S.C., Schmidt, C.C., Georgescu, H.I., Kwoh, C.K., Evans, C.H., and Woo, S.L-Y. 1997. Effect of growth factors on the proliferation of ligament fibroblasts from skeletally mature rabbits. *Connect Tissue Res* 36:1–8.

Schmidt, C.C., Georgescu, H.I., Kwoh, C.K., Blomstrom, G.L., Engle, C.P., Larkin, L.A., Evans, C.H., and Woo, S.L-Y. 1995. Effect of growth factors on the proliferation of fibroblasts from the medial collateral and anterior cruciate ligaments. *J Orthop Res* 13:184–90.

Sciore, P., Boykiw, R., and Hart, D. 1998. Semiquantitative reverse transcription–polymerase chain reaction analysis of mRNA for growth factors and growth factor receptors from normal and healing rabbit medial collateral ligament tissue. *J Orthop Res* 16:429–37.

Shelbourne, K.D. and Patel, D.V. 1995a. Management of combined injuries of the anterior cruciate and medial collateral ligaments. *J Bone Joint Surg* 77A:800–6.

Shelbourne, K.D. and Patel, D.V. 1995b. Timing of surgery in anterior cruciate ligament injured knees. *Knee Surg Sports Traumatol Arthrosc* 3:148–56.

Spindler, K.P., Dawson, J.M., Stahlman, G.C., and Davidson, J.M. 1996. Collagen synthesis and biomechanical response to TGF–b2 in the healing rabbit MCL. *Trans Orthop Res Soc* 21:793.

Spindler, K.P., Imro, A.K., Mayes, C.E., and Davidson, J.M. 1996. Patellar tendon and anterior cruciate ligament have different mitogenic responses to platelet derived growth factor and transforming growth factor b. *J Orthop Res* 14:542–6.

Spindler, K., Kenneth, D., Jeremy, T., Dawson, J., Lillian, N., and Jeff, D. 1998. The biomechanical response to doses of TGF–b2 in the healing rabbit MCL. *Trans Orthop Res Soc* 23:1023.

Stahlman, G., Spindler, K., Dawson, J., and Davidson, J. 1994. Do the growth factors TGF–b2 and PDGF-AB affect the structural properties of the healing medial collateral ligament? *Trans Am Orthop Soc Sports Med* 20:121–2.

Wakitani, S., Goto, T., Pineda, S.J., Young, R.G., Mansour, J.M., Caplan, A.I., and Goldberg, V.M. 1994. Mesenchymal cell based repair of large, full thickness defects of articular cartilage. *J Bone Joint Surg* 76A:579–92.

Watanabe, N., Takai, S., Morita, N., Kawata, M., and Hirasawa Y. 1998. New method of distinguishing between intrinsic cells in situ and extrinsic cells supplied by the autogeneic transplantation employing transgenic rats. *Trans Orthop Res Soc* 23:1035.

Weiss, J.A., Beck, C.L., Levine, R.E., and Greenwald, R.M. 1995. Effects of platelet derived growth factor on early medial collateral ligament healing. *Trans Orthop Res Soc* 20:159.

Weiss, J.A., Woo, S.L-Y., Ohland, K.J., Horibe, S., and Newton, P.O. 1991. Evaluation of a new injury model to study medial collateral ligament healing: primary repair vs. nonoperative treatment. *J Orthop Res* 9:516–28.

Wilder, R.L., Lafyatis, R., and Remmers, E.F. "Platelet derived growth factor and transforming growth factor beta in wound healing and repair." 1990. In W.B. Leadbetter, J.A. Buckwalter, and S.L. Gordon (eds.), *Sports Induced Inflammation*, 1st Ed., Vol. 1. 301–13. Park Ridge, IL: American Academy of Orthopaedic Surgeons.

Woo, S.L-Y., Inoue, M., McGurk–Burleson, E., and Gomez, M. 1987. Treatment of the medial collateral ligament injury II: structure and function of canine knees in response to differing treatment regimens. *Am J Sports Med* 15:22–9.

Woo, S.L-Y., Niyibizi, C., Matyas, J., Kavalkovich, K., Weaver-Green, C., and Fox, R.J. 1997. Medial collateral knee ligament healing: combined medial collateral and anterior cruciate ligament injuries studied in rabbits. *Acta Orthop Scand* 68:142–8.

Woo, S.L-Y., Suh, J.-K., Parsons, I., IM., Wang, J.-H., and Watanabe, N. 1998. Biological Intervention in Ligament Healing Effect of Growth Factors. *Sports Med Arthrosc Rev* 6:74–82.

Woo, S.L-Y., Yamaji, T., Taylor, B.J., Hildebrand, K.A., and Fox, R.J. 1996. The effects of dose levels of EGF and TGF–B on the healing of the rabbit medial collateral ligament. In *American Orthopaedic Society for Sports Medicine in Lake Buena Vista, Florida*, American Orthopaedic Society for Sports Medicine. 763.

Yang, Y., Nunes, F.A., Berencsi, K., Furth, E.E., Gonczol, E., and Wilson, J.M. 1994. Cellular immunity to viral antigens limits E1 deleted adenoviruses for gene therapy. *Proc Natl Acad Sci* 91:4407–11.

Young, R.P., Butler, D.L., Weber, W., Caplan, A.I., Gordon, S.L., and Fink, D.J. 1998. Use of mesenchymal stem cells in a collagen matrix for Achilles tendon repair. *J Orthop Res* 16:406–13.

11
Muscle-Based Tissue Engineering for Orthopaedic Applications

DOUGLAS S. MUSGRAVE, CHARLES S. DAY, PATRICK BOSCH,
JACQUES MÉNÉTREY, CHANNARONG KASEMKIJWATANNA,
MOREY S. MORELAND, FREDDIE H. FU, and JOHNNY HUARD

Introduction

The advent of gene therapy and tissue engineering has facilitated novel approaches to the treatment of orthopaedic disorders. The delivery of growth factors, cells, and therapeutic genes promises therapeutic possibilities not contemplated by previous generations of orthopaedic surgeons. Significant scientific contributions have been made in the last 3 decades toward the understanding of skeletal muscle biology and its potential therapeutic applications. However, despite tremendous progress, many questions remain unanswered. This chapter reviews the current status of muscle-based tissue engineering for musculoskeletal disorders and discusses the focus of ongoing research.

Skeletal muscle-based tissue engineering consists of muscle cell and/or muscle tissue manipulation to improve healing and restore function. Several different means may be utilized to achieve these ends. The simplest method is the direct delivery of growth factor proteins to an injured muscle. However, as discussed later, the direct delivery of proteins has limitations. Therefore, approaches to gene therapy that would improve growth factor delivery are attractive. Both direct (in vivo) gene therapy and ex vivo gene therapy approaches using skeletal muscle have been reported (Dhawan et al. 1991; Dai, Roman, Naviaux, and Verma 1992; Lynch, Clowes, M.M., Osborne, Clowes, A.W., and Miller 1992; Jiao, Williams, Safda, Schultz, and Wolff 1993; Lau, Yu, Fontana, and Stockert 1996; Simonsen, Groskreutz, Gorman, and MacDonald 1996; Bosch et al. 1998a,b; Musgrave et al. 1999). The direct gene therapy approach, albeit technically straightforward, presents theoretical risks of in vivo genetic manipulation and possible reversion to pathogenicity of attenuated viral vectors. Furthermore, the direct approach does not provide for the introduction of cells capable of participating in the healing process. The ex vivo approach addresses these issues by limiting genetic manipulation to the culture flask, thereby eliminating the potential risks of in vivo genetic manipulation and viral reversion (Figure 11.1). The ex vivo approach also allows for isolation

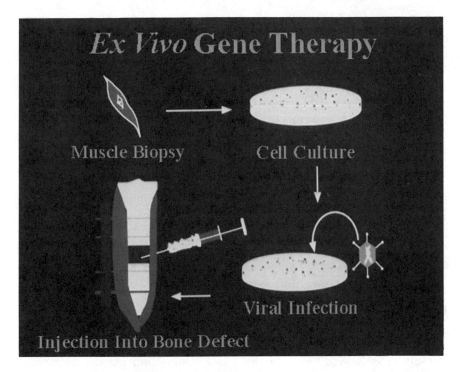

FIGURE 11.1. Basic approach of ex vivo gene therapy.

and expansion of muscle derived cells possibly capable of participating in the therapeutic process. Evidence suggests that muscle derived cells can participate in both muscle and bone healing (Urist 1965; Huard, Labrecque, Dansereau, Robitaille, and Tremblay 1991; Bosch et al. 1998b). Finally, myoblast transplantation delivers competent cells with complementary genomes to patients with skeletal muscle pathology. In addition to its clinical use for inherited myopathies such as Duchenne muscular dystrophy, recent studies also support its efficacy in treating acquired muscle injuries. Therefore, muscle-based tissue engineering combines gene therapy, cell transplantation, and tissue manipulation to effect an improved healing response.

The theory behind muscle-based tissue engineering is predicated on the unique biology of skeletal muscle derived cells. First, as discussed below, skeletal muscle contains satellite cells. These cells are resting mononucleated precursor cells capable of fusing to form postmitotic, multinucleated myotubes, and myofibers. The postmitotic, multinucleated myofibers are stable cells theoretically capable of persistent gene expression. Therefore, focusing tissue engineering approaches on the skeletal muscle satellite cell may make it possible to maximize the degree and persistence of gene expression. Second, as alluded to earlier, skeletal muscle may contain a population of mesenchymal stem cells. Mesenchymal stem cells are resting cells capable of differentiation into multiple tissue lineages (Caplan 1991). In vitro (Katagiri et al. 1994; Young et al. 1995; Warejcka, Harvey, Taylor, Young, and Lucas 1996; Kawasaki et al. 1998) and in vivo (Bosch et al. 1998b) data suggest that cells residing within skeletal muscle are also capable of differentiation into several different tissue lineages. Consequently, muscle-derived cells may be capable of regenerating many different tissues. Tissue engineering based on these cells may not only facilitate gene delivery, but also supply the needed stem cells for healing. Finally, muscle-derived cells are clinically accessible and reliably isolated. Skeletal muscle biopsies are of low morbidity and available on an outpatient basis. In vitro isolation of muscle-derived cells has been well described (Blau and Webster 1981; Rando and Blau 1994) and cryopreservation of the isolated cells facilitates flexibility in therapeutic applications. The burgeoning field of muscle based tissue engineering is based on these unique characteristics of skeletal muscle derived cells.

Isolation of Muscle-Derived Cells

Muscle regeneration occurs by release and activation of mononucleated satellite cells located between the basal lamina and plasma membrane (Figure 11.2) (Hurme, Kalima, Lehto, and Jarvinen 1992; Bischoff 1994). Disruption of the basal lamina and plasma membrane releases and activates the satellite cells which proliferate and differentiate into multinucleated myotubes and myofibers (Hurme et al. 1992; Bischoff 1994). This sequence partially repairs the injured muscle. The satellite cells, because of their unique biology, as discussed above, are the cells on which muscle-based tissue engineering is dependent.

Two basic approaches exist for obtaining myogenic cells for investigational and therapeutic applications: establishment of myogenic cell lines and primary myogenic cell isolation. The advantages of immortalized myogenic cell lines include establishment of a consistent and reliable population of pure myogenic cells and the relative technical ease of manipulation once the immortalized cell line is created. However, the cell lines are either allogeneic or xenogeneic and may have significant tumorigenic potential (Rando and Blau 1994). Therefore, myogenic cells lines, while useful research tools, are not applicable to clinical trials.

The alternative approach involves isolation of primary myogenic cells from the host. Isolated autogenous cells may be engineered in vitro prior to reimplantation into the host. Consequently, these autogenous cells have no tumorigenic potential and the immune reaction is minimized (Rando and Blau 1994; Qu et al. 1998). Two methods of primary muscle cell isolation exist. The first is fundamentally based on differential cell-adhesion characteristics (Rando and Blau 1994). A skeletal muscle biopsy is obtained

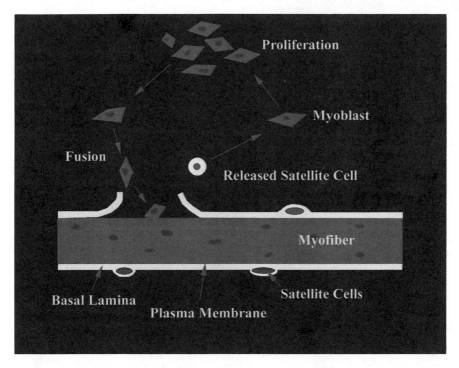

FIGURE 11.2. Schematic representation of skeletal myofiber injury and regeneration.

under sterile conditions. The muscle is placed in a balanced saline solution and finely minced with scalpel blades. The minced muscle is then enzymatically and mechanically dissociated to further release satellite cells from between the basal laminae and plasma membranes. The cell suspension is then placed into a collagen-coated flask, a technique referred to as "preplating." Different populations of cells (i.e., myoblasts, fibroblasts) adhere to the collagen-coated flask at varying intervals. The supernatent of cells not yet adhered to the first flask (preplate) is sequentially passed to subsequent flasks (preplates) (Qu et al. 1998). The proportion of myogenic cells in each of the flasks can be determined with immunohistochemical staining for desmin, a myogenic protein. Successive flasks contain progressively increasing percentages of mononucleated myogenic cells. These cells, believed to be satellite cells or possibly further undifferentiated stem cells, are expanded in cell culture until adequate numbers are obtained for the desired tissue engineering application. The second method of primary myogenic cell isolation, myoblast cell sorting, is predicated on the existence of myoblast specific cell surface proteins, such as NKH-1 antigen, an isoform of the neuronal cell adhesion molecule (NCAM). NKH-1, while expressed by myoblasts, is not expressed by fibroblasts. Therefore, these two cell populations can be distinguished using immunohistochemistry. A primary monoclonal antibody against NKH-1 binds myoblasts in cell culture. A secondary antibody, conjugated to a fluorescent protein, then binds the primary antibody. Fluorescent-labeled myoblasts can be identified and selected using a cytofluorometer. Consequently, a pure population of myoblasts may be created from which adequate cell numbers could be expanded.

Muscle Injuries

Muscle injuries make up a large percentage of recreational and competitive athletic injuries, and may result from both direct (contusions, lacerations) and indirect (strains, ischemia) trauma. Upon injury, satellite cells are released and activated in order to proliferate into myotubes and myofibers, thereby promoting muscle healing. However, this reparative process is usually incomplete and accompanied by a fibrous reaction, producing scar tissue that limits the muscle's potential for functional recovery (Hurme et al. 1991).

Investigations in animals identify possible clinical applications for muscle-based tissue engineering to treat muscle injuries (Taylor, Dalton, Jr., Seaber, and Garrett, Jr. 1991; Obremsky, Seaber, Ribbeck, and Garrett, Jr. 1994). Animal models of muscle laceration, contusion, and strain have been developed (Taylor et al. 1991; Obremsky et al. 1994; Kasemkijwattana et al. 1998). Laceration is performed in mice by incising 75% of the width and 50% of the thickness of the gastrocnemius muscle. Contusion is created by dropping a 16-gram ball from a height of 100 centimeters (cm) onto the mouse gastrocnemius muscle. Strain is created by elongating the mouse gastrocnemius muscle–tendon unit at a rate of 1 cm/min. Under these conditions, active muscle regeneration is found at 7 and 10 days after injury, but begins to decrease at 14 days and continues decreasing at 35 days. Concomitantly, fibrosis is observed beginning at 14 days and gradually increases until 35 days (Kasemkijwattana et al. 1998). Fibrosis appears at the time muscle regeneration diminishes and, therefore, may hinder the healing response.

Injured skeletal muscle releases numerous growth factors acting in autocrine and paracrine fashion to modulate muscle healing. These proteins stimulate satellite cells to proliferate and differentiate into myofibers (Schultz, Jaryszak, and Valiere 1985; Allamedine, Dehaupas, and Fardeau 1989; Schultz 1989; Hurme et al. 1991; Bischoff 1994). The delivery of exogenous growth factors, specifically selected to enhance myofiber regeneration, is an intuitive therapeutic approach to muscle injuries. In vitro experiments have identified several growth factors capable of enhancing myogenic proliferation and differentiation (Kasemkijwattana et al. 1998). Satellite cell activity in cell culture is assessed at 48 hours and 96 hours after incubation in prospective growth factors. Basic fibroblast growth factor (b-FGF), insulin-like growth factor-1 (IGF-1), and nerve growth factor (NGF) significantly enhance myoblast proliferation, whereas b-FGF, acidic fibroblast growth factor (a-FGF), IGF-1, and NGF increase myobast differentiation into myotubes. Consequently, bFGF, IGF-1, and NGF are the logical candidates for therapeutic applications to enhance muscle healing (Ménétrey et al. 2000).

The technique chosen to deliver prospective growth factors to injured muscle is of paramount importance to optimize therapeutic benefit. Options include direct injection of growth factors themselves, direct gene therapy, ex vivo gene therapy, and myoblast transplantation. Direct injections of human recombinant growth factors (b-FGF, IGF-1, and NGF) into injured muscle increase the number of regenerating myofibers in vivo and increase both muscle twitch and tetanic strength 15 days after injury (Ménétrey et al. 2000). However, secondary to rapid clearance and short half-lives, the effect of direct growth factor injections is probably transient and suboptimal. Gene therapy provides a mechanism to achieve persistent protein production and, thereby, theoretically improve muscle healing. Direct gene therapy to deliver genes to skeletal muscle involves the possibility of using naked DNA (Acsadi et al. 1991), retrovirus (Dunckley, Wells, Walsh, and Dickson 1993), adenovirus (Ascadi et al. 1994), herpes simplex virus (Huard, Goins, and Glorioso 1995), and adeno–associated virus (Xiao, Li, and Samulski 1996). Most of these vectors transfect/transduce relatively few adult myofibers. However, adenovirus is capable of tranducing a large number of regenerating muscle fibers, a condition present in injured muscle. Direct injection of adenovirus containing the β–galactosidase marker gene into lacerated, contused, and strained muscle results in many transduced myofibers at 5 days (Figure 11.3). Therefore, direct injection of adenovirus-carrying growth factor genes (i.e., bFGF, IGF-1, NGF) may result in sustained protein production in injured muscle. Recent data show that direct injection of adeno-associated virus (AAV) results in a high level of adult myofiber transduction in both injured and noninjured muscle. AAV may be the preferred vector for direct gene delivery to mature skeletal muscle, although it is only capable of carrying genes of less than 5 kilobase (kb) pairs.

Ex vivo gene therapy and myoblast transplantation are two closely related methods that require in vitro cell isolation and culture. Ex vivo techniques involve muscle biopsy and myogenic cell isolation as described earlier in this chapter (Rando and Blau 1994). The isolated satellite cells are transduced in vitro with the desired gene-carrying vector. The satellite cells are then reinjected into skeletal muscle where they fuse to form postmitotic myotubes and myofibers, and begin growth factor production. This technique is currently feasible with adenoviral (Huard, Verrault, Roy, Tremblay, G., and Tremblay, J.P. 1994), retroviral (Salvatori et al. 1993), and herpes simplex viral vectors (Booth et al. 1997). Ex vivo delivery of

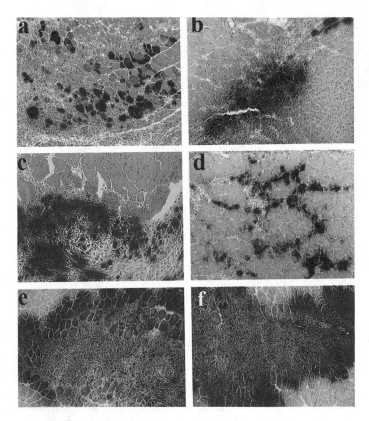

FIGURE 11.3. Adenovirus-mediated gene transfer to injured skeletal muscle. Direct B–galactosidase gene transfer to contused (a), lacerated (c), and strain-injured muscle (e). Myoblast mediated ex vivo β–galactosidase gene transfer to contused (b), lacerated (d), and strain-injured muscle (f).

the β–galactosidase marker gene to injured muscle produces many β–galactosidase positive myofibers (Figure 11.3). The ex vivo muscle cell-mediated approach provides not only an efficient method of delivering selected genes, but also provides cells capable of participating in the reparative process. Myoblast transplantation, a related approach also providing competent cells, lacks in vitro genetic manipulation. In addition to its historic application toward inherited muscle diseases, myoblast transplantation has been shown to improve myofiber regeneration in muscle experimentally injured with myonecrotic agents (Huard, Acsadi, Jani, Massie, and Karpati 1994). Furthermore, preliminary data regarding the use of autologous myoblast transplantation, as well as minced muscle tissue transplantation, into lacerated skeletal muscle is promising. Therefore, the closely related techniques of muscle cell mediated ex vivo gene therapy and myoblast transplantation are both applicable to muscle injuries.

Muscle-based tissue engineering offers exciting potential therapies for muscle disorders. A large number of recreational and professional athletic injuries involve skeletal muscle (Garrett 1990). Therapies to improve functional recovery and shorten rehabilitation may both optimize performance and minimize morbidity. Further research is ongoing to refine these muscle-based tissue engineering applications. The results of such investigations may provide revolutionary treatments for these common injuries.

Bone Healing

The need to improve bone healing permeates every orthopaedic subspecialty. Nonunions, implant stabilizations, arthrodeses, bone defect reconstructions, and leg length discrepencies all benefit from bone healing augmentation. Unfortunately, current techniques of autograft, allograft, bone transport, and electrical stimulation are often suboptimal. Therefore, tissue engineering approaches toward bone formation have immense implications.

Intramuscular bone formation is a poorly understood phenomenon. It can be present in the clinically pathologic states of heterotopic ossification, myositis ossificans, fibrodysplasia ossificans progressiva, and osteosarcoma. Radiation therapy and the antiinflammatory drug indomethicin are used clinically to suppress myositis ossificans. However, neither the mechanism of formation nor the suppression of ectopic bone is clearly understood. The first evidence of the existence of growth factors capable of stimulating intramuscular bone was gathered 30 years ago (Urist 1965). Now, a growing number of bone morphogenetic proteins (BMPs), members of the transforming growth factor-beta (TGF–B) superfamily, are recognized. Human BMP-2 is available in recombinant form (rhBMP-2) and induces bone formation when injected into skeletal muscle (Wang et al. 1990). Current applications focus on injecting rhBMP-2 directly into nonunions and bone defects. However, direct BMP implantation is hindered by the growth factors' short biological half-lives, their rapid dilution by extracellular fluid, and their requirement of a foreign body scaffold. The delivery of BMP cDNAs via gene therapy may circumvent these limitations. Therefore, muscle-based tissue engineering holds enormous promise in the arena of bone healing and, in the process, may shed light on the pathophysiology of ectopic bone formation.

Cells isolated from skeletal muscle are capable of responding to rhBMP-2 both in vitro and in vivo. Primary rodent and human myogenic cells in culture respond in a dose dependent fashion to rhBMP-2 by producing alkaline phosphatase, an osteogenic protein (Bosch et al. 1998b; Kawasaki et al. 1998). The purer the population of myogenic cells, as evidenced by desmin staining, the greater the alkaline phosphatase production (Bosch et al. 1998b). Recombinant human BMP-2 inhibits myogenic differentiation as it stimulates osteoblastic differentiation of muscle-derived cells (Yam-aguchi et al. 1991; Katagiri et al. 1994; Kawasaki et al. 1998). Therefore, the in vitro data suggest that myogenic cells are capable of responding to rhBMP-2 and entering an osteogenic lineage.

Primary rodent muscle-derived cells are capable of being engineered to produce intramuscular bone in vivo (Bosch et al. 1998a,b). The ex vivo approach is utilized to transduce the primary muscle-derived cells with an adenovirus carrying the BMP-2 cDNA. Intramuscular injection of as little as 300,000 transduced cells produces bone in severe combined immune deficient (SCID) mice (Figure 11.4). The resultant bone is mineralized, as evidenced by von Kossa's stain, and contains trabecular spaces. Not only do the transduced muscle cells produce BMP-2, but strong evidence suggests that the injected cells also respond to BMP-2 by producing bone (Bosch et al. 1998b). In addition to the ex vivo approach, an adenovirus mediated direct gene therapy approach in rodents produces large amounts of intra-muscular bone (Musgrave et al. In press). Consequently, both in vitro and in vivo data support the hypothesis that muscle cells may be engineered to become osteogenic cells. The ramifications of myogenic cells' capabilities to form bone are immense.

Muscle-based tissue engineering to produce bone may be applicable to multiple orthopaedic abnormalities. One example involves large bone defects resulting from trauma or oncologic resections. Muscle-derived cells capable of bone formation may be exploited to reconstruct the bone defect and minimize the use of autograft, allograft, and bone transport. Currently, we are investigating whether a muscle flap can be engineered to produce bone and, thereby, reconstruct an experimental bone defect. Both ex vivo and in vivo gene therapy techniques are being applied in this model. An alternative approach is to transform muscle, restricted to the confines of a silicone mold, into bone of desired geometry such as a proximal femur or midshaft tibia (Khouri, Khoudsi, and Reddi 1991). The muscle-based approach to bone defect reconstructions is especially appealing in light of the often poor vascularity of traumatic and oncologic bone defects (Khouri et al. 1996). The combination of vascularized muscle and osteocompetent muscle-derived cells offers revolutionary possibilities worthy of further investigation.

Intraarticular Disorders

Degenerative and traumatic joint disorders are encountered frequently as our population becomes more active and lives longer. These disorders include arthritis of various etiologies, ligament disruptions, meniscal tears, and osteochondral injuries. Currently, the orthopaedist's tools consist pri-marily of surgical procedures aimed at biomechanically alterating the joint (anterior cruciate ligament (ACL) reconstructions, total knee replacement, meniscal repair or excision, cartilage debridement, etc.) Tissue engineering

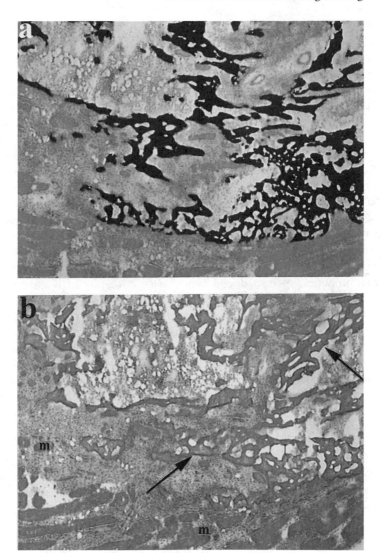

FIGURE 11.4a. Von Kossa's stain demonstrating mineralized intramuscular bone (black) formed via muscle cell-mediated ex vivo gene delivery of BMP-2. b. Corresponding H & E stain demonstrating intramuscular bone (black arrows) infiltrating skeletal myofibers (m).

applied to these intraarticular disease states theoretically offers a more biologic and less disruptive reparative process. Both direct (Nita et al. 1996) and ex vivo (Bandara et al. 1993) gene therapy approaches to arthritis models have been reported. The synovial cell-mediated ex vivo approach, while offering advantages of ex vivo gene tranfer such as the safety of in vitro genetic manipulation and precise cell selection, is hindered by a decline of gene expression after 5 to 6 weeks (Bandara et al. 1993). The satellite cell, because of its ability to form postmitotic myotubes and myofibers, offers theoretical advantages of longer term and more abundant protein production.

Muscle cell-mediated ex vivo gene delivery to numerous intraarticular structures is possible. Intraarticular injection of primary myoblasts, transduced by an adenovirus carrying the β–galactosidase marker gene, results in gene delivery to many intraarticular structures (Day et al. 1997). Tissues expressing β–galactosidase at 5 days after injection in the rabbit pup knee include the synovial lining, meniscal surface, and cruciate ligaments. Likewise, injection of transduced immortalized myoblasts results in gene delivery to various intraarticular structures, including the synovial lining and patellar ligament surface. However, purified immortalized myoblasts fused more readily and resulted in more de novo intraarticular myofibers than did primary myoblasts. This finding illustrates the importance of obtaining a pure population of myogenic cells, de void of fibroblast contamination. In contrast, injection of transduced synovial cells results in β–galactosidase expression only in the synovium (Day et al. 1997). Muscle cell-mediated ex vivo approaches are predicated on myoblast fusion to form myofibers, multicellular protein-producing factories. Intraarticular injection of transduced immortalized myoblasts into a SCID mouse results in gene expression and myofiber formation in multiple structures at 35 days. Therefore, intermediate term intraarticular gene expression resulting from muscle cell-mediated tissue engineering is feasible in animal models. Based on these data, muscle cell-based approaches are aimed at delivering genes specifically to the ACL and meniscus.

The second most frequently injured knee ligament, the ACL has a low healing capacity, possibly secondary to its encompassing synovial sheath or its exposure to synovial fluid when disrupted. Because complete tears of the ACL cannot undergo spontaneous healing, current treatment options are limited to surgical reconstruction using autograft or allograft. The replacement graft, often either patella ligament or hamstring tendon in origin, undergoes ligamentization with eventual collagen remodeling (Arnosczky, Tarvin, and Marshall 1982). Therefore, recent research is directed at augmentation of this ligamentization process using growth factors to affect fibroblast behavior. In vivo data suggest that PDGF, TGF-B, and EGF promote ligament healing (Conti and Dahners 1993). Transient, low levels of these growth factors resulting from their direct injection into the injured ligament are unlikely to produce a significant response. Therefore, an efficient delivery mechanism is essential to the development of a clinically applicable therapy. Muscle cell-mediated ex vivo gene therapy offers the potential to achieve persistent local gene expression and subsequent growth factor delivery to the ACL (Ménétrey, et al. 1999). The rabbit ACL, when injected with myoblasts transduced with the β–galactosidase reporter gene, contains β–galactosidase expressing cells beginning at 3 days postinjection. A high level of β–galactosidase expression is evident from 3 to 21 days after injection. Furthermore, de novo intraarticular myofibers expressing β–galactosidase are present beginning at 1 week. Colocalization of β–galactosidase and fluorescent microspheres used to label the myoblasts suggests that intraarticular myofibers result from fusion of injected myoblasts. Numerous transduced muscle cells are also present in the ligament's surrounding synovial sheath where gene expression persists for up to 6 weeks. In contrast, injection of transduced fibroblasts results in intraligament gene expression for only 2 weeks and synovial sheath gene expression for only 3 weeks. Like myoblast-mediated gene delivery, direct injection of adenovirus containing the β–galactosidase gene also results in gene expression for up to 6 weeks. However, as discussed previously, muscle cell-mediated ex vivo approaches have theoretical advantages over the direct technique, including safety and multinucleated myofiber formation. Therefore, muscle-based tissue engineering facilitates intermediate term (6 weeks) gene delivery to the ACL. This time period may be sufficient to effect a clinical response. Longer term expression may be achieved using other viral vectors to transduce the muscle cells. Investigations into the effect of muscle cell-mediated ex vivo gene therapy to enhance the healing of torn ACLs, reconstructed ACLs, and the bone ligament interface are currently ongoing and will bring this technique closer to clinical fruition.

The knee meniscus plays a critical role in maintaining normal knee biomechanics. Primary functions of the meniscus include load transmission, shock absorption, joint lubrication, and tibiofemoral stabilization in the ACL deficient knee. The historical treatment of menisectomy for meniscal tears has been replaced by meniscal repair when tears involve the meniscus' peripheral, vascular third. Growth factors, including platelet-derived growth factor (PDGF), are capable of enhancing meniscal healing (Spindler, Mayes, Miller, Imro, and Davidson 1995). In vitro data currently under review detail numerous growth factors' effects on fibroblast proliferation and collagen production. Regardless of which growth factor is proven optimal for meniscal healing, the cardinal issue of protein delivery must be addressed. Direct intrameniscal growth factor injections are unlikely to produce sustained levels without the need for multiple injections, a scenario not clinically appropriate. Efficient and sustained delivery of desired growth factors may be best accomplished by gene transfer. Muscle cell-mediated ex vivo gene delivery offers the possibility of sustained, high-level gene expression. Data currently submitted for review (Kasemkijwattana et al., in press) establish the feasibility of muscle cell-mediated and direct adenoviral-mediated gene delivery to the rabbit meniscus, with up to 6-week persistence. Further studies regarding muscle cell-mediated growth factor gene delivery may lead to novel therapies for meniscal injuries, preventing significant morbidity and chronic disability.

Muscle-Derived Stem Cells

The precise characterization of muscle-derived cells is an area of intense research. Because desmin positive cells, isolated using the preplate technique described earlier, are able to fuse into myotubes and myofibers, they are believed to be satellite cells or resting myoblasts. However, recent evidence calls into question the true identity and/or pluripotency of these cells. As discussed previously, in vitro and in vivo data strongly suggest that these desmin positive cells are capable of osteogenesis. Furthermore, in vitro dexamethasone stimulation of neonatal rat and adult rabbit muscle cells supports the existence of pluripotent stem cells residing within muscle (Warejcka et al. 1996). Three possibilities exist to explain these findings. First, the satellite cells may be true myoblasts which, for some reason, can transdifferentiate into another lineage, but such a hypothesis is contradictory to basic cell biology principles of differentiation. Second, the satellite cells may not be strict myoblasts, but pluripotent cells capable of differentiating into myogenic, osteogenic, and possibly other lineages. A population of cells residing within skeletal muscle capable of differentiating into multiple mesenchymal lineages would be, by definition, mesenchymal stem cells (Caplan 1991). Third, two or more populations of stem cells may exist in skeletal muscle. One population may consist of myoprogenitors

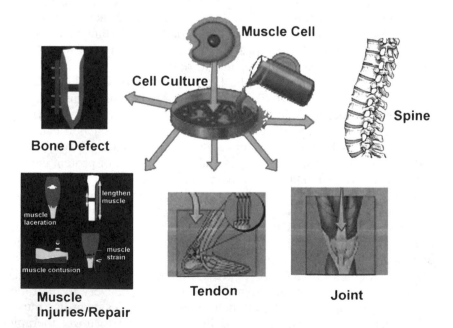

FIGURE 11.5. Schematic representation of multiple potential applications for muscle-based tissue engineering.

whereas other populations may consist of pluripotent stem cells. Which one of the above possibilities is true is the focus of ongoing research in our laboratory and others. These fundamental basic science issues have enormous significance for muscle-based tissue engineering and orthopaedic surgery. A population of pluripotent stem cells residing within skeletal muscle would facilitate innumerable orthopaedic tissue engineering applications.

Future Directions

Muscle-based tissue engineering is a burgeoning new discipline with unknown potential. Data gathered thus far challenge traditional scientific beliefs at many levels, from basic cell biology to clinical medicine. In addition to the thorough characterization of possible skeletal muscle-derived mesenchymal stem cells, investigators must aggressively pursue potential orthopaedic applications for muscle-based tissue engineering. The development of muscle-based tissue engineering approaches to the treatment of acquired muscle injuries, bone healing, and intraarticular disorders is underway. Furthermore, investigations have been initiated into the utility of muscle-based tissue engineering to heal cartilage defects, heal flexor tendon lacerations, and improve spinal fusions (Figure 11.5). An explosion of research, from basic science to clinical medicine, is mandated to fully elucidate the potential of muscle based tissue engineering for orthopaedic applications.

References

Acsadi, G., Dickson, G., Love, D., Jani, A., Gurusinghe, A., Walsh, F.S., Wolff, J.A., and Davies, K.E. 1991. Human dystrophin expression in mdx mice after intramuscular injection of DNA constructs. *Nature* 352:815–8.

Acsadi, G., Jani, A., Huard, J., Blaschuk, K., Massie, B., Holland, P., Lochmuller, H., and Karpati, G. 1994. Cultured human myoblasts and myotubes show markedly different transducibility by replication defective adenovirus recombinants. *Gene Ther* 1:338–40.

Allamedine, H.S., Dehaupas, M., and Fardeau, M. 1989. Regeneration of skeletal muscle fiber from autologous satellite cells multiplied in vitro. *Muscle Nerve* 12:544–5.

Arnoczky, S.P., Tarvin, G.B., and Marshall, J.L. 1982. Anterior cruciate ligament replacement using patellar tendon. *J Bone Joint Surg* 64-A:217–24.

Bandara, G., Mueller, G.M., Galea–Lauri, J., Tindal, M.H., Georgesco, H.J., Suchanek, M.K., Hung G.L., Glorioso, J.C., Rddoins, P.D., and Evans, C.H. 1993. Intraarticular expression of biologically active interleukin–1 receptor antagonist protein by ex vivo gene transfer. *Proc Natl Acad Sci USA* 90:10764–8.

Bischoff, R. 1994. The satellite cell and muscle regeneration. In: Engel AG, Franzini-Armstrong C. (eds.). *Myology*. 2nd ed. New York: McGraw–Hill 97–118.

Blau, H.M. and Webster, C. 1981. Isolation and characterization of human muscle cells. *Proc Natl Acad Sci USA* 78:5623–7.

Booth, D.K., Floyd, S.S., Day, C.S., Glorioso, J.C., Kovesdi, I., and Huard, J. 1997. Myoblast mediated ex vivo gene transfer to mature muscle. *J Tissue Eng* 3(2):125–33.

Bosch, P., Musgrave, D.S., Schuler, F.S., Ménétrey, J., Evans, C.H., Niyibizi, C., and Huard, J. 1998a. Myoblast mediated bone formation. *Tiss Eng* 4(4):470.

Bosch, P., Musgrave, D.S., Schuler, F.S., Ménétrey, J., Lowenstein, J.E., Evans, C.H., and Huard, J. 1998b. Osteoprogenitor cells in skeletal muscle. *Tiss Eng* 4(4):477.

Caplan, A.I. Mesenchymal stem cells. 1991. *J Orthop Res* 9:641–50.

Conti, N.A. and Dahners, L.E. 1993. The effect of exogenous growth factors on the healing of ligaments. *Trans Orthop Res Soc* 18:60.

Dai, Y., Roman, M., Naviaux, R.K., and Verma, I.M. 1992. Gene therapy via primary myoblasts: long term expression of factor IX protein following transplantation in vivo. *Proc Natl Acad Sci USA* 89:10892–5.

Day, C.S., Kasemkijwattana, C., Ménétrey, J., Floyd, S.S., Dooth, D., Moseland, M.S., Fu, F.H., and Huard, J. 1997. Myoblast-mediated gene transfer to the joint. *Journal of Orthopaedic Research* 15(6):894–903.

Dhawan, J., Pan, L.C., Pavlath, G.K., Tsavis, M.A., Lanctot, A.M., and Blau, H.M. 1991. Systemic delivery of human growth hormone by injection of genetically engineered myoblasts. *Science* 254:1509–12.

Dunckley, M.G., Wells, D.J., Walsh, F.S., and Dickson, G. 1993. Direct retroviral mediated transfer of a dystrophin minigene into mdx mouse in muscle in vivo. *Hum Mol Genet* 2(6):717–23.

Garrett, W.E. Jr. 1990. Muscle strain injuries: clinical and basic aspects (Review) *Medicine of Science in Sports & Exercise* 22(4):436–43.

Huard, J., Ascadi, G., Jani, A., Massie, B., and Karpati, G. 1994. Gene transfer into skeletal muscles by isogenic myoblasts. *Hum Gene Ther* 5:949–58.

Huard, J., Goins, B., and Glorioso, J.C. 1995. Herpes simplex virus type I vector mediated gene transfer to muscle. *Gene Ther* 2:1–9.

Huard, J., Labrecque, C., Dansereau, G., Robitaille, L., and Tremblay, J.P. 1991. Dystrophin expression in myotubes formed by the fusion of normal and dystrophic myoblasts. *Muscle Nerve* 14:178–82.

Huard, J., Verrault, S., Roy, R., Tremblay, G., and Tremblay, J.P. 1994. High efficiency of muscle regeneration following human myoblast clone transplantation in scid mice. *J Clin Invest* 93:586–99.

Hurme, T., Kalima, H., Lehto, H., and Jarvinen, M. 1991. Healing of skeletal muscle injury: an ultrastructural and immunohistochemical study. *Med Sci Sports Exerc* 23:801–810.

Jiao, S., Williams, P., Safda, N., Schultz, E., and Wolff, J.A. 1993. Cotransplantation of plasmid transfected myoblasts and myotubes into rat brains enables high levels of gene expression long-term. *Cell Transplant* 2:185–92.

Kasemkijwattana, C., Ménétrey, J., Day, C.S., Bosch, P., Buranapanitkit, B., Moreland, M.S., Fu, F.H., Watkins, S.C., and Huard, J. 1998. Biologic intervention in muscle healing and regeneration. *Sports Med Arthrose Rev* 6:95–102.

Kasemkijwattana, C., Ménétrey, J., Goto, H., Niyibizi, C., Fu, F.H., and Huard, J. (In Press). The use of grawth factors, gene therapy and tissue Engineering to improve meniscal healing. In Proceedings of the International Symposium on Advanced Biomaterials & Tissue Engineering, Tokyo, Japan.

Katagiri, T., Yamaguchi, A., Komaki, M., Abe, E., Takahashi, N., Ikeda, T., Rosen, V., Wozney, J.M., Fujisawá-Sehara, A., and Suda, T. 1994. Bone morphogenetic protein–2 converts the differentiation pathway of C2C12 myoblasts into the osteoblast lineage. *J Cell Biol* 127(6):1755–66.

Kawasaki, K., Aihara, M., Honmo, J., Sakurai, S., Fujimaki, Y., Sakamoto, K., Fujimaki, E., Wozney, J.M., and Yamaguchi, A. 1998. Effects of recombinant human bone morphogenetic protein–2 on differentiation of cells isolated from human bone, muscle, and skin. *Bone* 23(2):223–31.

Khouri, R.K., Brown, D.M., Koudsi, B., Deune, E.G., Gilula, L.A., Cooley, B.C., and Reddi, A.H. 1996. Repair of calvarial defects with flap tissue: role of bone morphogenetic proteins and competent responding tissues. *Plast Reconstr Surg* 98(1):103–9.

Khouri, R.K., Koudsi, B., and Reddi, H. 1991. Tissue transformation into bone in vivo: a potential practical application. JAMA 266(14):1953–5.

Lau, H.T., Yu, M., Fontana, A., and Stockert, C.J. 1996. Prevention of islet allograft rejection with engineered myoblasts expressing fasL in mice. *Science* 273:109–11.

Lynch, C.M., Clowes, M.M., Osborne, W.R.A., Clowes, A.W., and Miller, A.D. 1992. Long term expression of human adenosine deaminase in vascular smooth muscle cells of rats: a model for gene therapy. *Proc Natl Acad Sci USA* 89:1138–42.

Ménétrey, J., Kasemkijwattana, C., Day, C.S., Bosch, P., Fu, F.H., Moreland, M.S., and Huard, J. 1999. Direct, fibroblast and myoblast mediated gene transfer to the anterior cruciate ligament. *Tiss Eng* 5:435–441.

Ménétrey, J., Kasemkijwattana, C., Day, C.S., Bosch, P., Vogt, M., Fu, F.H., Moreland, M.S., and Huard, J. 2000. Growth factors improve muscle healing in vivo. *JBJS Journal of Bone a Joint Surgery*, 82B:131–7.

Musgrave, D.S., Bosch, P., Ghivazzani, S.C., Robbins, P.D., Evans, C.H., and Huard, J. 1999. Adenovirus mediated direct gene therapy with BMP–2 produces bone. *Bone* 24(6):541–7.

Nita, I., Ghivizzani, S.C., Galea–Lauri, J., Bandara, G., Georgescu, H.I., Robbins, P.D., and Evans, C.H. 1996. Direct gene delivery to synovium: an evaluation of potential vectors in vitro and in vivo. *Arthritis Rheum* 39:820–8.

Obremsky, W.T., Seaber, A.V., Ribbeck, B.M., and Garrett, Jr., W.E. 1994. Biomechanical and histological assessment of a controlled muscle strain injury treated with piroxicam. *Am J Sports Med* 22(4):558–61.

Qu, Z., Balkir, L., van Devtekom, J.C.T., Robbins, P.D., Pouchnic, R., and Huard, J. 1998. Development of approaches to improve cell survival in myoblast transfer therapy. *J of Cell Biology* 142:1257–67.

Rando, T.A. and Blau, H.M. 1994. Primary mouse myoblast purification, characterization, and transplantation for cell mediated gene therapy. *J Cell Biol* 125(6):1275–87.

Salvatori, G., Ferrari, G., Messogiorno, A., Servidel, S., Colette, M., Tonalli, P., Giarassi, R., Cosso, G., and Mavillo, F. 1993. Retroviral vector mediated gene transfer into human primary myogenic cells leads to expression in muscle fibers in vivo. *Hum Gene Ther* 4:713–23.

Schultz, E. 1989. Satellite cell behavior during skeletal muscle growth and regeneration. *Med Sci Sports Exerc* 21:181–6.

Schultz, E., Jaryszak, D.L., and Valiere, C.R. 1985. Response of satellite cells to focal skeletal muscle injury. *Muscle Nerve* 8:217–22.

Simonsen, G.D., Groskreutz, D.J., Gorman, C.M., and MacDonald, M.J. 1996. Synthesis and processing of genetically modified human proinsulin by rat myoblast primary cultures. *Hum Gene Ther* 7:71–8.

Spindler, K.P., Mayes, C.E., Miller, R.R., Imro, A.K., and Davidson, J.M. 1995. Regional mitogenic response to the meniscus to platelet derived growth factor (PDGF–AB). *J Orthop Res* 13(2):201–7.

Taylor, D.C., Dalton, Jr., J.D., Seaber, A.V., and Garrett, Jr., W.E. 1993. Experimental muscle strain injury: early functional and structural deficits and the increased risk for reinjury. *Am J Sports Med* 21(2):190–4.

Urist, M. 1965. Bone: formation by autoinduction. *Science* 150:893–9.

Wang, E.A., Rosen, V., D'Assandro, J.S., Bauduy, M., Cordes, P., Harada, T., Isreal, D.I., Hewick, R.M., Kerns, K.M., LaPan, P. et al. 1990. Recombinant human bone morphogenetic protein induces bone formation. *Proc Natl Acad Sci USA* 87:2220–4.

Warejcka, D.J., Harvey, R., Taylor, B.J., Young, H.E., and Lucas, P.A. 1996. A population of cells isolated from rat heart capable of differentiating into several mesodermal phenotypes. *J Surg Res* 62:232–42.

Xiao, X., Li, J., and Samulski, R.J. 1996. Efficient long term gene transfer into muscle tissue of immunocompetent mice by adeno associated virus vector. *J Virol* 70:8098–108.

Yamaguchi, A., Katagiri, T., Ideda, T., Wozney, J.M., Rosen, V., Wang, E.A., Kahn, A.J., Suda, T., and Yoshiki, S. 1991. Recombinant human bone morphogenetic protein–2 stimulates osteoblastic maturation and inhibits myogenic differentiation in vitro. *J Cell Biol* 113(3):681–7.

Young, H.E., Mancini, M.L., Wright, R.P., Smith, J.C., Black, Jr., A.C., Reagan, C.R., and Lucas, P.A. 1995. Mesenchymal stem cells reside within the connective tissues of many organs. *Dev Dynam* 202:137–44.

12
Application of Tissue Engineering to Cartilage Repair

JUN–KYO SUH and FREDDIE H. FU

Introduction

Once damaged, articular cartilage very rarely undergoes spontaneous healing, primarily because of the avascular nature of the tissue (Hunter 1743; Campbell 1969; Mankin 1982). Although many repair techniques have been proposed over the past 4 decades, none has successfully regenerated long-lasting hyaline cartilage tissue to replace damaged cartilage. In fact, most of the surgical interventions to repair damaged cartilage have been directed toward the treatment of clinical symptoms such as pain relief and functional restoration of joint structures and articulating surfaces, rather than toward the regeneration of hyaline cartilage.

An initial surgical attempt to restore the normal articulating surface of joint cartilage began with the introduction of Pridie's resurfacing technique (Pridie 1959; Insall 1974). This chondral repair technique utilized the disruption of subchondral bone by inducing bleeding from the bone marrow to promote the regular wound healing mechanism in the cartilage defect site. Since Pridie's abrasion arthroplasty, several surgical techniques of subchondral distruption with different variations have subsequently been introduced in an attempt to improve the healing mechanism of repaired tissue. They include subchondral drilling (Mitchell and Shepard 1976), arthroscopic abrasion (Brown et al. 1974; Johnson 1989; Bert 1993), and microfracture technique (Rodrigo, Steadman, Silliman, and Fullstone 1994). Several experimental animal studies on full-thickness cartilage repair have revealed that subchondral breaching techniques created a fibrin clot formation due to bleeding in the region of the cartilage defect. This clot was subsequently infiltrated by mesenchymal cells and acted like a 3–dimensional scaffold for migrating progenitor cells (Shapiro, Koide, and Glimcher 1993). These cells gradually differentiated into and within 6 weeks completely filled the defect region with a hyaline–like repair cartilage containing significantly less proteoglycan molecules than normal hyaline cartilage (Mitchell and Shepard 1976). Repair tissue at this stage was more cellular than the adjacent normal cartilage. Furthermore, the repair tissue showed no structural integration with residual cartilage. As a result, there was evidence of degeneration, fissuring in the cartilage tissue, and, at a later stage, even a subsequent failure to maintain normal appearance (O'Driscoll, Keeley, and Salter 1988; Scherping, Jr., Suh, Marui, Woo, and Steadman 1995). It appears that, in the full-thickness defect, a spontaneous repair mechanism was present for the first several weeks after surgical repair, but later failed due to inadequate mechanical and biochemical conditions in the repaired tissue (Suh, Aroen, Muzzonigro, Disilvestro, and Fu 1997).

Recently, tissue engineering concepts have been introduced to develop cell-based repair approaches for articular cartilage (Freed et al. 1993; Vacanti, C.A., Woo, Schloo, Upton, and Vacanti, J.P. 1994). Tissue engineering of articular cartilage involves the isolation of articular chondrocytes or their precursor cells that may be expanded in vitro and then seeded into a biocompatible matrix, or scaffold, for cultivation and subsequent implantation into the joint. Certainly, the type of cell used to engineer cartilage is critical to the long-term outcome. Different cell populations that have been investigated in experimental studies include matured articular chondrocytes (Green 1974; Robinson, Halperin, and Nevo 1990b), epiphyseal chondrocytes (Bentley and Greer 1971; Itay, Abramovici, and Nevo 1987) mesenchymal stem cells (Wakitani, Kimura, and Hirroka 1989), bone marrow stromal cells (Butnaria–Ephrat Robinson, Mendes, Halperin, and Nevo 1996; Martin, Padera, Vunjak-Novakovic, and Freed 1997; Wakitani et al. 1994), and perichondrocytes (Chu et al. 1995).

The choice of biomaterial is also critical to the success of such tissue engineering approaches in cartilage repair. A variety of biomaterials, naturally occurring and synthetic, biodegradable and nonbiodegradable, have been introduced as potential cell-carrier substances for cartilage repair (Grande, Halberstadt, Naughton, Schwartz, and Ryhana 1997). However, the most ideal cell-carrier substance would be the one which closely mimics the naturally occurring environment in the articular cartilage matrix. It has been shown that cartilage-specific extracellular matrix (ECM) components such as type II collagen and glycosaminoglycan (GAG) play a critical role in regulating expression of the chondrocytic phenotype and in supporting chondrogenesis both in vitro and in vivo (Kosher, Lash, and Minor 1973; Kosher and Church 1975). Otherwise, chondrocytes may undergo dedifferentiation and produce an inferior fibrocartilaginous matrix rich in type I collagen (von Der Mark, K., Gauss, von der Mark, H., and Muller 1977), which would lead to a failure of metaplastic changes to form hyaline cartilage in cartilage repair (Sokoloff 1974). The present chapter describes recent advances in tissue engineering approaches for repair of damaged articular cartilage, with particular emphasis on the development of biocompatible scaffold materials which are naturally occurring and biodegradable. Criteria for the choice of such biomaterial include biological friendliness and biomechanical strength in order to provide the biochemically and biomechanically appropriate environment necessary for engineered cells to regenerate a long-lasting hyaline cartilage in the defect site.

Biocompatible Cell Carrier Substances: Literature Review

Collagen-Based Scaffold Matrices

Collagen is the most abundant protein component in the human body. Most collagenous proteins can be solubilized in a strong acidic environment and reconstructed into a porous sponge structure by freeze drying. Due to their natural constituency and biodegradability, such reconstructed porous collagen sponges (of either type I or type II collagen) have been extensively studied for numerous applications in tissue repair and tissue engineering.

In cartilage tissue engineering, various forms, such as sponges (Speer, Chvapil, Volz, and Holmes 1979; Frenkel, Toolan, Menche, Pitman, and Pachence 1997; Nehrer et al. 1997) or gels (Kang et al. 1997; Kimura, Yasui, Ohsawa, and Ono 1984; Qi and Scully 1997; Wakitani et al. 1989) of type I and type II collagens, were tested for chondrogenic potentials of seeded chondrocytes in vitro as well as in vivo. While chondrocytes seeded in type I collagen sponge assumed an elongated fibroblastic cell morphology, those seeded in type II collagen sponge demonstrated a spherical cell morphology and increased GAG synthesis, which are the hallmarks of the chondrocyte-specific phenotype (Nehrer et al. 1997). Furthermore, chondrocytes cultured in type II collagen–alginate gel showed increased rates of cytokine-regulated proliferation and proteoglycan synthesiss in vitro, as compared to those cultured in type I collagen–alginate gel (Qi and Scully 1997). Since type II collagen is the major constituent of natural articular cartilage, a type II collagen-based material may be the most ideal substance for chondrocyte cultivation (Nehrer et al. 1997). A recent in vitro study (Shakibaei, De Souza, and Merker 1997) revealed that chondrocytes cultured in monolayer on a type II collagen-coated culture plate also produced an increased amount of fibronectin and specific membrane receptors (integrins of the beta 1 group). Due to the small amount present in the animal body and its limited supply, however, the type II collagen-based substance may not be a practically feasible scaffold material when creating clinically useful dimensions and structures of articular cartilage. Instead, type III collagen may have a good potential for practical use in cartilage tissue engineering and repair. Type I collagen sponge was also found to be an effective means to deliver chondrocytes into cartilage defects (Sams and Nixon 1995).

Synthetic Polymer Scaffolds

It has been suggested that animal collagen-based scaffold may not be an optimal choice because it cannot be manufactured with a high degree of reproducibility (Freed et al. 1995). As an alternative, a synthetic, biodegradable polymer scaffold can easily be manufactured with stringent reproducibility. In addition, its mechanical properties and degradation rate can be accurately controlled during the manufacturing process. When seeded in the biodegradable polymer scaffold, chondrocytes are expected to synthesize their native matrix proteins, deposit these proteins within the polymer scaffold, and thus organize and remodel it as the synthetic polymer substance slowly degrades away.

Polyglycolic acid (PGA) and poly-L-lactic acid (PLLA) are biodegradable polyesters belonging to the poly α-hydroxy acids group, are approved by the Food and Drug Administration (FDA) for implantation in humans, and are currently utilized as biodegradable surgical suture material and biodegradable fracture fixation devices (Agrawal, Niederaver, Micallef, and Athanasiou 1995). Their superb mechanical properties and well-studied degradation behavior under physiological conditions have made them a promising choice for scaffold substance in tissue engineering (Athanasiou, Niederauer, and Agrawal 1996). In cartilage tissue engineering, PGA (Freed et al. 1993, 1994; Vacanti 1994; Grande et al. 1997), PLLA (Freed et al. 1993; Chu et al. 1995) and PGA–PLLA copolymer (Woo et al. 1994; Athanasiou et al. 1996; Sittinger et al. 1996; Grande et al. 1997) have been studied for their efficacies as chondrocyte-delivering scaffolds in vitro and in vivo. When chondrocytes isolated from bovine or human articular cartilage were cultured on PGA and PLLA porous scaffolds in vitro and in vivo, the engineered cartilage constructs were reported to closely resemble normal cartilage, histologically as well as with respect to cell density and matrix composition (i.e., type II collagen and proteoglycan) (Freed et al. 1993). Chondrocytes–PGA (Freed et al. 1994) and perichondrocyte–PLLA scaffolds (Chu et al. 1995) were also tested for repair of full-thickness cartilage defects in rabbits. Although these studies demonstrated promising results of such cell–polymer scaffolds for cartilage repair using animal models, there are several questions to be addressed before they can be practically used for clinical applications. First, it has been suggested that a pH decrease in the vicinity of degrading PGA and PLLA results in adverse effects, which may be responsible for biocompatibility concerns recently raised about these polymers (Agrawal and Athanasiou 1997). Further studies are needed to evaluate the possibility that such adverse effects may cause synovitis (Messner and Gilquist 1993). Second, one of the major difficulties in using cell–polymer scaffolds for cartilage repair is the stable fixation of the cell–polymer scaffold to the cartilage defect (Freed et al. 1994). A failure of the stable structural integrity of a scaffold and its surrounding cartilage would cause a severe mechanical defect in the graft, particularly at the scaffold–cartilage interface.

Investigators have also found that some nonbiodegradable polymer substances, such as polytetrafluoroethylene (Hanff, Sollerman, Abrahamsson, and Lundborg 1990), polyethylmethacrylate (Reisses, Downes, and Bentley 1994), and hydroxyapatite/Dacron composites (Messner and Gillquist 1993), would facilitate the restoration of an articular surface. However, further studies are needed to evaluate the relative merits of these biomaterials in comparison with others.

Development of New Hydrogel Substances

Polysacchride-Based Hydrogels

One of the major macromolecules in articular cartilage is proteoglycan, which consists of a core protein and extended polysaccharide (glycosaminoglycan, or GAG) chains. The cartilage specific GAGs include chondroitin 4-sulfate (N–acetyl–galactosamine with an SO_4^- on the 4-carbon position and glucuronic acid), chondroitin 6-sulfate (N-acetyl-galactosamine with an SO_4^- on the 6–carbon position and glucuronic acid) and keratan sulfate (N–acetyl–glucosamine with an SO_4^- on the 6–carbon position and galactose) (Rosenberg 1975). Another important GAG is the hyaluronic acid protein complex (glucuronic acid and glucosamine), one of the major macromolecules in synovial fluid. Hyaluronic acid molecules are also present in cartilage matrix in the form of a backbone structure in proteoglycan aggregate, to which the proteoglycan monomers are attached through a link protein.

Hyaluronic acid has been used as a therapeutic intervention in the treatment of osteoarthritis to improve the lubrication mechanism of articulating surfaces and thus reduce joint pain (Iwata 1993; Frizziero, Govoni, and Bacchini 1998). Several in vitro culture studies have demonstrated that hyaluronic acid has a beneficial effect, inhibiting chondrocytic chondrolysis mediated by fibronectin fragments (Homandberg, Hui, Wen, Kuettner and Williams 1997). Hyaluronic acid has also been shown to have antiinflammatory effects (Hakansson, Haligren, and Verge 1980; Pisko, Turner, Soderstrom, and Panetti 1983; Forrester and Balazs 1989), and to inhibit prostaglandin synthesis (Punzi et al. 1989; Akatsuka, Yamamoto, Tobetto, Yasui, and Ando 1993), and the release (Shimazu et al. 1993) and degradation (Morris, Wilcon, and Treadwell 1992) of proteoglycan. Based on these findings, it has been suggested that hyaluronic acid is a good candidate for cell-carrier substance in chondrocyte transplantation therapy (Robinson, Halperin, and Nevo 1990a). Bone marrow-derived mesenchymal cells, delivered in hyaluronic acid hydrogel, demonstrated a good potential for cartilage resurfacing in animal models (Butnaria–Ephrat et al. 1996;

Solchaga et al. 1998). However, further studies are needed in order to determine the efficacy of the hyaluronic acid-based hydrogel as a cell-carrier substance for cartilage repair as well as to evaluate the cell–material interaction and the mechanical characteristics of the cell–material construct.

GAG-Augmented Polysacchride Hydrogel

In light of the supportive influence of tissue specific matrix molecules on the chondrocytic phenotype in vitro as described above, a GAG and a GAG analog appear to be logical means to develop a novel biomaterial to provide chondrogenesis. One such candidate is chitosan, a partially deacetylated derivative of chitin, found in the exoskeletons of most arthropods. Structurally, chitosan is a polycationic linear polysaccharide, a repeating monosaccharide of β-1,4 linked glucosamine monomers with randomly located N-acetyl glucosamine units, and thus shares characteristics similar to GAG and hyaluronan (Chandy and Sharma 1990). Chemical modifications and processing methods have produced a variety of forms of chitosan (solutions, powders, flakes, gel matrices, microcapsules, semipermeable membranes, and bulk porous scaffolds) for various biomedical applications (Allen et al. 1984).

Through an ionical cross-linking mechanism between chitosan and chondroitin sulfate-A (CSA), one of the major GAGs found in cartilage, a hydrogel material can be formed to mimic the GAG-rich ECM of the articular chondrocyte. Recent preliminary studies have demonstrated the biocompatability and the chondrogenic characteristics of GAG-augmented chitosan hydrogel (Sechriest, Miao, et al. 1998; Sechriest, Shuler, et al. 2000; Suh et al. 1998).

Experimental Procedures

Preparation of Chitosan-Based Hydrogel with GAG Augmentation

CSA-augmented chitosan hydrogel–membrane surfaces were prepared for cell-culture experiments. First, a 1% chitosan solution was prepared by dissolving chitosan (CarboMer Inc., Westborough, MA) into a 2% acetic acid solution (pH = 2.0). Standard 24-well polystyrene tissue culture plates were then coated with $500\,\mu L$ of the chitosan solution, and heat-dried for 6 hours. To form CSA-augmented chitosan membranes, a 10% CSA solution was applied to the chitosan residue and allowed 18 hours for cross-linking. The membranes were rinsed 5 times with phosphate-buffered saline (PBS) prior to cell seeding.

Cell Growth Kinetics, Morphology, and Biosynthetic Activities

Articular chondrocytes, isolated from fresh bovine metacarpophalangeal joints, were seeded in equal density (6×10^4 cells/mL) onto the CSA-chitosan surfaces as well as onto the standard polystyrene culture plates, and maintained in standard culture medium (low-glucose DMEM) with 10% fetal bovine serum in an incubator at 37°C with 5% CO_2. Degree of cell proliferation and cell morphology were then assessed after 1 week using hemocytometer and scanning electron microscope, respectively. After 1 week of preculture in FBS supplemented low-glucose DMEM, the culture systems were incubated for an additional 48 hours in fresh media either in the presence of ^{35}S-sulfate for the assay of proteoglycan synthesis or in the presence of ^{3}H-proline for sodium-dodecyl-sulfate-polyacrylamide-gel-electrophoresis (SDS-PAGE) for collagen typing.

Experimental Results

Cell Growth Kinetics

Chondrocytes seeded at low density (6×10^4 cells/mL) onto either the standard polystyrene culture plates or the CSA-chitosan membrane surfaces attached to the substrate within 2 hours. Over the course of 1 week, the majority of cells attached to the polystyrene surface assumed a fibroblast-like morphology, and the cell population rapidly expanded (Figure 12.1). In contrast, primary chondrocytes seeded at the same low density onto the CSA–chitosan surface assumed a spherical morphology after attachment. After 1 week, chondrocytes attached to the CSA–chitosan had largely maintained a rounded or polygonal morphology and undergone only a modest degree of mitosis (Figure 12.2). The increase in cell population for chondrocytes cultured on polystyrene and CSA–chitosan surfaces were 313 ± 61% and 66 ± 28 %, respectively (Figure 12.3).

Scanning Electron Microscopic Morphology

Scanning electron microscopic evaluation of the CSA–chitosan membrane revealed a matrix material with an irregular surface (Figure 12.4). We found no evidence that chondrocytes were embedded or partially embedded in the hydrogel membrane. Rather, chondrocytes adherent to CSA–chitosan had formed discrete points of attachment with the biomaterial through pseudopodial extensions (Figure 12.5). These microscopic findings suggest that maintenance of the phenotypic morphology is due to an interaction between cell-membrane receptors and the CSA-rich ECM. A similar phenomenon was previously described with primary murine articular chondrocytes cultured on a collagen type II membrane (Shakibaei et al. 1997).

Proteoglycan Synthesis

The total ^{35}S-sulfate-incorporation rate produced by primary chondrocytes cultured on the CSA–chitosan surface (2580 ± 460 cpm/10^6 cells) was approximately equal to that produced by chondrocytes cultured on the

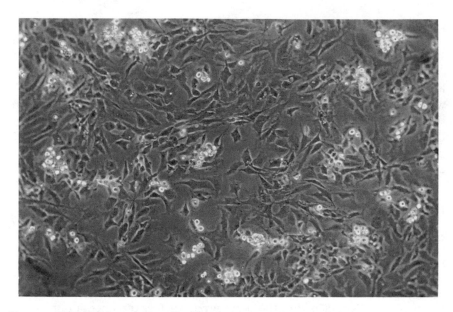

FIGURE 12.1. (Light micrograph, 100×) At 1 week, most primary chondrocytes cultured on the polystyrene surface had assumed a fibroblast-like morphology and undergone a significant degree of proliferation in monolayer. Adapted from Sechriest, V.F. et al., *J Biomed Mat Res* 49:534–541, 2000.

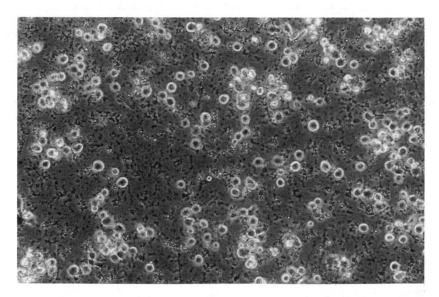

FIGURE 12.2. (Light micrograph, 200×) After 1 week, primary chondrocytes cultured on the CSA–chitosan surface maintained a spherical morphology and demonstrated a limited rate of mitosis. Adapted from Sechriest, V.F. et al., *J Biomed Mat Res* 49:534–541, 2000.

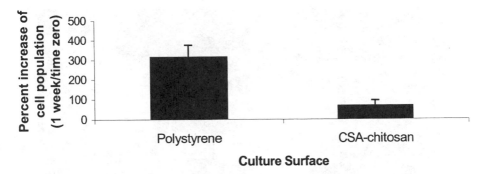

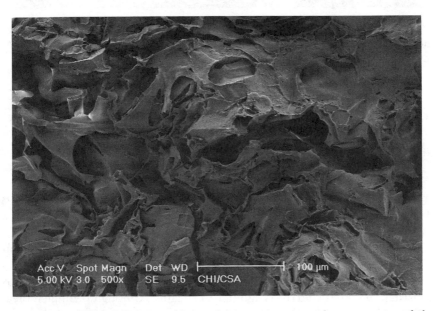

FIGURE 12.3. Proliferation of primary chondrocytes. Primary chondrocytes cultured on the polystyrene surface demonstrated a 313 ± 61% increase in cell number. Chondrocytes cultured on CSA–chitosan were more stable, with a 66 ± 28% increase in cell number after 7 days in monolayer. The data represent the mean ± SD, n = 6. Adapted from Sechriest, V.F. et al., *J Biomed Mat Res* 49:534–541, 2000.

FIGURE 12.4. (SEM, 500×) Irregular and heterogeneous surface geometry of the CSA–chitosan membrane.

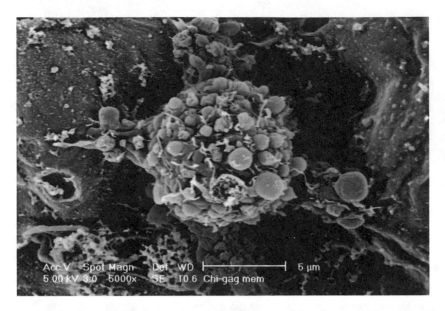

FIGURE 12.5. (SEM, 5000×) An isolated chondrocyte with a discrete point of attachment (arrow) to the CSA–chitosan surface. Note that a rounded morphology is maintained.

standard polystyrene surface (2810 ± 320 cpm/10^6 cells). Thus, CSA–chitosan supports, but does not appear to enhance the total amount of proteoalycan synthesized by primary articular chondrocytes (Figure 12.6).

Collagen Synthesis

After 1 week, chondrocytes cultured on the control polystyrene surface (P) demonstrated a complex collagen phenotype. The protein migration pattern shown in Figure 12.7 indicates that the collagens produced by cells in this culture system were of types I and II, type I collagen being the predominant collagen synthesized. The presence of type I collagen is indicated by the presence of α2(I) chains. Also present in smaller amounts were protein bands consistent with α1(V)/α1(XI) and α2(XI) chains. Although the high molecular weight protein bands were not identified, they may represent chains of type III collagen, type IX collagen, and/or β-chains of type I collagen.

For chondrocytes cultured on the CSA–chitosan, the migration pattern shown in Figure 12.7 signifies that the predominant collagen produced by chondrocytes was type II, indicated by the presence of the strongly stained protein band in the α1(II) position with only faint protein bands in the α2(I) position. Chondrocytes in this culture system also demonstrated protein bands consistent with α1(V)/α1(XI) and α2(XI) chains. The high molecular weight protein bands in the CSA–chitosan group were not identified, but may represent pepsin-derived fragments of type IX collagen. These data

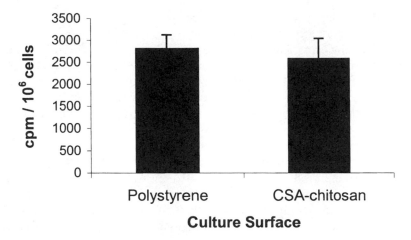

FIGURE 12.6. ^{35}S-Sulfate incorporation. CSA–chitosan supports PG synthesis by primary bovine chondrocytes in vitro. The data represent the mean ± SD, n = 10. Adapted from Sechriest, V.F. et al., *J Biomed Mat Res* 49:534–541, 2000.

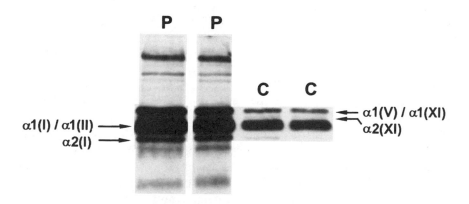

FIGURE 12.7. SDS–PAGE analysis for collagen typing (duplicate sample lanes). SDS–PAGE shows that primary bovine chondrocytes cultured on polystyrene (P)-produced collagens type I and II as indicated by the presence of protein bands in the α1(I) and α2(I) positions. Adapted from Sechriest, V.F. et al., *J Biomed Mat Res* 49:534–541, 2000.

indicate that the primary chondrocytes cultured on CSA–chitosan maintain the synthesis of cartilage-specific collagens.

Future Directions and Discussion

Despite the extensive research on articular cartilage conducted over the past 4 decades, we still do not have a definitive answer for repairing damaged cartilage. Although many new repair techniques have recently emerged and demonstrated promising results in experimental models, few have exhibited long-term clinical efficacy.

Tissue engineering has recently emerged as a new interdisciplinary science to repair injured body parts and restore their functions by using laboratory-grown tissues, molecules or artificial implants. It is our hope that tissue engineering with novel applications will provide a method for the study of articular cartilage and associated pathologies. This understanding of the pathogenesis of articular cartilage will help us develop new methods to enhance the repair of damaged cartilage.

The biochemical and morphologic results of this study of primary chondrocytes cultured on CSA–chitosan reemphasize the favorable influence of cartilage-specific matrix components on the chondrocytic phenotype in vitro. Our findings, coupled with the favorable material properties of chitosan, suggest that CSA–chitosan may be well suited as a carrier material for the transplant of autologous chondrocytes as described by Brittberg et al. (1994) and/or as a scaffold for the tissue engineering of cartilage–like tissue as originally described by Vacanti, C.A., Langer, Schloo, and Vacanti, J.P. (1991). Future studies using chondrocytes that have undergone proliferative expansion in vitro will explore this potential. Further development of a chitosan-based material that can support chondrogenesis may be significant not only in terms of the quality of neocartilage produced, but also in terms of the ability of that tissue to integrate with the host matrix.

Along with tissue engineering as a newly emerging science for various tissue repairs, gene therapy has been suggested as a means of delivering sustained therapeutic levels of antiarthritic gene products to the transplanted cells in diseased joints (Forrester and Balazs 1989; Evans and Robbins 1994, 1997). Genes have been successfully introduced locally to the target cells in the knee joints of rabbits by both ex vivo and in vivo methods, with gene expression lasting for several weeks (Bandara et al. 1993; Roessler, Allen, Wilson, Hartman, and Davidson 1993). Such gene therapy has been successfully combined with chondrocyte transplantation to investigate the feasibility of ex vivo gene transfer to chondrocytes in full thickness cartilage defects (Kang et al. 1997). A combination of tissue engineering for cartilage scaffolds, gene therapy, use of growth factors, and improved postoperative protocols, will hopefully improve the long-term outcome of cartilage repair in the clinical setting.

References

Agrawal, C.M. and Athanasiou, K.A. 1997. Technique to control pH in vicinity of biodegrading PLA PGA implants. *J Biomed Mat Res* 38(2):105–14.

Agrawal, C., Niederauer, G., Micallef, D., and Athanasiou, K. 1995. The use of PLA-PGA polymers in orthopaedics. In D.L. Wise, D.J. Trantolo, D.E. Altobelli, M.J.

Yaszemski, J.D. Gresser, and E.R. Schwart, eds. *Encyclopedic Handbook of Biomaterials and Bioengineering.* 2081–115. New York: Marcel Dekker.

Akatsuka, M., Yamamoto, Y., Tobetto, K., Yasui, T., and Ando, T. 1993. In virtro effects of hyaluronan on prostaglandin E2 induction by interleukin–1 in rabbit articular chondrocytes. *Agent Actions* 38:122–5.

Allen, G., Altman, L., Bensinger, R., Ghosh, D., Hirabayashi, Y., and Neogi, A. 1984. *Biomedical applications of chitin and chitosan.* In J.P. Zikakis (ed.), *Chitin, Chitosan, and Related Enzymes.* 19–134. New York: Academic Press.

Athanasiou, K.A., Niederauer, G.G., and Agrawal, C.M. 1996. Sterilization, toxicity, biocompatibility, and clinical applications of polylactic acid/polyglycolic acid copolymers. *Biomaterial* 17:93–102.

Bandara, G., Mueller, G.M., Galea–Lauri, J., Tindal, M.H., Georgescu, H.I., Suchanek, M.K., Hung, G.L., Glorioso, J.C., Robbins, P.D., and Evans, C.H. 1993. Intraarticular expression of biologically active interleukin–1–receptor–antagonist protein by ex vivo gene transfer. *Proc Nat Acad Sci* USA 90:10764–8.

Bently, G. and Greer, R.B. 1971. Homotransplantation of isolated epiphyseal and articular cartilage chondrocytes into joint surfaces of rabbits. *Nature* 230:385–8.

Bert, J.M. 1993. Role of abrasion arthroplasty and debridement in the management of arthritis of the knee. *Rheum Dis Clin N Am* (19):725–39.

Brittberg, M., Lindahl, A., Nilsson, A., Ohlsson, C., Isaksson, O., and Peterson, L. 1994. Treatment of deep cartilage defects in the knee with autologous chondrocyte implantation. *N Engl J Med* 331(14):889–95.

Brown, R., Blazina, M.E., Kerlan, R.K., Carter, V.S., Jobe, F.W., and Carlson, G.J. 1974. Osteochondritis of the capitellum. *J Sports Med* 2:27–46.

Butnaria–Ephrat, M., Robinson, D., Mendes, D., Halperin, N., and Nevo, Z. 1996. Resurfacing of goat articular cartilage by chondrocytes derived from bone marrow. *Clin Orthop Rel Res* 330:234–43.

Campbell, C. 1969. The healing of cartilage defects. *Clin Orthop* 64:45–63.

Chandy T. and Sharma C. 1990. Chitosan as a biomaterial. *Biomat Art Cell Art Org* 18(1):1–24.

Chu, C.R., Coutts, R.D., Yoshioka, M., Harwood, F.L., Monosov, A.Z., and Amiel, D. 1995. Articular cartilage repair using allogenic perichondrocyte seeded biodegradable porous polylactic acid (PLA): a tissue engineering study. *J Biomed Mat Res* 29:1147–54.

Evans, C.H. and Robbins, P.D. 1994. Prospects for treating arthritis by gene therapy. *J Rheumatol* 21:779–82.

Evans, C.H. and Robbins, P.D. 1997. Possible orthopaedic applications of gene therapy. *J Bone Joint Surg* 77–A(7):1103–14.

Forrester, J. and Balazs, E.A. 1989. Inhibition of phagocytosis by high molecular weight hyaluronate. *Immunology* 40:435–46.

Freed, L.E., Grande, D.A., Lingbin, Z., Emmanuel, J., Marquis, J.C., and Langer, R. 1994. Joint resurfacing using allograft chondrocytes and synthetic biodegradable polymer scaffolds. *J Biomed Mat Res* 28:891–9.

Freed, L.E., Marquis, J.C., Nohria, A., Emmanual, J., Mikos, A.G., and Langer, R. 1993. Neocartilage formation in vitro and in vivo using cells cultured on synthetic biodegradable polymers. *J Biomed Mat Res* 27:11–23.

Freed, L.E., Vunjak–Novakovic, G., Biron, R.J., Eagles, D.B., Lesnoy, D.C., Barlow, S.K., and Langer, R. 1995. Biodegradable polymer scaffolds for tissue engineering. *Biotechnology* 12:689–93.

Frenkel, S.R., Toolan, B., Menche, D., Pitman, M., and Pachence, J.M. 1997. Chondrocyte transplantation using a collagen bilayer matrix for cartilage repair. *J Bone Joint Surg* 79-B:831–6.

Frizziero, L., Govoni, E., and Bacchini, P. 1998. Intraarticular hyaluronic acid in the treatment of osteoarthritis of the knee: clinical and morphological study. *Clin Exp Rheum* 16:441–9.

Grande, D.A., Halberstadt, C., Naughton, G., Schwartz, R., and Ryhana, M. 1997. Evaluation of matrix scaffolds for the tissue engineering of articular cartilage grafts. *J Biomed Mat Res* 34:211–20.

Green, W.T. 1977. Articular cartilage repair: behavior of rabbit chondrocytes during tissue culture and subsequent allografting. *Clin Orthop Rel Res* 124:237–50.

Hakansson, L., Haligren, R., and Verge, P. 1980. Regulation of granulocyte function by hyaluronic acid. In vitro and in vivo effect on phagocytosis, locomotion and metabolism. *J Clin Invest* 66:298–305.

Hanff, G., Sollerman, C., Abrahamsson, S.O., and Lundborg, G. 1990. Repair of osteochondral defects in the rabbit knee with Gore–Tex (expanded polytetrafluoreethylene). An experimental study. *Scand J Plast and Reconstr Surg* 24:217–23.

Homandberg, G.A., Hui, F., Wen, C., Kuettner, K.E., and Williams, J.M. 1997. Hyaluronic acid suppresses fibronectin fragment mediated cartilage chondrolysis: I. in vitro. *Osteoarthritis Cartilage* 5:309–19.

Hunter, W. Of the structure and diseases of articulating cartilages. 1743. *Phil Trans* 470:514–21.

Insall, J. 1974. The Pridie debridement operation for osteoarthritis of the knee. *Clin Orthop Rel Res* 101:61–7.

Itay, S., Abramovici, A., and Nevo, Z. 1987. Use of cultured embryonal chick epiphyseal chondrocytes as grafts for defects in chick articular cartilage. *Clin Orthop* 220:284–301.

Iwata, S. 1993. Pharmacological and clinical aspects of intraarticular injection of hyaluronate. *Clin Orthop Rel Res* 289:285–91.

Johnson, L.L. 1989. Arthroscopic abrasion arthroplasty historical and pathological perspective: present status. *Arthroscopy* 2:54–69.

Kang, R., Marui, T., Nita, I., Georgescu, H.I., Suh, J.–K., Robbins, R.D., and Evans, C.H. 1997. Ex vivo gene transfer to chondrocytes in full–thickness articular cartilage defects—a feasibility study. *Osteoarthritis Cartilage* 5:139–43.

Kimura, T., Yasui, N., Ohsawa, S., and Ono, K. 1984. Chondrocytes embedded in collagen gel maintain cartilage phenotype during long term cultures. *Clin Orthop Rel Res* 186:231–9.

Kosher, R.A. and Church, R.L. 1975. Stimulation of in vitro somite chondrogenesis by procollagen and collagen. *Nature* 258:327–30.

Kosher, R.A., Lash, J.W., and Minor, R.R. 1973. Environmental enhancement of in vitro chondrogenesis. Stimulation of somite chondrogenesis by exogenous chondromucoprotein. *Dev Biol* 35(2):210–20.

Mankin, H.J. 1982. The response of articular cartilage to mechanical injury. *J Bone Joint Surg* 64-A:460–6.

Martin, I., Padera, R.F., Vunjak-Novakovic, G., and Freed, L.E. 1997. In vitro differentiation of chick embryo bone marrow stromal cells into cartilaginous and bone-like tissues. *J Orthop Res* 16:181–9.

Messner, K. and Gillquist, J. 1993. Synthetic implants for the repair of osteochondral defects of the medial femoral condyle: a biomechanical and histological evaluation in the rabbit knee. *Biomaterials* 14:513–21.

Mitchell, N. and Shepard, N. 1976. The resurfacing of adult articular cartilage by multiple perforations through the subchondral bone. *J Bone Joint Surg* 58A(2): 230–3.

Morris, E.A., Wilcon, S., and Treadwell, B.V. 1992. Inhibition of interleukin–1 mediated proteoglycan degradation in bovine articular cartilage explants by addition of sodium hyaluronate. *Am J Vet Res* 53:1977–82.

Nehrer, S., Breinan, H.A., Ramappa, A., Shortkroff, S., Young, G., Minas, T., Sledge, C.B., Yannas, I.V., and Spector, M. 1997. Canine chondrocytes seeded in type I and type II collagen implants investigated in vitro. *J Biomed Mat Res* 38:95–104.

O'Driscoll, S.W., Keeley, F.W., and Salter, R.B. 1988. Durability of regenerated articular cartilage produced by free autologous periosteal grafts in major full thickness defects in joint surface under the influence of continuous passive motion. A follow–up report at one year. *J Bone Joint Surgery* 70–A(4):595–606.

Pisko, E., Turner, R.A., Soderstrom, L.P., and Panetti, M. 1983. Inhibition of neutrophil phagocytosis and enzyme release by hyaluronic acid. *Clin Exp Rheum* 1:41–4.

Pridie, K.H. 1959. A method of resurfacing osteoarthritic knee joints. In Proceedings of the *British Orthopaedic Association. (J Bone Joint Surgery)* 41B:618–9.

Punzi, L., Schiavon, F., Cavasin, F., Ramonda, R., Gambari, P.F., and Todesco, S. 1989. The influence of intraarticular hyaluronic acid on PGE2 and cAMP of synovial fluid. *Clin Exp Rheum* 7:247–50.

Qi, W.N. and Scully, S.P. 1997. Extracellular collagen modulates the regulation of chondrocytes by transforming growth factor–beta 1. *J Orthop Rel Res* 15(4):483–90.

Reisses, N., Downes, S., and Bentley, G. 1994. Cartilage repair using a new synthetic polymer. in *Trans 40th Orthop Res Soc* New Orleans, LA, p. 482.

Robinson, D., Halperin, N., and Nevo, Z. 1990a. Regenerating hyaline cartilage in articular defects in old chickens using implants of embryonal chick chondrocytes embedded in a new natural delvery substance. *Calcif Tissue Int* 46:246–53.

Robinson, D., Halperin, N., and Nevo, Z. 1990b. Use of cultured chondrocytes as implants for repairing cartilage defects. In A. Maroudos and K. Kuettner (eds.), *Methods In Cartilage Research* 327–30. London: Academic Press.

Rodrigo, J.J., Steadman, R.J., Silliman, J.F., and Fullstone, H.A. 1994. Improvement of full thickness chondral defect healing in the human knee after debridement and microfracture using continous passive motion. *Am J Knee Surg* 7(3):109–16.

Roessler, B.J., Allen, E.D., Wilson, J.M., Hartman, J.W., and Davidson, B.L. 1993. Adenoviral mediated gene transfer to rabbit synovium in vivo. *J Clin Invest* 92:1085–92.

Rosenberg, L.C. 1975. Structure of cartilage proteoglycans. In P.M.C. Burgleigh and A.R. Poole (eds.), *Dynamics of Connective Tissue Macromolecules*. 105–28. Amsterdam: North Holland.

Sams, A.E. and Nixon, A.J. 1995. Chondrocyte laden collagen scaffolds for resurfacing extensive articular cartilage defects. *Osteoarthritis Cartilage* 3:47–59.

Scherping, Jr., S., Suh, J.-K., Marui, T., Woo, S.L.-Y., and Steadman, R. 1995. Effect of subchondral plate preservation on repair of large cartilage defects. In *American Orthopaedic Society for Sports Medicine*. Toronto, ONT, Canada.

Sechriest, V.F., Miao, Y.J., Westerhausen–Larson, A., Niyibizi, C., Fu, F.H., and Suh, J.–K. 1998. GAG aumented polysachharide hydrogel for chondrocyte culture. In *Annual Meeting, Biomed Eng Society*, Cloveland, OH.

Sechriest, V.F., Shuler, F.D., Miao, Y.J., Niyibizi, C., Evans, C.H., Westerhausen–Larson, A., Matthew, H.W., Fu, F.H., and Suh, J.–K. 2000. GAG aumented poly-sachharide membrane stabilizes differentiated phenotype of primary chondro-cytes in monolayer. *J Biomed Mat Res* 49:534–41.

Shakibaei, M., De Souza, P., and Merker, H.J. 1997. Integrin expression and colla-gen type II implicated in maintenance of chondrocyte shape in monolayer culture: an immunomorphological study. *Cell Biol Int* 21(2):115–25.

Shapiro, F., Koide, S., and Glimcher, M.J. 1993. Cell origin and differentiation in the repair of full thickness defects of articular cartilage. *J Bone Joint Surg* 75–A(4):532–53.

Shimazu, A., Jikko, A., Iwamoto, M., Koike, T., Yan, W., Okada, Y., Shinmei, M., Nakamura, S., and Kato, Y. 1993. Effects of hyaluronic acid on the release of pro-teoglycans from the cell matrix in rabbit chondrocyte culture in the presence and absence of cytokines. *Arthritis Rheum* 36:247–53.

Sittinger, M., Reitzel, D., Dauner, M., Hierlemann, H., Hammer, C., Kastenbauer, E., Planck, H., Burmester, G.R., and Bujia, J. 1996. Resorbable polyesters in car-tilage engineering: affinity and biocompatibility of polymer fiber structures to chondrocytes. *J Biomed Mat Res* 33:57–63.

Sokoloff, L. 1974. Cell biology and the repair of articular cartilage. *J Rheumatol* 1:1–10.

Solchaga, L.A., Arm, D., Johnstone, B., Yoo, J.U., Awadallah, A., Goldberg, V.M., and Caplan, A.I. 1998. HYAFF-11: A new cell delivery vehicle for tissue engineering. in *Trans 44th Orthop Res Soc* New Orleans, p. 800.

Speer, D.P., Chvapil, M., Volz, R.G., and Holmes, M.D. 1979. Enhancement of healing in osteochondral defects by collagen sponge implants. *Clin Orthop Rel Res* 144:326–35.

Suh, J.-K., Aroen, A., Muzzonigro, T.S., DiSilvestro, M., and Fu, F.H. 1997. Injury and repair of articular cartilage: related scientific issues. *Operative Techniques in Orthopaedics*, 7:270–8.

Suh, J.-K., Sechriest, V.F., Miao, Y.J., Shuler, F.D., Manson, T.T., Niyibizi, C., Evans, C.H., Matthew, H.W., and Fu, F.H. 1998. Glycosaminoglycan aumented polysach-haride hydrogel: A potential carrier for chondrocyte transplantation. in *2nd Bi–Annual Meeting of Tissue Engineering Society*, Orlando, FL, 1–39.

Vacanti, C.A., Langer, R., Schloo, B., and Vacanti, J.P. 1991. Synthetic polymers seeded with chondrocytes provide a template for new cartilage formation. *Plast Reconstr Surg* 88(5):753–9.

Vacanti, C.A., Woo, S.K., Schloo, B., Upton, J., and Vacanti, J.P. 1994. Joint resurfac-ing with cartilage grown in situ from cell polymer structures. *Am J Sports Med* 22(4):485–8.

von Der Mark, K., Gauss, V., von Der Mark, H., and Muller, P. 1977. Relationship between cell shape and type of collagen synthesized as chondrocytes lose their cartilage phenotype in culture. *Nature* 267:531–2.

Wakitani, S., Kimura, T., and Hirroka, A. 1989. Repair of rabbit articular surfaces with allograft chondrocytes embedded in collagen gel. *J Bone Joint Surg* 71-B:74–80.

Wakitani, S., Tatsuhiko, G., Pineda, S., Young, G.Y., Mansour, J.M., Caplan, A.J., and Goldberg, V. 1994. Mesenchymal cell–based repair of large, full thickness defects of articular cartilage. *J Bone Joint Surg* 76–A(4):579–92.

Woo, S.K., Vacanti, J.P., Cima, L., Mooney, D., Upton, J., Puelacher, W.C., and Vacanti, C.A. 1994. Cartilage engineered in predetermined shapes employing cell transplantation on synthetic biodegradable polymers. *Plast Reconstr Surg* 94(2):233–7.

Part IV
Hurdles and Development of New Vectors for Gene Therapy and Tissue Engineering

13
Immune Reaction Following Cell and Gene Therapy

JACQUES P. TREMBLAY

Introduction

The progress of medicine in the past relied on the development of pharmacological agents, most of which were empirically extracted from plants. A second important phase of medical progress was the discovery of vaccination, which not only led to an understanding of the complexity of the immune response, but also to a systematic method of treating a variety of transmissible diseases. The third important step in the development of medicine was the transplantation of organs where, again, the main progress came from the understanding and control of the immune response. The usefulness of organ transplantation is, however, currently limited by the availability of organs. A solution to this problem may reside in the transplantation of cells proliferated in culture, a possibility that has triggered a burst of research activity. Cell transplantation, that is the use of myoblasts, could potentially be used to treat several dystrophies (Law, Goodwin, and Wang 1988; Morgan, Watt, Sloper, and Partridge 1988; Partridge, Morgan, Coulton, Hoffman, and Kunkel 1989). Gene therapy treatment of hereditary diseases as well as many acquired diseases and conditions will be another important development in the medicine of the next century. Both of the new solutions for the future of medicine are, however, faced with a very old problem: our limited understanding and even more limited control of the immune response. Both cell transplantation and gene therapy can also be used to treat orthopaedic problems, as recently suggested by Dr. Johnny Huard (Day et al. 1997). Indeed, as indicated in several chapters of this book, these techniques can be used to improve the healing of several tissues of the musculoskeletal system. The basic problems facing cell transplantation and gene therapy are the same for treatment of dystrophies as for treatments of orthopaedic problems.

Immune Reaction Following Cell Transplantation

As with organ transplantation, cell transplantation triggers immune responses against xenotransplantation, the major histocompatibility complex (MHC), and minor antigens. Moreover, cell transplantation can also induce immune responses against components of the culture medium. Although the practical experience of the author is in the field of myoblast transplantation, problems observed in that procedure will also occur in other types of cell transplantation.

Cell Xenotransplantation

Few experiments of cell xenotransplantation have been published. Although it was initially reported that xenotransplantation of myoblasts, in mice human, was possible due to the low immunogenicity of the cells which were thought not to express MHCs (Karpati et al. 1989), this report was denied a few years later by my research group (Guérette, Asselin, Vilquin, Roy, and Tremblay 1994). We indeed observed a strong cellular immune response following the transplantation of human myoblasts in mice. This result demonstrated that, even with cell transplantation, the species barrier could not be easily overcome. Xenotransplantation of myoblasts is, however, possible with great success in severe combined immunodeficient (SCID) mice, confirming that the lack of previous success was actually due to the immune response (Huard, Verreault, Roy, Tremblay, M., and Tremblay, J.P. 1994).

Against Major Histocompatibility Complex (MHC)

Our experiments in mice have also clearly demonstrated that MHC incompatible myoblasts are rapidly rejected. Without immunosuppression, the transplantation of myoblasts obtained from a MHC-incompatible mouse leads to rapid rejection due to a strong cellular immune response. The presence of a potential immune response against the MHCs was underestimated in initial clinical trials of myoblast transplantation. Most research groups selected the father (or an unrelated person) as the donor despite an incompatibility for MHC antigens in most cases (Tremblay and Guerette 1997). The immune reaction against the MHCs can, however, be controlled with adequate immuno suppression.

Against Minor Antigens

Syngeneic myoblast transplantation can be performed with excellent success without any immunosuppression. Indeed, our group transplanted myoblasts obtained from C57BL10J +/+ into dystrophic mice, C57BL10J mdx/mdx (Vilquin, Wagner, Kinoshita, Roy, and Tremblay 1995). Since the host muscles were lacking dystrophin, success was assessed by the presence of a high percentage of dystrophin positive fibers up to 9 months post-transplantation. This result indicated that the transplantation of perfectly compatible myoblasts (i.e., compatible for MHCs and minor antigens) was possible without any immunosuppressive treatment.

Our group has also transplanted normal myoblasts obtained from different mouse strains that were MHC compatible with the host (C57BL10J mdx/mdx) without any immunosuppressive treatment. For some donor strains, the presence of an increased number of dystrophin positive fibers was observed 1 month after transplantation, but in all cases such increases were not detected a few months after the transplantation (Boulanger, Asselin, Roy, and Tremblay 1997). This observation was attributed to immune responses against minor antigens. Such responses may have caused the rejection of MHC compatible myoblasts in the clinical trial conducted by our group (Huard et al. 1992; Tremblay et al. 1993), since the transplantations were done without immunosuppression.

Against Components of the Culture Medium

The transplantation of cells has an important particularity relative to organ transplantation: the cells have to be proliferated for a few days or even a few weeks before their transplantation. Culture medium usually contains animal serum. So far, it has been difficult, if not impossible, to completely eliminate the serum requirement, most likely because serum contains many growth factors, some of which remain unidentified. Before transplantation, the cells are detached from the culture vials and washed by three suspensions and centrifugations with culture medium but without serum. Despite this precaution, most of the mice transplanted with syngeneic mouse myoblasts develop antibodies against calf serum (Tremblay, unpublished results). The presence of antibodies against serum does not prevent the success of the transplantation or induce a rejection of the muscle fibers formed by donor myoblasts (Vilquin, Wagner, Kinoshita, Roy, and Tremblay 1995a). The presence of such antibodies may, however, reduce the success of a second syngeneic transplantation or of autotransplantation done without immunosuppression. Thus, it may be necessary to transiently immunosuppress the host even for an autologous transplantation.

Against the Normal Gene Product Introduced by Transplantation

Cell transplantation will in some cases be used to correct a genetic defect by introducing the normal gene coding for a protein missing in the patient through the implantation of normal cell transplantation. This is the case for myoblast transplantation in Duchenne muscular dystrophy (DMD) patients. Indeed, one of the aims of such transplantation is to restore dystrophin, a protein located under the membrane of the normal muscle fiber but absent in these patients. Our research group was the first to detect an immune response against dystrophin following the transplantation of normal myoblasts in DMD patients (Huard et al. 1992; Tremblay et al. 1993). The presence of antibodies against dystrophin was later confirmed following syngeneic myoblast transplantation in mdx mice, an animal model of DMD that also lacks dystrophin (Vilquin, Wagner, et al. 1995). The presence of antibody against dystrophin did not lead to a rejection of the muscle fibers containing this protein even when the mdx mice were maintained alive for 8 months following the detection of such antibodies. Because dystrophin is located under the sarcolemma, it is inaccessible to antibodies. However, when the new protein introduced by the cellular transplantation is either located on the external face of the cell membrane or secreted, the presence of antibody against the new antigen may lead to the rejection of the transplanted cells or to an elimination of the secreted protein from the extracellular milieu.

Immune Reaction Following Gene Therapy

Gene therapy is certainly the most exciting new therapeutic avenue for the next millennium. The introduction of a normal gene or the correction of a mutated gene has extremely large potential for therapy in both hereditary diseases and acquired diseases, including orthopedic disorders. Many different viruses have been used for gene therapy: retrovirus, adenovirus, herpes virus, lentivirus, and adeno–associated virus (AAV). The potential immune response following gene therapy, however, was initially underestimated. Since the early clinical trials, the presence of specific immune responses and inflammatory responses have often been demonstrated. Immune responses have been detected against the viral coat, proteins produced by the vector genome itself, the therapeutic gene product, the product of the selection gene, and the marker gene.

Against the Viral Coat

Every viral vector contains DNA or RNA and is surrounded by a coat or a capsid made of proteins that represent a new antigenic source for the patient. Immune reaction against these proteins has been detected following the injection of viral vectors (Gahery–Segard, Juillard, et al. 1997; Gahery–Segard, Farace, et al. 1998). Although the development of an immune response against the components of the viral coat is now well established, it does not seem to affect the success of the first therapeutic intervention if the host is not already preimmune to the vector. However, this type of immune response will reduce the effectiveness of a second administration of the same vector. In the case of the adenoviral vector, a potential solution to this problem is to change the serotype of the vector for the following administration. However, the recent death of a patient following administration of a high dose of an adenoviral vector is probably due to an immune reaction against the viral coat since the first symptoms appeared after only 3 hours (Lehrman 1999).

Against the Viral Vector Proteins

The first generation of adenoviral vectors and Herpes viral vectors contained genes coding for viral proteins that were necessary for their reproduction and cellular activity. The proteins encoded by these genes were digested in peptides and presented on the cell surface by the MHC where they triggered a cellular immune reaction. This led to a rapid rejection of the infected cells. The reduction of the cells expressing the transgene was initially attributed to a shut off of the promoter controlling their expression. In the following years, many research groups have demonstrated the presence of immune responses against the remaining viral gene products (Yang et al. 1994; Vilquin, Guérette, et al. 1995). This has led to the development of a new generation of adenoviral vectors devoid of all viral genes (Kochanek et al. 1996; Hauser, Amalfitano, Kumar-Singh, Hauschka, and Chamberlain 1997). This new generation of vectors will resolve the problem of the immune response against viral gene products, but immune responses against viral coat proteins will persist. The potential for immune response against the viral coat also exists for other viral vectors, such as retrovirus, AAA, and lentivirus.

Against the Therapeutic Gene Product

The aim of using a viral vector is usually to introduce gene coding for a protein whose own encoding is missing or mutated in the genome of the patient. Expression of the transgene will therefore lead to the presence of a protein that was either completely absent or truncated in the patient. However, since this new or different protein has been shown to trigger an immune reaction that limits the success of gene therapy, the only solutions to this fundamental problem are either immunosuppression or the development of tolerance to the specific protein involved (see below).

Against the Selection Gene

The production of viral vector is usually performed with plasmid reproduced in bacteria and transferred in special cells that have genes incorporated into their genome in order to insure reproduction of the replication defective viral vector. Bacteria are chosen on the basis of a selection gene usually present in the plasmid. Often, this selection gene codes for a protein conferring resistance to an antibiotic and representing a new antigen that will be recognized by the patient's immune system, thus possibly triggering an immune response. Unfortunately, this potential immune response against the selected antigen was not taken into consideration in development of the first generation of gene therapy vectors.

Against the Marker Gene

In the early versions, a gene coding for an easily detectable product was inserted into the vectors. This gene often coded for an easily detectable enzyme, such as the bacteria β–galactosidase. Since the product of the marker gene was usually a new protein for the treated animals, it thus was also capable of triggering immune reactions that were not anticipated by molecular biologists (Yang, Kaecker, Su, and Wilson 1996). This problem can now be overcome by introducing a marker gene expressed in a different tissue of the treated animal. As the protein coded by that gene is, thus, not new to the host's immune system, it does not trigger an immune response.

Solution to the Immune Problems

Both cell transplantation and gene therapy trigger immune responses, a result which has so far limited the usefulness of these new therapeutic approaches. As mentioned above, there are two major solutions to this problem. The first, using drugs to impair the effectiveness of the immune system, has already been used with success for organ transplantation. The second, involving the development of specific tolerance, has not often been used for organ transplantation, but may be an interesting solution, especially in gene therapy.

Immunosuppression

Immunosuppression with Drugs

In routine immunosuppression for organ transplantation, the most commonly used drugs are cyclosporine and tacrolimus, both of which inhibit the secretion of the interleukin-2 (IL–2) cytokine. Cyclosporine and tacrolimus have also been used to prevent the immune reaction following cell transplantation. Our research group has obtained good transplantation results in mice (Kinoshita et al. 1994) and in monkeys (Kinoshita, Vilquin, Gravel, Roy, and Tremblay 1995; Skuk, Roy, Goulet, and Tremblay 1999), even when the donor was incompatible for MHC with the host. These immunosuppressive drugs can also follow gene therapy to prevent the immune reaction against vector proteins or the transgene product. The best results following gene therapy in skeletal muscle with adenovirus were obtained with tacrolimus (Vilquin, Guérette, et al. 1995).

Immunosuppression with Antibodies

The immune response involves a complex interaction between helper T lymphocytes and antigen presenting cells. However, as this interaction can be blocked by antibodies reacting specifically with the proteins involved, immunoreaction can be prevented just as effectively with such antibodies as with tacrolimus. For example, by using a combination of anti-CD4 (cell determinant 4), anti-CD8 (cell determinant 8) and anti-LFA-1 (leukocyte function associated molecule-1), our research group was able to suppress the rejection of myoblasts injected into skeletal muscle (Guérette, Gingras, Wood, Roy, and Tremblay 1996).

Immunosuppression with Recombinant Proteins

Interactions between the surface proteins involved in triggering the immune response can also be blocked using recombinant proteins of two types: an unmodified protein involved in the interaction or a genetically modified protein. The genetic modification is usually necessary to make the protein more soluble, for example, in the case of a receptor protein, by coupling it to an antibody. Thus, CTLA4-Ig (cytotopic T lymphocyte antigen-4 coupled with IgG1) a recombinant protein which blocks the interaction between CD28 present on the T lymphocyte and B7 present on antigen presenting cells, has been shown to prolong the effect of adenovirus based gene therapy to skeletal muscle (Kay et al. 1995; Guérette et al. 1996).

Introduction of an Immunosuppressing Gene with a Vector

Gene coding for the protein that blocks the interaction between cells involved in the immune response can be introduced in the transplanted cells by using a vector. Thus, immunosuppression is produced by proteins secreted directly by the transplanted cells themselves (Chahine et al. 1995). A similar approach could also be used for gene therapy. For example, the vector introducing the therapeutic gene might also contain a gene to immunosuppress the host or to make the cell immunoprivileged for the immune system. Some caution should be used with this last strategy since it may create regions of tissue that are vulnerable to bacterial and viral infection.

Development of Tolerance

Tolerance is the absence of an active rejection against transplanted tissue without any immunosuppressive agent. This is the best possible situation following transplantation since there are no potential side effects of an immunosuppressive drug. Several research groups in immunology have therefore focused all of their attention toward this ultimate goal. Many different strategies have been investigated, including injection of antigen in the thymus, oral antigen presentation, treatments with various antibodies, and recombinant proteins.

Injection of Antigen or Vector in the Thymus

Newly formed lymphocytes initially migrate to the thymus where those which might react with the patient's own antigens will die. Many experiments have, therefore, shown that the injection of antigens into the thymus leads to the removal of any new lymphocytes which could react with foreign substances (Oluwole et al. 1995; Naji 1996). Intrathymic injection needs to be accompanied, though, by the destruction of existing lymphocytes in the body, which is usually accomplished with an antilymphocyte serum. Antigens injected in the thymus have included pieces of tissue, purified proteins, or cells infected with virus containing the gene coding for the antigenic protein.

Development of Oral Tolerance

An intriguing way to develop tolerance to a substance is to feed it to the subject animal (MacDonald 1994; Weiner et al. 1994). Oral presentation of antigen induces tolerance or anergy (i.e., inactivation) of the lymphocytes which recognize the antigenic substance. This approach has been successful mostly for pure antigen as well as some types of cell transplantation. It has also been used to develop tolerance to a new protein introduced by a viral vector, and may, therefore, be an interesting approach to aid in developing tolerance for gene therapy.

Treatments with Antibodies

As mentioned above, the onset of the immune response involves many interactions between cell surface proteins. The antigen is normally presented to the T lymphocytes by the MHC complex. Activation of the lymphocytes, however, requires interactions between other cell surface proteins providing necessary secondary signals. Other reinforcing signals are also provided by cytokines. If the antigen is presented to lymphocytes while the second or third signal is blocked, they will enter into anergy, meaning that they will not be able to react later with this antigen even if the secondary signals are no longer blocked. Several monoclonal antibodies have, therefore, been used to block the interactions between antigen presenting cells (APC) and helper T lymphocytes (Cosimi 1995). Long-term tolerance has recently been obtained in mice and in monkeys using anti-CD154 treatment following islet transplantation (Rossini, Greiner, and Mardes 1999).

Treatments with Recombinant Proteins

The interaction between APC and lymphocytes can also be blocked with recombinant proteins that react with the surface proteins involved. Again, this transient interference in the period following the introduction of a new antigen causes anergy of the lymphocytes (Cosimi 1995).

Immunoprotection of the Transplanted Cells

When cells are transplanted with the sole aim of secreting a protein, they may be hidden from the immune system by encapsulation. In animal experiments, various polymers have been used to enrobe the cells. The basic requirement is that the polymer forms a pore too narrow for infiltration by macrophages and lymphocytes. Pore size should, however, be large enough to permit diffusion of the secreted protein (see review by van Schelfgaarde and de Vos 1999).

Autologous Transplantation of Cells

One way of avoiding part of the immune problem is to transplant their own cells back into patients undergoing treatment. This approach prevents any immune reaction against the MHCs and the minor antigens. However, if the cells have been grown in culture before transplantation, there could be an immune reaction against components of the culture medium, as mentioned above. In addition, if the cells have been genetically modified before their transplantation, there could be a possible reaction against the new recombinant protein. Therefore, even following autologous cell transplantation, transient and perhaps even sustained immunosuppression may be required if the recombinant protein is presented to the immune system. As mentioned above, the development of tolerance toward the new recombinant protein may be an interesting way to avoid sustained immunosuppressive treatment.

In the special case of myoblast transplantation in a DMD patient, one of the problems is the senescence of myoblasts obtained from older patients, which prevents these myoblasts from being expanded in culture. A potential solution to this problem would be to immortalize these cells by introducing a T antigen or a telomerase gene (Deschênes, Chahine, Tremblay, Paulin, and Puymirat 1997).

Conclusion

Both cell transplantation and gene therapy are faced with the same formidable barrier, the immune system. Although immunosuppression with drugs represents an immediate control of this problem, this is far from an ideal solution. Development of tolerance toward the new antigens would be an ideal solution, and the recent encouraging results obtained with an anti-CD154 antibody suggests that this approach may be a reality soon.

References

Boulanger, A., Asselin, I., Roy, R., and Tremblay, J.P. 1997. Role of nonmajor histocompatibility complex antigens in the rejection of transplanted myoblasts. *Transplantation* 63:893–9.

Chahine, A.A., Yu, M., McKernan, M.M., Stoeckert, C., and Lau, H.T. 1995. Immunomodulation of pancreatic islet allografts in mice with CTLA–4Ig secreting muscle cells. *Transplantation* 59:1313–18.

Cosimi, A.B. 1995. Future of monoclonal antibodies in solid organ transplantation. *Dig Dis Sci* 40:65–72.

Day, C.S., Kasemkijwattana, C., Ménétrey, J., Floyd, S.S., Booth, D.K., Moreland, M.S., Fu, F.H., and Huard, J. 1997. Myoblast mediated gene transfer to the joint. *J Orthop Res* 15:894–903.

Deschênes, I., Chahine, M., Tremblay, J., Paulin, D., and Puymirat, J. 1997. Increase in the proliferative capacity of human myoblasts by using the T antigen under the vimentin promoter control. *Muscle Nerve* 20:437–45.

Gahery-Segard, H., Farace, F., Godfrin, D., Gaston, J., Lengagne, R., Tursz, T., Boulanger, P., and Guillet, J.G. 1998. Immune response to recombinant capsid proteins of adenovirus in humans: antifiber and antipenton base antibodies have a synergistic effect on neutralizing activity. *J Virol* 72(3):2388–97.

Gahery-Segard, H., Juillard, V., Gaston, J., Lengagne, R., Pavirani, A., Boulanger, P., and Guillet, J.G. 1997. Humoral immune response to the capsid components of recombinant adenoviruses: routes of immunization modulate virus induced Ig subclass sifts. *Eur J Immunol* 27(3):653–9.

Guérette, B., Asselin, I., Vilquin, J.T., Roy, R., and Tremblay, J.P. 1994. Lymphocyte infiltration following allo and xeno myoblast transplantation in mice. *Transp Proc* 26:4061–2.

Guérette, B., Gingras, M., Wood, K., Roy, R., and Tremblay, J.P. 1996. Immunosuppression with monoclonal antibodies and CTLA4–Ig after myoblast transplantation in mice. *Transplantation* 62(7):962–7.

Hauser, M.A., Amalfitano, A., Kumar–Singh, R., Hauschka, S.D., and Chamberlain, J.S. 1997. Improved adenoviral vectors for gene therapy of Duchenne muscular dystrophy. *Neuromusc Disord* 7:277–83.

Huard, J., Bouchard, J.P., Roy, R., Malouin, F., Dansereau, G., Labrecque, C., Albert, N., Richards, C.L., Lemieux, B., and Tremblay, J.P. 1992. Human myoblast transplantation: preliminary results of 4 cases. *Muscle Nerve* 15:550–60.

Huard, J., Verreault, S., Roy, R., Tremblay, M., and Tremblay, J.P. 1994. High efficiency of muscle regeneration after human myoblast clone transplantation in SCID mice. *J Clin Invest* 93:586–99.

Karpati, G., Pouliot, Y., Zubrzycka–Gaan, E., Carpenter, S., Ray, P.N., Worton, R.G., and Holl, P. 1989. Dystrophin is expressed in mdx skeletal myscle fibers after normal myoblast inmplantation. *Am J Path* 135:27–32.

Kay, M.A., Holterman, Z.X., Meuse, L., Gown, A., Ochs, H.D., Linsley, P.S., and Wilson, C.B. 1995. Long term hepatic adenovirus mediated gene expression in mice following CTLA4–Ig administration. *Nat Genet* 11:191–7.

Kinoshita, I., Vilquin, J.T., Gravel, C., Roy, R., and Tremblay, J.P. 1995. Myoblast allotransplantation in primates. *Muscle Nerve* 18:1217–18.

Kinoshita, I., Vilquin, J.T., Guérette, B., Asselin, I., Roy, R., and Tremblay, J.P. 1994. Very efficient myoblast allotransplantation in mice under FK506 immunosuppression. *Muscle Nerve* 17:1407–15.

Kochanek, S., Clemens, Mitani, K., Chen, H.H., Chan, S., and Caskey, C.T. 1996. A new adenoviral vector: replacement of all viral coding sequences with 28 kb of DNA independently expressing both full length dystrophin and B–galactosidase. *Proc Natl Acad Sci USA* 93(12):5731–6.

Law, P.K., Goodwin, T.G., and Wang, M. 1988. Normal myoblast injections provide genetic treatment for murine dystrophy. *Muscle Nerve* 11(6):525–33.

MacDonald, T.T. 1994. Oral tolerance: eating your way towards immunosuppression. *Curr Biol* 4:178–81.

Lehrman S. 1999. Virus treatment questioned after gene therapy. *Death Nature* 401:517–18.

Morgan, J.E., Watt, D.J., Sloper, J.C., and Partridge, T.A. 1988. Partial correction of an inherited biochemical defect of skeletal muscle by grafts of normal muscle precursor cells. *J Neurol Sci* 86:137–47.

Naji, A. 1996. Induction of tolerance by intrathymic inoculation of alloantigen. *Curr Opin Immunol* 8:704–9.

Oluwole, S.F., Jin, M.X., Chowdhury, N.C., Engelstad, K., Ohajekwe, O.A., and James, T. 1995. Induction of peripheral tolerance by intrathymic inoculation of soluble alloantigens: evidence for the role of host antigen presenting cells and suppressor cell mechanism. *Cell Immunol* 162:33–41.

Partridge, T.A., Morgan, J.E., Coulton, G.R., Hoffman, E.P., and Kunkel, L.M. 1989. Conversion of mdx myofibres from dystrophin negative to positive by injection of normal myoblasts. *Nature* 337:176–9.

Rossini, A.A., Greiner, D.L., and Mardes, J.P. 1999. Induction of immunologic tolerance for transplantation. *Physiol Rev* 79(1):99–141.

Skuk, D., Roy, B., Goulet, M., and Tremblay, J.P. 1999. Successful myoblast transplantation in primates depends on appropriate cell delivery and induction of regeneration in host muscle. *Exp Neurol* 155(1):22–30.

Tremblay, J.P. and Guérette, B. 1997. Myoblast transplantation: identification of the problems and some solutions. *Basic Appl Myol* 7:221–30.

Tremblay, J.P., Malouin, F., Roy, R., Huard, J., Bouchard, J.P., Satoh, A., and Richard, C.L. 1993. Results of triple blind clinical study of myoblast transplantations without immunosuppressive treatment in young boys with Duchenne muscular dystrophy. *Cell Trans* 2:99–112.

van Schilfgaarde R. and de Vos P. 1999. Factors influencing the properties and preformance of microcapsules for immunoprotection of pancreatic islets. *J Mol Med* 77(1):199–205.

Vilquin, J.T., Guérette, B., Kinoshita, I., Roy, B., Goulet, M., Gravel, C., Roy, R., and Tremblay, J.P. 1995. FK506 immunosuppression to control the immune reactions triggered by first generation adenovirus–mediated gene transfer. *Hum Gene Ther* 6:1391–401.

Vilquin, J.T., Wagner, E., Kinoshita, I., Roy, R., and Tremblay, J.P. 1995. Successful histocompatible myoblast transplantation in dystrophin deficient mdx mouse despite the production of antibodies against dystrophin. *J Cell Biol* 131:975–88.

Weiner, H.L., Friedman, A., Miller, A., Khoury, S.J., Al–Sabbagh, A., Santos, L., Sayegh, M., Nussenblatt, R.B., Trentham, D.E., and Hafler, D.A. 1994. Oral tolerance: immunologic mechanisms and treatment of animal and human organ–specific autoimmune diseases by oral administration of autoantigens. *Ann Rev Immunol* 12:809–37.

Yang, Y., Kaecker, S.E., Su, Q., and Wilson, J.M. 1996. Immunology of gene therapy with adenoviral vectors in mouse skeletal muscle. *Hum Mol Genet* 5(11):1703–12.

Yang, Y., Nunes, F.A., Berencsi, K., Furth, E.E., Gönczöl, E., and Wilson, J.M. 1994. Cellular immunity to viral antigens limits E1 deleted adenoviruses for gene therapy. *Proc Natl Acad Sci USA* 91:4407–11.

14
Vectors for Gene Transfer to Joints

PAUL D. ROBBINS, STEVEN C. GHIVIZZANI, JOSEPH C. GLORIOSO, and CHRISTOPHER H. EVANS

Introduction

Gene therapy represents a novel approach for treating joint and bone disorders (Evans et al. 1997; Evans, Ghivizzani, and Robbins 1998; Lattermann et al. 1998). As discussed in other chapters, gene transfer can be used for delivering therapeutic agents to synovium, cartilage, ligaments, tendons, meniscus, intervertebral disc, and bone to block disease progression or to promote repair. In addition, the recent completion of the first gene therapy trial for rheumatoid arthritis (RA) has demonstrated the feasibility of using gene transfer for the treatment of orthopaedic and rheumatologic disorders. The use of gene transfer to deliver a therapeutic agent offers certain advantages over the use of recombinant protein. In particular, the use of genes as therapeutic agents can results in persistent expression locally at the site of disease, bypassing the need for multiple injections and preventing possible side effects associated with systemic administration.

There are two general methods for gene transfer, direct in vivo and indirect ex vivo. The direct method involves injection of the vector at the appropriate site, whereas the ex vivo method involves the genetic modification of cells in culture followed by injection at the diseased site. Although there are many components necessary for successful gene therapy, including identification of the therapeutic gene(s) and a good understanding of the pathology to be treated, the rate limiting step for successful gene therapy is the efficiency of gene transfer, especially in vivo gene transfer. There are two general types of vectors for gene transfer, viral and nonviral. Viruses have evolved to be highly efficient at the delivery of nucleic acid, either DNA or RNA, to specific cell types while avoiding the host immune response. If the pathogenicity of a virus can be eliminated while still retaining the efficiency of gene transfer and expression, then it may be well suited for gene therapy applications. Nonviral vectors such as liposomes have the advantage of being nonpathogenic, but are also less efficient in the transfer of nucleic acid to cells. A list of viral and nonviral vectors that can be used for both direct and indirect gene transfer to joints is summarized in Figure 14.1. Each

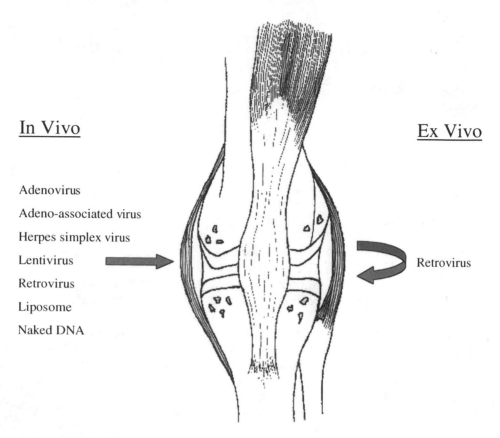

In Vivo

Adenovirus

Adeno-associated virus

Herpes simplex virus

Lentivirus

Retrovirus

Liposome

Naked DNA

Ex Vivo

Retrovirus

FIGURE 14.1. Vectors for in vivo and ex vivo gene delivery to joints.

of these vector systems, retrovirus, adenovirus, herpes simplex virus, adeno-associated virus, and nonviral DNA formulations, has its strengths as well as weaknesses that require further modification to make it suited to general gene therapy applications. The focus of this chapter is on the type of vectors that have been used in animal models for gene transfer to joints. A description of each vector system and how it has been applied to intraarticular gene transfer is provided.

Retroviral Vectors

Murine Leukemia Virus

The majority of gene therapy clinical trials to date have utilized retroviral vectors based on murine leukemia virus (MLV). MLV has been developed as a vector because it has been extensively characterized, has no homology with human retroviruses, and causes no direct pathology in mice. However, the replication competent retrovirus can cause leukemia after a long latency period by inserting next to and activating a set of oncogenes. Retroviruses are RNA viruses that replicate through a double-stranded DNA intermediate, integrating into the host DNA where they are able to express a viral RNA for the life of the cell. Thus, retroviruses have the advantage of permanently modifying cells.

The MLV genome encodes for three polyproteins, Gag, Pol, and Env, which are required in trans for viral replication and packaging. However, all that is required for viral replication in cis are the 5′ and 3′ long–terminal repeats (LTRs) that contain promoter, polyadenylation, and integration sequences, a packaging site termed ψ, and a tRNA binding site as well as several additional sequences involved in reverse transcription. The genes encoding the three viral proteins can be removed and heterologous genes and transcriptional regulatory sequences inserted. In order to produce infectious, replication-defective retroviral vectors, cell lines that express the three viral genes, Gag, Pol, and Env, have been generated and are termed packaging cells. The vector, in the proviral form of double-stranded DNA, can then be introduced into the packaging cells either transiently or stably by transfection, resulting in the production of infectious virus. However, following infection of a target cell and integration of the vector into the host DNA, the virus is unable to replicate due to the absence of the viral trans-acting proteins.

There are three general types of retroviral vectors (Guild, Finer, Housman, and Mulligan 1988; Dranoff et al. 1993; Miller 1992; Miller, A.D., Miller, D.G., Garcia, and Lynch 1993; Byun et al. 1996; Kim et al. 1997). The first class of vector is the LTR-based vector where the gene of interest is expressed directly from the LTR. In the LTR-based vectors, two or more genes can be expressed by either differential splicing or through the use of an internal ribosome entry site or IRES. The second type of vector is the internal promoter vector where the gene of interest is driven from an internal promoter. In this type of vector, the 3′ LTR can be mutated so that after infection, LTR-based expression is abolished or the LTR can be maintained or modified. For instance, it is possible to insert heterologous enhancer elements into the U3 region of the 3′ LTR, resulting in better or cell-type specific gene expression. The presence of a wild type or enhancer-modified LTR allows for expression of a second gene from the LTR. Finally, there is a so-called reverse orientation vector where the gene of interest is expressed from its own promoter in reverse orientation. This type of vector is used for insertion of genomic sequences with introns, allowing for more appropriate regulation of gene expression. If the genomic sequences are placed in the sense orientation, the introns will be removed from the full-length RNA by splicing.

Retroviral vectors have the advantage of stably integrating into the host DNA in the infected cell and, in theory, expressing a therapeutic gene for the life of that cell. Furthermore, the provirus is maintained during subsequent mitotic division; thus a retrovirally-transduced cell can be clonally expanded. Packaging lines, which can produce moderate titers (10^6 to 10^7) of the replication-defective vectors, without giving rise to replication-competent helper virus through recombination, have been generated and used clinically. The disadvantage to the MLV-based retroviral vectors is that they require cell division, in particular mitosis, for integration. Thus the current retroviral vectors are better suited for ex vivo gene therapy where isolated cells can be propagated in culture, genetically modified by retroviral infection, and then transplanted into the recipient. Retroviral vectors also may be used in vivo, to infect certain rapidly dividing cells such as hepatocytes following partial hepatechtomy, rapidly dividing tumor cells, or proliferating synovial cells lining inflamed joints.

Lentiviruses

Lentiviruses are a class of retroviruses, including human immunodeficiency virus (HIV), simian immunodeficiency virus (SIV), feline immunodeficiency virus (FIV) and equine infectious anemia virus (EIAV), that are able to infect nondividing cells. Lentiviruses are more complex than MLV-based retroviruses since they encode for accessory proteins in addition to Gag, Pol, and Env encoded by multiply spliced viral RNAs. Recently, vectors based on HIV, SIV, FIV, and EIAV have been developed that allow for infection of slowly or nondividing cell types like β-cells from pancreatic islets, hepatocytes, and neurons (Naldini, Blomer, Gage, Trono, and Verma 1996; Naldini, Blomer, Gallay, et al. 1996). Recent advances in producing higher titer viruses and the further development of lentivirus vectors, which can infect nondividing cells, may allow for the general application of retroviral vectors to in vivo gene therapy.

Retroviral Gene Transfer to Joints

Although MLV-based vectors are able to stably modify cells by integrating into the genome of the host cells, they are only able to infect dividing cells. Thus MLV-based vectors are more appropriate for ex vivo applications where cells are modified in culture prior to reintroduction. We have used retroviral vectors extensively to modify in culture rabbit synovial cells that were then injected intraarticularly into the rabbit knee (Bandara et al. 1993; Hung et al. 1994). The injected cells are able to colonize the synovial lining and express the transgene for extended periods of time. A retroviral vector has been used to deliver the gene for human interleukin 1 receptor antagonist, an inhibitor of IL-1, to rabbit synovial cells. Following injection of the modified cells into a rabbit knee with antigen-induced arthritis, a significant reduction in inflammation as well as chondroprotective effect was observed

(Otani et al. 1996). Similarly, we have modified meniscal fibroblasts, ligament and tendon cells, and chondrocytes in culture with retroviral vectors prior to implantation at the appropriate site in the rabbit knee.

The results in the rabbit knee model using genetically modified synovial cells have led to a clinical trial to assess the safety and feasibility of transferring a gene to human joints (Evans et al. 1996). In a recently completed trial, the IL-1Ra gene was transferred to autologous synovial cells, isolated during a scheduled surgical procedure, that were then injected back into metacarpophalangeal (knuckle) joints of patients with rheumatoid arthritis (RA). The knuckle joints that were treated with the modified cells were joints scheduled to undergo joint replacement surgery 1 week posttreatment. The removal of the genetically modified joints allowed for analysis of gene expression and changes in the pathophysiology of the joints. Injection of retrovirally transduced synovial cells has been well tolerated with no apparent toxicity, and expression of the transgene has been observed in the treated joints by RT-PCR. These results should lead to a Phase II trial to examine the efficacy of intraarticular IL-1ra gene transfer in patients with earlier stage RA.

Although retroviral vectors have been used extensively for ex vivo gene transfer to joints, they have also been shown to be potentially useful for in vivo delivery (Evans and Robbins 1997). Injection of a high titer stock of amphotropic virus into a naïve rabbit joint results in only a low level of injection. However, injection of virus into an inflamed joint results in a significant level of infection due to the number of cells undergoing cell division (Ghivizzani et al. 1997). Similar results have been obtained in a rat knee model (Nguyen et al. 1998). Interestingly, expression of the transgene seems to persist longer following delivery in vivo compared to the ex vivo method, possibly due to differences in the life span of the infected or injected cells or due to differences in the immune response to the expressed transgene proteins. The ability to infect human synovial cells in vivo has been demonstrated following transplantation of human synovial cells into a severe combined immunodeficient (SCID) mouse (Jorgensen et al. 1997). Although lentiviruses have not been tested for in vivo gene delivery, it is possible that they may be able to infect slowly or nondividing cells in the joint.

Adenoviral Vectors

Adenoviruses are double stranded DNA viruses that cause respiratory and eye infections in a number of mammalian species including humans. Although there are many serotypes of adenovirus (Ad), it is serotype 5 that is the best characterized and is used predominantly as a vector for gene transfer. The viruses have four early genes that encode for multiple polypeptides important for the virus life cycle. The proteins encoded by the E1 region of the virus are essential to activate transcription of the E2, E3, and E4 genes that encode for additional regulatory proteins as well as polypeptides important for viral replication. Late in infection, the major late promoter (MLP) is activated, resulting in the expression of protein important for packaging of the viral DNA.

In order to generate a replication defective virus from Ad–5 (Graham and Prevec 1995), the E1 gene is deleted from the viral genome and an expression cassette carrying the gene of interested inserted. The E1 virus is unable to replicate unless the E1 polypeptides, E1A and E1B, are provided in trans. The human embryonic kidney cell line 293 stably expresses E1A and E1B, allowing for replication and propagation of an E1 virus. However, after infection of a target cell, the virus is unable to replicate because E1 proteins are not provided in trans. However, the transgene carried by the vector can potentially be expressed for as long as the viral DNA is maintained as an episome in the nucleus.

The advantages of adenoviral vectors is that they can be grown to a high titer, greater than 10^{14} particles/ml, and are able to infect a wide variety of dividing and nondividing cell types. However, the viral DNA that is unable to replicate in the infected cell is eventually degraded, resulting in loss of gene expression. In addition, the "first generation" E1 vectors still express a low level of viral proteins, resulting in stimulation of an immune response to the virally infected cell in vivo. The first generation adenoviral vectors may be well suited for cancer gene therapy where a stimulation of a local immune response to the tumor is needed, but, for other applications, the vector needs to be further developed.

One approach to reduce the immunogenicity of the virus after infection in vivo is to remove some or all of the open reading frames (ORFs) of the E4 gene that encode trans-acting regulatory proteins (Wang and Finer 1996). The "second generation" vectors are grown on 293 cells modified to express the polypeptides encoded by the E4 gene (Krougliak and Graham 1995). These vectors not only have a reduced immunogenicity in vivo, but also appear to have a corresponding decrease in transgene expression since E4 polypeptides may play a role in enhancing gene expression from the viral genome. An alternative approach for making a more defective, nonimmunogenic virus is to completely remove all the viral genes, leaving only the inverted terminal repeats (ITRs) and the packaging site (Kochanek et al. 1996). This "gutted" or "gutless" virus is grown using a first or second generation virus as a helper virus in 293 cells. The only difficulty with this approach is how to enrich for the gutted vector in relation to the helper virus. The insertion of sequences into the helper virus that hinder its growth specifically in modified 293 cells has resulted in 1000-fold better production of the gutted vectors compared to the helper virus. However, it is still unclear whether the virus is able to express high levels of the transgene for longer periods of time in vivo.

Adenoviral Gene Transfer to Joints

Adenoviral vectors have been used to deliver a number of different marker or therapeutic genes to joints in rabbit, rat, and murine models (Roessler, Allen, Wilson, Hartman, and Davidson 1993; Nita et al. 1996; Ghivizzani et al. 1998). Injection of 10^7 to 10^8 infectious particles results in good infection of the synovial layer with expression persisting 1 to 2 weeks. However, expression is eventually lost, presumably due to the immune response to the virally infected cells. Intraarticular injection of higher doses of virus results in more efficient infection and in higher levels of transgene expression, but also in a stronger immune response that abolishes expression in 3 to 7 days. In addition, there is a humoral response to the injected virus, reducing the extent of infection achieved after a second injection. The ability to infect cells in vivo in human joints may also be reduced by circulating antiadenoviral antibodies. Thus, it is unclear if adenoviral vectors will be useful for gene transfer to synovium following intraarticular injection. It is important to note that injection of adenovirus into tendons (Gerich, Kang, Fu, Robbins, and Evans 1997), ligaments (Hildebrand et al. In press), and bone (Baltzer et al. In press) has resulted in longer term gene expression, most likely due to the poor immune response in these tissues. Thus, it is possible that local, transient delivery of genes that can promote repair of certain intraarticular tissues as well as repair of bone fractures can be facilitated by first, second, or third generation adenoviral vectors.

Although adenoviral vectors may not be useful for treatment of arthritis by intraarticular gene delivery, they have proven highly useful for testing the efficacy of specific genes in animal models of arthritis. We have shown that delivery of the soluble receptors for interleukin-1 beta (IL-1β) and TNF-α (tumor necrosis factor) results in a significant therapeutic effect in a rabbit knee model of antigen-induced arthritis (Ghivizzani et al. 1998). Similarly, intra or periarticular injection of an adenoviral vector expressing a virally encoded form of interleukin–10 (vIL-10) conferred an antiarthritis effect in both murine (Apparailly et al. 1998; Ma et al. 1998; Whalen et al. 1999) and rabbit models of arthritis (Lechman et al. 1999). Interestingly, following injection of Ad–vIL–10, a therapeutic effect was observed in the contralateral rabbit knees or in untreated mouse paws (Whalen et al., 1999; Lechman et al., 1999). Although the mechanism of the therapeutic effect observed in untreated knees or paws following treatment of a single joint is unclear, the trafficking of infected cells to lymph nodes as well as to other joints has been observed (Lechman et al. 1999). It is possible that a subset of adenovirally infected cells such as dendritic cells is able to migrate to certain immunoregulatory sites to suppress the immune response. Given the ability of adenovirus in joints to infect a cell type that is then able to confer a "contralateral effect," it is possible that adenoviral vectors may have certain clinical applications in polyarticular disease.

Herpes Simplex Virus Vectors

Herpes simplex viruses (HSV) are large, linear DNA viruses of approximately 150 kb that are able to infect cells both lytically and, for certain cell types such as neurons, latently. Similarly to adenovirus, HSV can infect a wide variety of cell types including muscle, tumor, and pancreatic islets. In addition, HSV vectors are able to incorporate exogenous DNA of up to 35 kb, allowing for insertion of large fragments of DNA. The normal life cycle of HSV following uptake into cells involves expression of a set of immediate early genes (e.g. ICP0, ICP4, ICP22, ICP27, and ICP47). ICP4 and ICP27 are essential for the expression of early and late genes important for virus morphogenesis and infectious particle production. In order to generate a vector based on HSV, the immediate early genes important for subsequently activating the early and late genes are deleted or inactivated (Fink et al. 1995; Marconi et al. 1996; Fink and Glorioso 1997a, 1997b). The resulting virus can be grown on complementing cell lines expressing the appropriate immediate early genes. The first HSV vector had a deletion in the ICP4 gene, but was still able to express the other immediate early gene products that are toxic to certain cell types. Recently, further deletion of ICP0, ICP22, and ICP27 has resulted in a nontoxic virus that can express a transgene for extended periods in culture and in vivo (Wu, Watkins, Schaffer, and Deluca 1996). Alternatively, it is possible to package plasmid vectors, termed amplicons, using defective HSV helper virus or plasmids encoding helper functions (Geller 1997). To generate helper-free amplicon vectors, a plasmid carrying the HSV origin of replication and packaging signals is cotransfected with a series of plasmids that contain all of the HSV genes, but are missing the packaging sequence. The vector plasmid is packaged into infectious HSV particles and used for gene transfer. The problem with the amplicon system is that the titers are relatively low and that recombination can give rise to a low level of HSV genomes that then can be packaged.

HSV Gene Transfer to Joints

We have evaluated the use of HSV-based vectors for gene transfer to joints (Nita et al. 1996; Glorioso et al. 1997; Oligino, Ghivizzani, Evans, et al. 1999). Injection of an ICP4-deficient virus into a rabbit knee resulted in efficient infection but short term expression (Nita et al. 1996; Oligino et al. 1999). However, the use of a more defective HSV vector containing mutations in ICP4, 22, and 27, resulted in more prolonged gene expression of up to 14 days (Oligino et al. 1999). Interestingly, the number of infectious particles of HSV vector needed to express significant levels of a sTNF-α receptor marker intraarticularly was substantially less than that needed to obtain similar levels of expression with adenoviral vectors. The ability of HSV vectors to confer a therapeutic effect in the rabbit knee model of arthritis was also demonstrated using a ICP4, 22, and 27 virus expressing IL-1ra. Thus, if HSV vectors are more fully developed, they may be useful for intraarticular gene expression.

Recent data suggest that intraarticular delivery of HSV-based vectors may provide substantial gene delivery to neurons innervating the joint (Wolfe, Goins, Kaplan et al. submitted for publication). The infected neurons appear to serve as a repository for transgene production in the infected joint with expression of a nerve growth factor (NGF) transgene detected at sustained, high intraarticular levels for many months. This novel application of HSV vectors, using neurons as depots for expression of antiarthritic gene products, also may be highly useful for clinical applications in the future.

Adeno-Associated Virus

Adeno-associated virus (AAV) is a member of the parvovirus family of single-stranded small DNA viruses that require a helper virus such as adenovirus or herpes simplex virus for replication. Although the majority of the population is seropositive for AAV, the virus is not associated with any pathology in humans. The virus contains two genes, Rep and Cap, that encode polypeptides important for viral replication and encapsidation. Similarly to retroviruses, rep and cap can be supplied in trans with only the ITR sequences required in cis for the virus life cycle. Thus, therapeutic genes can be inserted between the ITRs and infectious virus generated by cotransfection of the 293 cell line with the vector and a Rep and Cap expression plasmid, followed by infection with a first generation adenoviral vector (Rolling and Samulski 1995). The resulting virus can be purified away from the contaminating adenovirus by centrifugation. However, to avoid possible adenoviral contamination, infectious recombinant AAV also can be generated by cotransfection of a plasmid expressing the complementing adenoviral genes.

Wild-type AAV can be grown to high titers and infect a wide variety of cell types. Following infection, the virus forms a double stranded DNA intermediate that can be integrated at a certain frequency into chromosome 19 through a Rep-dependent mechanism (Samulski 1993). However, it appears as if recombinant AAV vectors are unable to integrate specifically and may actually persist as episomal DNA in nondividing cells. One of the rate limiting steps is the virus life cycle is the rate of second strand synthesis, necessary for expression of the transgene. Normally, adenoviral gene products are able to stimulate the rate of second strand synthesis, but in the absence of adenoviral helper functions, AAV requires cellular factors (Ferrari, Samulski, T., Shenk, and Samulski, R.J. 1996). These unknown factors are either high in certain cell types or can be induced by DNA-damaging agents (Russell, Alexander, and Miller 1995). In particular, AAV has been shown to stably infect muscle and liver following delivery of recombinant AAV produced using adenoviral complementing plasmids (Fisher et al. 1997; Snyder et al. 1999).

AAV Gene Transfer to Joints

Because of significant differences in the ability of AAV to infect stably different cell types, it is unclear whether AAV will be able to infect many of the intraarticular cell types. Recent unpublished analysis of AAV-mediated gene transfer to joints has provided mixed results. In a rat model, intraarticular injection of an AAV–LacZ vector only resulted in detectable expression of the marker gene in synovium following induction of inflammation (Pan, Xiao et al. 1999; Oligino et al. submitted for publication). Interestingly, induction of inflammation at later intervals resulted in stimulation of gene expression, suggesting that inflammation was stimulating gene expression, not second strand synthesis. In a rabbit model, we have shown that injection of virus into a naïve knee joint resulted in significant expression of a secreted therapeutic protein, sIL-1 receptor, after 3 days, but expression was abolished after 2 weeks. Subsequently, induction of inflammation did not result in further transgene expression. In addition, a repeat injection of an AAV vector carrying a different marker gene did not result in transgene expression, suggesting generation of a neutralizing antibody to AAV that blocks viral infection. Whether AAV is able to transduce cartilage, ligaments, or tendons in vivo is unknown, but is currently under investigation. Clearly, AAV needs to be evaluated further as a vector for gene transfer to joints.

Nonviral Vectors

Although nonviral gene transfer to joints is less efficient than viral-mediated DNA delivery, it has several advantages. First, it is considerably cheaper to produce large amounts of DNA than to produce viruses for clinical application. Second, there is no possibility of recombination giving rise to replication competent virus that can result in significant pathology. Third, it may be possible to repeat doses with DNA formulations that are non or weakly immunogenic. For these reasons, many laboratories and biotechnology companies have focused on improving nonviral gene delivery methods. Although certain cells such as myoblasts are able to take up naked DNA at some efficiency, complexing of the DNA to different chemical formulations increases DNA uptake and subsequent gene expression. In particular, cationic and anionic liposomes are able to increase the uptake of DNA in tissue culture cells with cationic liposomes improving the efficiency in vivo (Gao and Huang 1995; Liu et al. 1996). It is also possible to couple the DNA to polypeptides in order to target the DNA to specific cell surface proteins. Condensation of the DNA using different polycations is able to improve DNA uptake and prevent degradation.

Nonviral Gene Transfer to Joint

Similar to intramuscular injection of DNA, injection of naked DNA into rabbit and rat joints has resulted in detectable levels of transgene expression in synovial cells (Yovandich, O'Malley, Sikes, and Ledley 1995; Sant et al. 1998). However, delivery of a secreted marker gene did not result in a therapeutic level of expression. Cationic liposomes also have been used to deliver the gene for IL-1Ra intraarticularly with significant levels of expression observed (unpublished observations). However, expression of the transgene was highly transient and there was intraarticular inflammation associated with injection of the DNA complex. Double-stranded oligonucleotide decoys for the transcription factor NK-kB also have been delivered to rat ankles with arthritis using cationic liposomes at a level sufficient to confer a therapeutic effect (Miagkov et al. 1998). One group has demonstrated that intraarticular injection of DNA/liposome complex with an inactivated virus, hemagglutinating virus of Japan (HVJ), resulted in efficient gene transfer to synoviocytes as well as chondrocytes (Tomita et al. 1997). This interesting observation needs further examination to determine if HVJ liposomes will indeed be useful for treating intraarticular pathologies.

Summary

As described above, many different viral and nonviral vectors have been used for in vivo and ex vivo gene transfer to joints. Currently, there is no one vector that is appropriate for all gene therapies for bone and joint disorders; it is likely that certain vectors will be used for specific applications. For instance, a vector that may be used to transfer a gene encoding a protein able to promote new bone growth will differ from a vector encoding an antiinflammatory agent for treatment of RA. In addition to vector development for treatment of specific intraarticular pathologies, there also will be a need to optimize as well as regulate gene expression following delivery. Clearly, gene therapy approaches for treating orthopaedic and rheumatologic disorders will be developed as the technology to deliver genes to the appropriate target tissues improves.

Acknowledgments. This work was supported in part by public health service grants AR62225 to Paul D. Robbins and AR44526 to Joseph C. Glorioso from the National Institutes of Health.

256   P.D. Robbins *et al.*

References

Apparailly, F., Verwaerde, C., Jacquest, C., Auriault, C., Sany, J., and Jorgensen, C. 1998. Adenovirus–mediated transfer of viral IL–10 gene inhibits murine collagen–induced arthritis. *J Rheumatol* 160:5213–20.

Baltzer, A.W.A., Lattermann, C., Weiss, K., Wooley, P., Grimm, M., Ghivizzani, S.C., Robbins, P.D., and Evans, C.H. In press. Genetic enhancement of fracture repair: Evaluation of Ad-BMP-2 in a lapine femoral defect model. *Gene Ther.*

Bandara, G., Mueller, G.M., Galea–Lauri, J., Tyndall, M.H., Georgescu, H.I., Suchanek, M.K., Hung, G.L., Glorioso, J.C., Robbins, P.D., and Evans., C.H. 1993. Intraarticular expression of the interleukin–1 receptor antagonist protein by ex vivo gene transfer. *Proc Natl Acad Sci USA* 90:10764–8.

Byun, J., Kim, S.-H., Kim, J.M., Robbins, P.D., Yim, J., and Kim, S. 1996. Analysis of the relative level of gene expression from different retroviral vectors used for gene therapy. *Gene Ther* 3:780–8.

Dranoff, G., Jaffee, E., Lazenby, A., Golumbek, P., Levitsky, H., Brose, K., Jackson, V., Hamada, H., Pardoll, D., and Mulligan, R.C. 1993. Vaccination with irradiated tumor cells engineered to secrete murine granulocyte-macrophage colony-stimulating factor stimulates potent, specific, and long-lasting anti-tumor immunity. *Proc Natl Acad Sci USA* 90:3539–43.

Evans, C.H., Ghivizzani, S.C., Kang, R., Muzzonigro, T., Wasko, M.C., Herndon, J.H., and Robbins, P.D. 1997. Gene Therapy in Rheumatic Diseases. *Arthritis Rheum* 42:1–16.

Evans, C.H., Ghivizzani S.C., and Robbins, P.D. 1998. Blocking cytokines with genes. *J Leuk Biol* 64:55–61.

Evans, C.H. and Robbins, P.D. 1997. Getting genes into human synovium. *J Rheumatol* 24:2061–3.

Evans, C.H., Robbins, P.D., Ghivizzani, S.C., Herndon, J.H., Kang, R., Bahnson, A.B., Barranger, J.A., Elders, E.M., Gay, S., Tomaino, M.M., Wasko, C., Watkins, S.C., Whiteside, T.L., Glorioso, J.C., Lotze, M.T., and Wright, T.M. 1996. Clinical trial to assess the safety, feasibility, and efficacy of transferring a potentially antiarthritic cytokine gene to human joints with rheumatoid arthritis. *Hum Gene Ther* 7:1261–80.

Ferrari, F.K., Samulski, T., Shenk, T., and Samulski, R.J. 1996. Second strand synthesis is a rate limiting step for efficient transduction by recombinant adeno-associated virus vectors. *J Virol* 70:3227–34.

Fink, D.J. and Glorioso, J.C. 1997a. Engineering herpes simplex virus vectors for gene transfer to neurons. *Nat Med* 3:357–9.

Fink, D.J. and Glorioso, J.C. 1997b. Herpes simplex virus based vectors: problems and some solutions. *Adv Neurol* 72:149–56.

Fink, D.J., Ramakrishnan, R., Marconi, P., Goins, W.F., Holland, T.C., and Glorioso, J.C. 1995. Advances in the development of herpes simplex virus based gene transfer vectors for the nervous system. *Clin Neurosci* 3:284–91.

Fisher, K.J., Jooss, K., Alston, J., Yang, Y., Haecker, S.E., High, K., Pathak, R., Raper, S.E., and Wilson, J.M. 1997. Recombinant adeno–associated virus for muscle directed gene therapy. *Nat Med* 3:306–12.

Gao, X. and Huang, L. 1995. Cationic liposome mediated gene transfer. *Gene Ther* 2:710–22.

Geller, A.I. 1997. Herpes simplex virus–1 plasmid vectors for gene transfer into neurons. *Adv Neurol* 72: 143–8.

Gerich, T.G., Kang, R., Fu, F.H., Robbins, P.D., and Evans, C.H. 1997. Gene transfer to the patellar tendon. *Knee Surg Sports Traumatol Arthrosc* 5:118–23.

Ghivizzani, S.C., Lechman, E.R., Kang, R., Tio, C., Kolls, J., Evans, C.H., and Robbins, P.D. 1998. Direct adenoviral mediated gene transfer of IL–1 and TNF–alpha soluble receptors to rabbit knees with experimental arthritis has local and distal antiarthritic effects. *Proc Natl Acad Sci USA* 95:4613–8.

Ghivizzani, S.C., Lechman, E.R., Tio, C., McCormick, J., Chada, S., Kang, R., Mule, K.M., Evans, C.H., and Robbins, P.D. 1997. Direct retroviral mediated gene transfer to the synovium of the rabbit knee. *Gene Ther* 4:977–82.

Glorioso, J.C., Robbins, P.D., Krisky, D., Marconi, P., Goins, W.F., Schmidt, M.C., and Evans, C.H. 1997. Progress in the development of herpes simplex viurs gene vectors for treatment of rheumatoid arthritis. In B. Davidson (ed.), *Advanced Drug Delivery Reviews* Vol. 27:41–57.

Graham, F.L. and Prevec, L. 1995. Methods for construction of adenovirus vectors. *Mol Biotechnol* 3:207–20.

Guild, B.C., Finer, M.H., Housman, D.E., and Mulligan, R.C. 1988. Development of retrovirus vectors useful for expressing genes in cultured murine embryonal cells and hematopoietic cells in vivo. *J Virol* 62:3795–801.

Hildegrand, K.A., Dele, M., Allen, C.R., Smith, D.W., Georgescu, H.I., Robbins, P.D., Evans, C.H., and Woo, S.L.-Y. In Press. The expression of marker genes in the medial collateral and anterior cruciate ligaments: the use of different viral vectors and the effects of injury. *J Orth Res.*

Hung, G.L., Bandara, G., Galea-Lauri, J., Mueller, G.M., Georgescu, H.I., Larkin, L.A., Tyndall, M.H., Robbins, P.D., and Evans, C.H. 1994. Suppression of intraarticular response to interleukin–1 by gene transfer to synovium. *Gene Ther* 1:64–9.

Jorgensen, C., Demoly, P., Noel, D., Mathieu, M., Piecharczyc, M., Gougat, C., Bousquet, J., and Sany, J. 1997. Gene transfer to human rheumatoid synovial tissue engrafted in SCID mice. *J Rheumatol* 24:2076–9.

Kim, S.H., Yu, S.S., Park, J.S., Robbins, P.D., An, C.S., and Kim, S. 1997. Construction of retroviral vectors with improved safety, gene expression, and versatility. *J Virol* 72:994–1004.

Kochanek, S., Clemens, P.R., Mitani, K., Chen, H.H., Chan, S., and Caskey, C.T. 1996. A new adenoviral vector: replacement of all viral coding sequences with 28 kb of DNA independently expressing both full length dystrophin and β–galactosidase. *Proc Natl Acad Sci USA* 93:5731–6.

Krougliak, V. and Graham, F.L. 1995. Development of cell lines capable of complementing E1, E4, and protein IX defective adenovirus type 5 mutants. *Hum Gene Ther* 6:1575–86.

Lattermann, C., Baltzer, A.W.A., Whalen, J.D., Evans, C.H., Evans, C.H., Robbins, P.D., and Fu, F.H. 1998. Gene therapy in sports medicine. *Sports Med Arthrosc Rev* 6:83–8.

Lechman, E.R., Jaffurs, D., Ghivizzani, S.C., Gambotto, A., Kovesdi, I., Mi, Z., Evans, C.H., and Robbins, P.D. 1999. Direct adenoviral mediated gene transfer of vIL-10 to rabbit knees with experimental arthritis halts disease progression both locally and systemically. *J Immunol* 163:2202–8.

Liu, F., Yang, J., Huang, L., and Liu, D. 1996. New cationic lipid formulations for gene transfer. *Pharm Res* 13:1856–60.

Ma, Y., Thornton, S., Duwel, L.E., Boivin, G.P., Giannini, E.H., Leiden, J.M., Bluestone, J.A., and Hirsch, R. 1998. Inhibition of collagen-induced arthritis in mice by viral IL-10 gene transfer. *J Immunol* 161:1516–24.

Marconi, P., Krisky, D., Oligino, T., Poliani, P.L., Ramakrishnan, R., Goins, W.F., Fink, D.J., and Glorioso, J.C. 1996. Replication defective herpes simplex virus vectors for gene transfer in vivo. *Proc Natl Acad Sci USA* 93:11319–20.

Miagkov, A.V., Kovelenko, D.V., Brown, C.E., Didsbury, J.R., Cogswell, J.P., Stimpson, S.A., Bladwin, A.S., and Makarov, S.S. 1998. NF–kB activation provides the potential link between inflammation and hyperplasia in the arthritic joint. *Proc Natl Acad Sci USA* 95:13859–64.

Miller, A.D. 1992. Retroviral vectors. *Curr Topics Micro Immunol* 158:1–24.

Miller, A.D., Miller, D.G., Garcia, J.V., and Lynch, C.M. 1993. Use of retroviral vectors for gene transfer and expression. *Meth Enzym* 217:581–99.

Naldini, L., Blomer, U., Gage, F.H., Trono, D., and Verma, I.M. 1996. Efficient transfer, integration, and sustained long term expression of the transgene in adult rat brains injected with a lentiviral vector. *Proc Natl Acad Sci USA* 93:11382–8.

Naldini, L., Blomer, U., Gallay, P., Ory, D., Mulligan, R., Gage, F.H., Verma, I.M., and Trono, D. 1996. In vivo gene delivery and stable transduction of nondividing cells by a lentiviral vector. *Science* 272:263–7.

Nguyen, K.H., Boyle, D.L., McCormack, J.E., Chada, S., Jolly, D.J., and Firestein, G.S. 1998. Direct synovial gene transfer with retroviral vectors in rat adjuvant arthritis. *J Rheumatol* 25:1118–25.

Nita, I., Ghivizzani, S., Galea–Lauri, J., Bandara, G., Georgescu, H.I., Robbins, P.D., and Evans, C.H. 1996. Direct gene delivery to synovium: an evaluation of potential vectors in vitro and in vivo. *Arthritis Rheum* 39:820–8.

Oligino, T., Ghivizzani, S.C., Wolfe, D., Lechman, E.R., Krisky, D., Mi, Z., Evans, C.H., Robbins, P.D., and Glorioso, J.C. 1999. Herpes virus viral vector mediated delivery of the IL-1Ra gene to rabbit joints reduces the pathophysiology of IL-1 induced arthritis. *Gene Therapy* 6:1713–20.

Oligino, T., Yao, O., Xiao, X., Glorioso, J.C., Evans, C.H., Robbins, P.D., and Ghivizzani, S.C. Under revision. Intra-articular expression of the IL-1Ra gene folllowing direct delivery with an adeno-associated virus vector. *Gene Ther.*

Otani, K., Nita, I., Macaulay, W., Georgescu, H.I., Mueller, G.M., Robbins, P.D., and Evans, C.H. 1996. Suppression of antigen induced arthritis in rabbits by ex vivo gene therapy. *J Immunol* 156:3558–62.

Pan, R.Y., Xiao, X., Chen, S.L., Li, J., Lin, L.C., Wang, H.J., and Tsao, Y.P. 1999. Disease-inducible transgene expression from a recombinant adeno-associated virus vector in a rat arthritis model. *J Virol* 73(4):3410–7.

Roessler, B.J., Allen, E.D., Wilson, J.M., Hartman, J.W., and Davidson, B.L. 1993. Adenoviral gene transfer to rabbit synovium in vivo. *J Clin Invest* 92:1085–92.

Rolling, F. and Samulski, R.J. 1995. AAV as a viral vector for human gene therapy. Generation of recombinant virus. *Mol Biotechnol* 3:9–15.

Russell, D.W., Alexander, I.E., and Miller, A.D. 1995. DNA synthesis and topoisomerase inhibitors increase transduction by adeno–associated virus vectors. *Proc Natl Acad Sci USA* 92: 5719–23.

Samulski, R.J. 1993. Adeno–associated virus: integration at a specific chromosomal locus. *Curr Opin Genet Dev* 3:74–80.

Sant, S.M., Suarez, T.M., Moalli, M.R., Wu, B.Y., Blaivas, M., Laing, T.J., and Roessler, B.J. 1998. Molecular lysis of synovial lining cells by in vivo herpes simplex virus–thymidine kinase gene transfer. *Hum Gene Ther* 9:2735–43.

Snyder, R.O., Miao, C., Meuse, L., Tubb, J., Donahue, B.A., Lin, H.F., Stafford, D.W., Patel, S., Thompson, A.R., Nichols, T., Read, M.S., Bellinger, D.A., Brinkhous, K.M., and Kay, M.A. 1999. Correction of hemophilia B in canine and murine models using recombinant adeno–associated viral vectors. *Nat Med* 5:64–70.

Tomita, T., Hashimoto, H., Tomitaq, N., Morishita, R., Lee, S.B., Hayashida, K., Nakamura, N., Yonenobu, K., Kaneda, Y., and Ochi, T. 1997. In vivo direct gene transfer into articular cartilage by intraarticular injection mediated by HVJ (Sevdai Virus) and liposomes. *Arthritis Rheum* 40:901–6.

Wang, Q. and Finer, M.H. 1996. Second generation adenovirus vectors. *Nat Med* 2:714–6.

Whalen, J.D., Lechman, E.R., Carlos, C.A., Weiss, K., Kovesdi, I., Robbins, P.D., and Evans, C.H. 1999. Adenoviral transfer of the viral interleukin 10 gene periarticularly to mouse paws suppresses development of collagen induced arthritis in both injected and uninjected paws. *J Immunol* 162:3625–32.

Wolfe, D., Goins, W.F., Kaplan, T.J., Capuano, S., Murphey-Corb, M., Cohen, J.B., Robbins, P.D., and Glorioso, J.C. Submitted for publication. Systemic accumulation of biologically active nerve growth factor following intra-articular herpesvirus gene transfer.

Wu, N., Watkins, S.C., Schaffer, P.A., and DeLuca, N.A. 1996. Prolonged gene expression and cell survival after infection by a herpes simplex virus mutant defective in the immediate-early genes encoding ICP4, ICP27, and ICP22. *J Virol* 70:6358–69.

Yovandich, J., O'Malley, B., Sikes, M., and Ledley, F.D. 1995. Gene transfer to synovial cells by intraarticular adminstration of plasmid DNA. *Hum Gene Ther* 6:603–10.

15
Adeno–Associated Virus (AAV) Vectors for Musculoskeletal Gene Transfer

Xiao Xiao, Ryan Pruchnic, Juan Li, and Johnny Huard

Effective gene delivery techniques into musculoskeletal tissues for gene therapy will revolutionize molecular medicine and pave the way for advances in orthopaedic treatment. However, numerous gene transfer approaches, including direct (in vivo), indirect (ex vivo), and systemic delivery of nonviral and viral vectors, have been hampered by limitations such as low transfection efficiency, immunologic responses, cytotoxic effects, and maturation-dependent factors that preclude the vector from transducing postmitotic and/or mature tissues. As a result, general application of gene therapy for treatments of muscular diseases and orthopedic injuries and repair will not be realized until improvement is made in vector development (see Chapter 14). Recent progress in the development adeno-associated virus (AAV), a small, nonpathogenic human DNA virus, has shed new light on this frontier.

Muscle-Directed Gene Delivery and Its Hurdles

Before discussion of AAV vector, previous development in other vectors and background in gene delivery in the musculoskeletal system will be briefly reviewed. For example, direct injection of nonviral vector (plasmid DNA) as a simple means for muscle gene delivery has demonstrated low cytotoxicity and low immunogenicity. However, the transfection efficiency is relatively poor, even with very large quantities of DNA (Acsadi et al. 1991; Wolff et al. 1992). The efficiency of gene transfer by nonviral vectors can be somewhat improved by the use of nontargeted liposomes, polylysine condensed plasmid DNA, pretreatment of the tissues with agents to induce tissue regeneration, and injection at the myotendinous junction (Davis, Demeneix, Quantin, Coulombe, and Whalen 1993; Davis, Whalen, and Demeneix 1993; Vitiello, Chonn, Wasserman, Duff, and Worton 1996; Doh, Vahlsing, Hartikka, Liang, and Manthorpe 1997) (see Chapter 6). Nonetheless, overall gene transfer efficiency and duration must still be improved.

Alternative approaches for intramuscular gene delivery involve the direct injection of viral vectors such as those derived from retrovirus (RV), adenovirus (Ad), and herpes simplex virus (HSV-1). Much recent research has been focused on these approaches. Retroviral vectors infect dividing cells with high efficiency and are able to achieve long-term persistence by stably integrating into the host genome (Barr and Leiden 1991; Dhawan et al. 1991; Dunckley, Davies, Walsh, Morris, and Dixon 1992; Salvatori et al. 1993). A major hurdle of this vector system for muscle-directed gene therapy, however, is the requirement for cell division, which is very limited in mature muscle. New retroviral vectors based on lentivirus, such as human immunodeficiency virus (HIV), can infect both dividing and nondividing cells, but their efficiency in muscle tissue remains to be examined.

Adenoviral (Ad) vectors can infect both dividing and nondividing muscle cells, and can be readily prepared to high titer stocks (10^9 to 10^{11} pfu/ml) (Quantin, Perricaudet, Tajkakhsh, and Mandel 1992; Ragot et al. 1993; Vincent et al. 1993; Acsadi et al. 1994, 1996). However, these advantages are overshadowed by immunological problems, especially those stemming from the early generation Ad vectors, that severely hinder long term in vivo transduction. Third-generation Ad vector, with most of its genes removed, has shown much reduced immunogenicity (see Chapter 2).

Herpes simplex viral type 1 (HSV-1)-based vectors display some advantages for use in gene transfer, including reasonably high titers (10^7 to 10^9 pfu/ml) and the capacity to carry large DNA fragments such as full-length dystrophin. However, first generation HSV-1 vectors suffer from significant cytotoxicity and immunogenicity which limit their in vivo application (Huard et al. 1996; Huard, Akkaraju, Watkins, Cavalcoli, and Glorioso 1997; Huard, Krisky, et al. 1997). Newly improved vectors (Huard, Krisky, et al. 1997) have been engineered to attenuate cytotoxicity by eliminating multiple viral immediate-early (IE) genes (see Chapter 2).

The common and foremost limitation of the aforementioned viral vectors for muscle gene transfer is the maturation-dependent vector transduction of myofibers. Recent results suggest that several factors contribute to reduced viral transducibility in mature skeletal muscle: 1) muscle cells become postmitotic in an early stage of development; 2) the extracellular matrix develops into a physical barrier; and 3) myoblasts are lost due to progression into a quiescent stage during muscle maturation (Huard et al. 1996; Feero et al. 1997; Huard, Akkaraju, et al. 1997; Huard, Krisky, et al. 1997; van Deutekom, Floyd, et al. 1998; van Deutekom, Hoffman, and Huard 1998). Approaches are being developed to improve viral gene transfer into skeletal muscle. They include the use of particular enzymes to increase the permeability of the extracellular matrix, pretreatment of muscle tissue with myonecrotic agents to induce an artificial release of myoblasts, and application of myoblast mediated ex vivo gene transfer (van Deutekom, Floyd, et al. 1998; van Deutekom, Hoffman, et al. 1998). The extracellular matrix has been demonstrated to act as a physical barrier that prevents viral particles from contacting and infecting the myofibers (Huard et al. 1996; Feero et al. 1997). Estimated pore size of the extracellular matrix is approximately 40 nm (Yurchenco 1990). On the other hand, the particle size of HSV-1 is about 120 to 300 nm in diameter, and AdV and RV, including lentivirus, are about 100 nm in diameter. Obviously, large viral particle size prevents efficient diffusion and penetration through the extracellular matrix. This observation is further supported by the fact that Ad and HSV-1 vectors can achieve higher transduction efficiency in mature muscle of dy/dy mouse, which has defects in the extracellular matrix due to merosin deficiency (Huard et al. 1996; Feero et al. 1997). Based on those previous observations, identifying new strategies and new vectors to bypass the barriers of transduction in mature muscle has become a top priority in muscle gene therapy research. One such strategy is to use various enzymes to permeate the extracellular matrix (van Deutekom, Floyd, et al. 1998; van Deutekom, Hoffman, et al. 1998) and facilitate viral penetration, although this has achieved limited success. Another strategy is to identify new vectors that can readily bypass these barriers. Adeno-associated virus (AAV)-based vector is a gene vehicle that fulfills this requirement.

Biology of AAV Vector

AAV is a nonpathogenic and defective parvovirus that requires essential helper functions from other unrelated viruses such as Ad to efficiently reproduce (Berns and Linden 1995; Berns 1996). AAV has no etiologic association with any known diseases (Berns and Bohenzky 1987). This safety feature should minimize the liability to gene therapy patients. AAV is capable of infecting a broad range of mammalian cell types and tissues, including muscle (Muzyczka 1992). Upon infection, the virus enters the cell nucleus, integrates into the host chromosome, and persists. The only cis-acting sequence required for efficient integration into the host chromosome is the 145-base-pair inverted terminal repeat (ITR) at each end of the AAV genome. This sequence also serves as the origin for DNA replication and viral DNA packaging (Xiao, X., Xiao, W., Li, and Samulski 1997). In recombinant AAV (rAAV), vectors have all viral sequences eliminated except the 145 bp ITRs. The removal of all viral genes adds another safety feature that not only prevents the generation of wild-type helper virus via homologous recombination (Samulski, Chang, and Shenk 1989), but also mitigates the possibility of immune reactions caused by undesired viral gene expression (Afione et al. 1996; Kessler et al. 1996; Xiao, Li, and Samulski 1996; Clark, Sferra, and Johnson 1997; Fisher et al. 1997; Snyder et al. 1997; Monahan et al. 1998), a significant problem seen with other viral vector systems (Yang et al. 1994; Yang, Li, Ertl, and Wilson 1995; Yang and Wilson 1995).

The commonly used production method for rAAV is the 2 plasmid (vector/packaging) transient cotransfection system (Samulski et al. 1989;

Snyder, Xiao, and Samulski 1996; Li, Samulski, and Xiao, 1997; Xiao, Li, et al. 1998). The vector plasmid contains the foreign transgene(s) that is flanked by 145 bp ITRs. The packaging plasmid contains all the AAV genes (Rep gene for viral DNA replication and Cap gene for encapsidation) but without ITRs. From this packaging plasmid, AAV gene products, and thus Rep and Cap proteins, are provided in trans to replicate and package the vector DNA. Such produced vectors are basically without packaging of wild-type coding sequences, despite some reports documenting emergence of low-level wild-type AAV by nonhomologous recombination (Alexander, Russell, and Miller 1997). Although generally successful, the above system has at least two major drawbacks: the inconvenience of the transient cotransfection regime and the requirement of helper Ad infection. Even though Ad can be removed during purification, a risk of contamination persists. Generation of efficient packaging cells harboring both functional AAV genes and the vector DNA will offer solutions to transient cotransfection (Clark, Voulgaropoulou, Fraley, and Johnson 1995; Tamayose, Hirai, and Shimada 1996). Employment of mini Ad plasmid containing only the necessary helper functions, rather than the entire Ad virus, should prevent Ad contamination. Recently, we have reported a high-titer rAAV production method completely free of Ad virus by using a mini Ad plasmid containing only the essential Ad helper genes (Ferrari, Xiao, McCarty, and Samulski 1997; Xiao, Li, et al. 1998). Using such 3-plasmid transfection methods, Ad-free, high-titer, and high purity rAAV viral stocks of up to 10^{10} to 10^{11} transducing units (t.u.)/ml (up to 10^{13} to 10^{14} viral particles/ml) can be obtained, approaching wild-type AAV yields. Given recent improvements in current methodology (Li et al. 1997; Xiao, Li, et al. 1998), rAAV vector production should no longer represent the bottleneck in this system in preclinical studies (Verma and Somia 1997).

AAV vectors can also efficiently transduce both dividing and nondividing cells. Transduction of proliferating cells has been well documented (Muzyczka 1992; Xiao, DeVlaminick, and Monahan 1993; Kotin 1994; Berns and Linden 1995; Flotte and Carter 1995; Xiao, Li, McCown, and Samulski 1997). However, despite earlier speculation and observations (Flotte et al. 1993; Xiao et al. 1993), the first solid evidence of transducing nondividing cells by rAAV vectors came from our in vivo studies in rat brains (Kaplitt et al. 1994). We have unequivocally demonstrated long-term transduction of neuronal cells in different areas of adult rat brain using rAAV vectors, containing either a LacZ reporter gene or a therapeutic tyrosine hydroxylase gene in a Parkinsonian rat model (Kaplitt et al. 1994). Subsequently, we have also shown that a rAAV–LacZ vector delivered through coronary artery achieved efficient gene transfer in pig heart muscle for over 6 months without significant decline of the transgene expression (Kaplitt et al. 1996).

Recently, we and others have demonstrated that AAV vectors, when injected in the mature skeletal muscle of mice, achieved sustained expression of LacZ gene up to 19 months, with molecular characterization of the vector suggesting integration as a mechanism of persistence (Xiao et al. 1996). Besides brain and muscle, other predominantly nondividing tissues efficiently transduced by AAV vectors include liver (Koeberl, Alexander, Halbert, Russell, and Miller 1997; Snyder et al. 1997; Xiao, Berta, Lu, Tazelaar, and Wilson 1998), spinal cord (Peel, Zolotukhin, Schrimsher, Muzyczka, and Reier 1997), and eye (Flannery et al. 1997). A very important factor behind the success of AAV vector in muscle tissue is its small particle size (20 nm in diameter), which makes it easy to bypass capillary blood vessel and/or basal lamina (extracellular matrix) of myofibers to consequently infect mature muscle cells. By contrast, other viral vectors, such as adenovirus (70 to 100 nm), retrovirus (~100 nm) and herpes viruses (120 to 300 nm), poorly transduce mature muscle cells because of the physical barrier generated by the basal lamina and other factors (Huard et al. 1996; van Deutekom, Floyd, et al. 1998; van Deutekom, Hoffman, et al. 1998).

The lack of a cellular immune response against vector-transduced cells and the lack of cytotoxicity are common characteristics of in vivo transduction experiments with rAAV vectors (Flotte et al. 1993; Kaplitt et al. 1994, 1996; Afione et al. 1996; Kessler et al. 1996; Xiao et al. 1996; Clark et al. 1997; Fisher et al. 1997; Flannery et al. 1997; Koeberl et al. 1997; Peel et al. 1997; Ponnazhagan, Erikson, et al. 1997; Snyder et al. 1997; Xiao, McCown, Li, et al. 1997). These results are in major contrast to in vivo transduction with Ad-based vectors, particularly the early generation ones, which induce severe immune reactions that result in the clearance of the transduced cells and destruction of the target tissue. Apparently, the cellular immune reaction is caused by the expression of the remaining Ad genes as well as transgenes themselves in the Ad vectors (Yang et al. 1994; Yang, Li, et al. 1995; Yang and Wilson 1995; Tripathy, Black, Goldwasser, and Leiden 1996). In contrast, rAAV vectors do not contain any viral genes or gene sequences. Moreover, the phenomena of reactions against vector transgenes generally do not hold true for AAV vectors. Foreign proteins expressed from AAV vectors, regardless of the nature of the transgenes, whether intracellular proteins, such as LacZ, GFP, Neo, and cellular enzymes (Fisher et al. 1997), or secreted proteins, such as factor IX (Koeberl et al. 1997; Snyder et al. 1997; Monahan et al. 1998), Epo (Kessler et al. 1996), or α–1 antitrypsin (Xiao, Berta, et al. 1998), do not cause CTL response to the transduced cells, even though the secreted proteins are recognized by antibodies in the humoral immune mechanism. This lack of CTL response has been attributed to the inability of rAAV vectors to infect professional antigen presenting cells (APCs), such as dendritic cells. It has been demonstrated (Jooss, Yang, Fisher, and Wilson 1998) that administration of dendritic cells preinfected with Ad–LacZ to the mice bearing AAV–LacZ transduced muscle cells can cause the elimination of those LacZ positive muscle cells, whereas administration of dentritic cells preinfected with AAV–LacZ did not have the same effect. In addition, the simple viral shell structure of AAV, with only one major viral protein (Berns and Bohenzy 1987), may also minimize the potential adjuvant effect.

Integration is an intrinsic part of the wild-type AAV latent life cycle (Berns and Bohenzy 1987; Berns and Linden 1995). A substantial body of work on rAAV vectors confirms integration in vitro. The discovery of site specific integration of wild-type AAV into human chromosome 19 has inspired additional interest in research concerning this vector (Kotin et al. 1990; Samulski et al. 1991; Berns and Linden 1995). Further characterization of the integration process revealed the involvement of AAV replication proteins (rep) in the site-specific event by binding to both AAV ITR and a consensus sequence on chromosome 19 (Weitzman, Kyostio, Kotin, and Owens 1994; Giraud, Winocour, and Berns 1995). However, in the absence of rep gene products, rAAV integration has lost its preference for chromosome 19 (Walsh et al. 1992; Samulski 1993; Kearns et al. 1996; Ponnazhagan, Mukherjee, et al. 1997). Although well documented in tissue culture cells, integration in vivo has not been well characterized. Wild-type AAV–2 DNA was reportedly detected by polymerase chain reaction (PCR) in peripheral blood leukocytes in 2 of 55 healthy blood donors, and 2 of 16 hemophiliac patients (Grossman et al. 1992). However, integration events were not examined in the report, despite the fact that wild-type AAV and rAAV vector integrations in human hematopoietic progenitor cells in vitro were documented (Goodman et al. 1994; Fisher–Adams, Wong, Podsakoff, Forman, and Chatterjee 1996). Recently, we and others have reported genomic integration by rAAV vector DNA after intramuscular injection into nondividing muscle tissues (Fisher–Adams et al. 1996; Xiao et al. 1996; Clark et al. 1997). Besides genomic integration other mechanisms for vector DNA persistence in vivo, such as the episomal circular form, also exist.

Despite the advantages of small viral particle size, however, a major inherent disadvantage of the rAAV system is its 5 kb packaging constraint that precludes the use of this vector for certain large genes, such as full-length dystrophin gene (>11 kb) for Duchenne muscular dystrophy (DMD). Even the truncated Becker like dystrophin gene (>7 kb) is beyond the packaging limit capacity for rAAV. Also, like retroviral vectors, AAV vector DNA integration into the host cell DNA may pose the potential risk of mutagenesis, although the probability of this is extremely low.

AAV Vector for Muscle Gene Therapy

Despite the success of rAAV injection in muscle tissue using numerous reporter genes as well as therapeutic genes for metabolic disease and secreted proteins, the utility of this vector system to treat muscle disease itself and to rescue muscle function loss caused by genetic deficiencies has been largely untested. Previous restrictions were either due to the extra large size of certain muscle disease genes, dystrophin and merosin, for example, or to the lack of animal disease models for small genes of muscular diseases, sarcoglycans (<2 kb) for limb girdle muscular dystrophy (LGMD), for example. However, recent advances have made it possible to test the feasibility of using rAAV vectors in treating muscle diseases. One such advance is the identification of a mutation in δ–sarcoglycan gene (cDNA <1 kb) in a dystrophic hamster model (Bio14.6) (Nigro et al. 1997; Sakamoto et al. 1997). The engineering of rAAV to carry the δ-sarcoglycan gene and the development of intramuscular injection of viral vectors have recently led to an efficient gene transfer of δ–sarcoglycan gene into the dystrophic muscles (Li et al. 1999) and a full functional recovery of the muscle deficit caused by the genetic mutation.

Muscle injuries present a formidable challenge in traumatology and are the most frequently occurring injuries in orthopaedics and sports medicine (see Chapter 5). Despite the natural ability of muscle to regenerate following injury, the healing process is often slow and incomplete. Healing is further hampered by the formation of fibrotic scar tissue. During the healing process, injured muscle fibers develop a hematoma, and then undergo necrosis, at which time the damaged myofibers are removed by macrophages (Kalimo, Rantanen, and Jarvinen 1997). New muscle fibers regenerate within the connective tissue framework of the damaged muscle. This natural healing process following muscle injuries subsequently leads to morbidity, possible reinjury, muscle atrophy, contracture, and pain (Jackson and Feagin 1973; Ryan, Wheeler, Hoppkinson, Arciero, and Kolakowski 1991; Aronen and Chronister 1992). Gene transfer of therapeutic growth factors to the injured muscle site can enhance muscle regeneration and healing, and help in the prevention of fibrotic scar tissue postinjury (see Chapter 5). AAV has the potential to incorporate the genes that encode for these growth factors and cytokines, and may allow for an efficient delivery into the injured muscle. Because growth factors are usually small peptides that bind to membrane receptors to influence various stages of cell development and cell functions (Grounds 1991; Chambers and McDermott 1996), these small genes can be readily inserted into recombinant AAV vectors. It is presumed that during muscle regeneration trophic substances released by the injured muscle activate satellite cells. Studies have shown that basic fibroblast growth factor (bFGF), insulin-like growth factor (IGF-1), and nerve growth factor (NGF) are potent stimulators of myoblast proliferation and fusion in vitro and of improved muscle regeneration in vivo (see Chapter 5).

The use of rAAV to encode for small genes could provide long-term persistence of these growth factors at the injured muscle site. Although other vectors are capable of carrying the genes for these growth factors, AAV is perhaps the best candidate. The advantages of rAAV transduction are its ability to transduce muscle at different maturities and its low

Newborn

Adult

Injured muscle

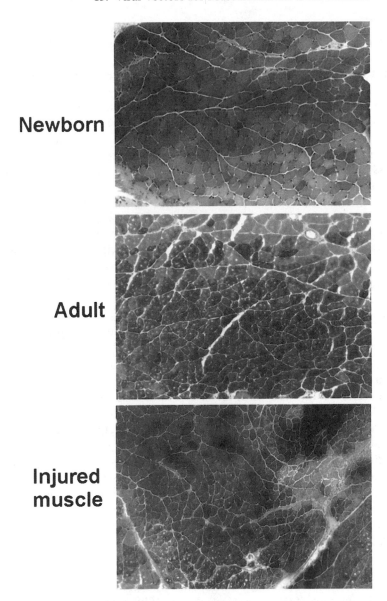

FIGURE 15.1. Efficient transduction with recombinant adeno-associated virus (rAAV) in muscle at different maturities. The injection of rAAV in newborn, adult, and cardiotoxin-injured muscle leads to an efficient gene transfer of β galactosidase at 3 weeks postinjection. Magnification: newborn = 20×. Adult and injured muscles = 10×.

immunogenicity, which consequently leads to long term persistence of expression of these growth factors at the injured site. A recent report has demonstrated that an AAV vector carrying the IGF-I gene can efficiently prevent age-related muscle injury and loss of muscle mass (Barton–Davis et al. 1998).

We have observed that rAAV can highly transduce both newborn and adult skeletal muscle (Figure 15.1). More important, rAAV also transduces regenerating muscle following injuries (cardiotoxin) with a high efficiency, suggesting that this vector can eventually deliver a high level and persistent expression of growth factors capable of improving muscle healing following injuries (Figure 15.1).

Since it has been observed that irradiation of the adult muscle prior to rAAV injection does not prevent the virus from transducing mature myofibers, we have concluded that rAAV does not require myoblasts as intermediates for transducing myofibers. Such a finding suggests that this vector can transduce skeletal muscle despite the level of muscle maturation. The ability of rAAV to highly transduce mature myofibers without the requirement of myoblasts, in contrast to other viral vectors (van Deutekom, Floyd, et al. 1998; van Deutekom, Hoffman, et al. 1998), makes rAAV ideal for the delivery of specific growth factors to improve muscle healing following injury.

In conclusion, rAAV has numerous unique advantages over other vectors used in gene therapy. It is not cytotoxic, does not trigger a cellular immunological response, and can achieve long-term transgene expression. Moreover, rAAV infects postmitotic cells, is small enough (20 nm) to penetrate the pores of the extracellular matrix, and does not require the presence of myoblasts to transduce myofibers. Therefore, rAAV is an attractive vector for delivering genes encoding for therapeutic growth factors and cytokines that could enhance the healing process following muscle injuries. The ability of rAAV to deliver genes in different tissues of the musculoskeletal system is under investigation.

References

Acsadi, G., Dickson, G., Love, D.L., Jani, A., Walsh, F.S., Gurusinghe, A., Wolff, J.A., and Davies, K.E. 1991. Human dystrophin expression in mdx mice after intramuscular injection of DNA constructs. *Nature* 352:815–18.
Acsadi, G., Jani, A., Massie, B., Simoneau, M., Holland, P., Blaschuk, K., and Karpati, G. 1994. A differential efficiency of adenovirus mediated in vivo gene transfer into skeletal muscle cells of different maturity. *Hum Mol Genet* 3:579–84.
Acsadi, G., Lochmueller, H., Jani, A., Huard, J., Massie, B., Prescott, S., Simoneau, M., Petrof, B., and Karpati, G. 1996. Dystrophin expression in muscles of mdx mice after adenovirus mediated in vivo gene transfer. *Hum Gene Ther* 7:129–40.
Afione, S.A., Conrad, C.K., Kearns, W.G., Chunduru, S., Adams, R., Reynolds, T.C., Guggino, W.B., Cutting, G.R., Carter, B.J., and Flotte, T.R. 1996. In vivo model of adeno–associated virus vector persistence and rescue. *J Virol* 70:3235–41.

Alexander, I.E., Russell, D.W., and Miller, A.D. 1997. Transfer of contaminants in adeno–associated virus vector stocks can mimic transduction and lead to artifactual results. *Hum Gene Ther* 8:1911–20.

Aronen, J.G. and Chronister, R. 1992. Quadriceps contusions: hastening the return to play. *Phys Sports Med* 20:130–6.

Barr, E. and Leiden, J.M. 1991. Systemic delivery of recombinant proteins by genetically modified myoblasts. *Science* 254:1507–9.

Barton–Davis, E.R., Shoturma, D.I., Musaro, A., Rosenthal, N., and Sweeney, H.L. 1998. Viral mediated expression of insulin–like growth factor I blocks the aging related loss of skeletal muscle function. *Proc Natl Acad Sci USA* 95:15603–7

Berns, K.I. 1996. Parvoviridae: the viruses and their replication. In B.N. Fields, D.M. Knipe, and P.M. Howley (eds.). Philadelphia/New York: Lippincott-Raven.

Berns, K.I. and Bohenzky, R.A. 1987. Adeno–associated viruses: an update. [review]. *Advances in Virus Research* 32:243–306.

Berns, K.I. and Linden, R.M. 1995. The cryptic life style of adeno–associated virus. *Bioessays* 17:237–45.

Chambers, R.L. and McDermott, J.C. 1996. Molecular basis of skeletal muscle regeneration. *Can J Appl Physiol* 21:155–84.

Clark, K.R., Sferra, T.J., and Johnson, P.R. 1997. Recombinant adeno–associated viral vectors mediate long–term transgene expression in muscle. *Hum Gene Ther* 8:659–69.

Clark, K.R., Voulgaropoulou, F., Fraley, D.M., and Johnson, P.R. 1995. Cell lines for the production of recombinant adeno–associated virus. *Hum Gene Ther* 6:1329–41.

Davis, H.L., Demeneix, B.A., Quantin, B., Coulombe, J., and Whalen, R.G. 1993. Plasmid DNA is superior to viral vectors for direct gene transfer into adult mouse skeletal muscle. *Hum Gene Ther* 4:733–40.

Davis, H.L., Whalen, R.G., and Demeneix, B.A. 1993. Direct gene transfer into skeletal muscle in vivo: factors affecting efficiency of transfer and stability of expression. *Hum Gene Ther* 4:151–9.

Dhawan, J., Pan, L.C., Pavlath, G.K., Travis, M.A., Lanctot, A.M., and Blau, H.M. 1991. Systemic delivery of human growth hormone by injection of genetically engineered myoblasts. *Science* 254:1509–12.

Doh, S.G., Vahlsing, J., Hartikka, J., Liang, X., and Manthorpe, M. 1997. Spatial–temporal patterns of gene expression in mouse skeletal muscle after injection of LacZ plasmid DNA. *Gene Ther* 4:648–63.

Dunckley, M.G., Davies, K.E., Walsh, F.S., Morris, G.E., and Dickson, G. 1992. Retroviral mediated transfer of a dystrophin minigene into mdx mouse myoblasts in vitro. *FEBS Lett* 296:128–34.

Feero, W.G., Rosenblatt, J.D., Huard, J., Watkins, S.C., Epperly, M., Clemens, P.R., Kochanek, S., Glorioso, J.C., Partridge, T.A., and Hoffman, E.P. 1997. Viral gene delivery to skeletal muscle: insights on maturation dependent loss of fiber infectivity for adenovirus and herpes simplex type 1 viral vectors. *Hum Gene Ther* 8:371–80.

Ferrari, F.K., Xiao, X., McCarty, D., and Samulski, R.J. 1997. New developments in the generation of Ad free, high titer rAAV gene therapy vectors. *Nat Med* 3:1295–7.

Fisher, K.J., Jooss, K., Alston, J., Yang, Y., Haecker, S.E., High, K., Pathak, R., Raper, S.E., and Wilson, J.M. 1997. Recombinant adeno–associated virus for muscle directed gene therapy. *Nat Med* 3:306–12.

Fisher–Adams, G., Wong Jr., K.K., Podsakoff, G., Forman, S.J., and Chatterjee, S. 1996. Integration of adeno–associated virus vectors in CD34+ human hematopoietic progenitor cells after transduction. *Blood* 88:492–504.

Flannery, J.G., Zolotukhin, S., Vaquero, M.I., LaVail, M.M., Muzyczka, N., and Hauswirth, W.W. 1997. Efficient photoreceptor targeted gene expression in vivo by recombinant adeno–associated virus. *Proc Natl Acad Sci USA* 94:6916–21.

Flotte, T.R., Afione, S.A., Conrad, C., McGrath, S.A., Solow, R., Oka, H., Zeitlin, P.L., Guggino, W.B., and Carter, B.J. 1993. Stable in vivo expression of the cystic fibrosis transmembrane conductance regulator with an adeno–associated virus vector. *Proc Natl Acad Sci USA.* 90:10613–17.

Flotte, T.R. and Carter, B.J. 1995. Adeno–associated virus vectors for gene therapy. [review]. *Gene Ther* 2:357–62.

Giraud, C., Winocour, E., and Berns, K.I. 1995. Recombinant junctions formed by site specific integration of adeno–associated virus into an episome. *J Virol* 69:6917–24.

Goodman, S., Xiao, X., Donahue, R.E., Moulton, A., Miller, J., Walsh, C., Young, N.S., Samulski, R.J., and Nienhuis, A.W. 1994. Recombinant adeno–associated virus mediated gene transfer into hematopoietic progenitor cells [published erratum appears in 1995 Blood 1:85(3):862]. *Blood* 84:1492–500.

Grossman, Z., Mendelson, E., Brok–Simoni, F., Mileguir, F., Leitner, Y., Rechavi, G., and Ramot, B. 1992. Detection of adeno–associated virus type 2 in human peripheral blood cells. *J Gen Virol* 73:961–6.

Grounds, M.D. 1991. Towards understanding skeletal muscle regeneration. *Path Res Pract* 187:1–22.

Huard, J., Akkaraju, G., Watkins, S.C., Cavalcoli, M.P., and Glorioso, J.C. 1997. LacZ gene transfer to skeletal muscle using a replication defective herpes virus type 1 mutant vector. *Hum Gene Ther* 8:439–52.

Huard, J., Feero, W.G., Watkins, S.C., Hoffman, E.P., Rosenblatt, D.J., and Glorioso, J.C. 1996. The basal lamina is a physical barrier to herpes simplex virus mediated gene delivery to mature muscle fibers. *J Virol* 70:8117–23.

Huard, J., Krisky, D., Oligino, T., Marconi, P., Day, C.S., Watkins, S.C., and Glorioso, J.C. 1997. Gene transfer to muscle using herpes simplex virus based vectors. *Neuromusc Disor* 7:1–15.

Jackson, D.W. and Feagin, J.A. 1973. Quadriceps contusions in young athletes: relation of severity of injury to treatment and prognosis. *J Bone Joint Surg (Am)* 55A:95–105.

Jooss, K., Yang Y., Fisher, K.J., and Wilson, J.M. 1998. Transduction of dendritic cells by DNA viral vectors directs the immune response to transgene products in muscle fibers. *J Virol* 72:4212–23.

Kalimo, H., Rantanen, J., and Jarvinen, M. 1997. Muscle injuries in Sports. *Balliere's Clin Orthop* 2(1):1–24.

Kaplitt, M.G., Leone, P., Samulski, R.J., Xiao, X., Pfaff, D.W., O'Malley, K.L., and During, M.J. 1994. Long term gene expression and phenotypic correction using adeno–associated virus vectors in the mammalian brain. *Nat Genet* 8:148–54.

Kaplitt, M.G., Xiao, X., Samulski, R.J., Li, J., Ojamaa, K., Klein, I.L., Makimura, H., Kaplitt, M.J., Strumpf, R.K., and Diethrich, E.B. 1996. Long term gene transfer in porcine myocardium after coronary infusion of an adeno–associated virus vector. *Ann Thorac Surg* 62:1669–76.

Kearns, W.G., Afione, S.A., Fulmer, S.B., Pang, M.C., Erikson, D., Egan, M., Landrum, M.J., Flotte, T.R., and G.R. Cutting, T.R. 1996. Recombinant adeno–associated virus (AAV–CFTR) vectors do not integrate in a site specific fashion in an immortalized epithelial cell line. *Gene Ther* 3:748–55.

Kessler, P.D., Podsakoff, G.M., Chen, X., McQuiston, S.A., Colosi, P.C., Matelis, L.A., Kurtzman, G.J., and Byrne, B.J. 1996. Gene delivery to skeletal muscle results in sustained expression and systemic delivery of a therapeutic protein. *Proc Natl Acad Sci USA* 93:14082–7.

Koeberl, D.D., Alexander, I.E., Halbert, C.L., Russell, D.W., and Miller, A.D. 1997. Persistent expression of human clotting factor IX from mouse liver after intravenous injection of adeno–associated virus vectors. *Proc Natl Acad Sci USA* 94:1426–31.

Kotin, R.M. 1994. Prospects for the use of adeno–associated virus as a vector for human gene therapy. *Hum Gene Ther* 5:793–801.

Kotin, R.M., Siniscalco, M., Samulski, R.J., Zhu, X.D., Hunter, L., Laughlin, C.A., McLaughlin, S., Muzyczka, N., Rocchi, M., and Berns, K.I. 1990. Site specific integration by adeno–associated virus *Proc Natl Acad Sci USA* 87:2211–5.

Li, J., Dressman, D., Tsao, Y.P., Sakamoto, A., Hoffman, E.P., and Xiao, X. 1999. *Gene Ther* 6:74–82.

Li, J., Samulski, R.J., and Xiao, X. 1997. Role for highly regulated rep gene expression in adeno–associated virus vector production. *J Virol* 71:5236–43.

Monahan, P.E., Samulski, R.J., Tazelaar, J., Xiao, X., Nichols, T.C., Bellinger, D.A., Read, M.S., and Walsh, C.E. 1998. Direct intramuscular injection with recombinant AAV vectors results in sustained expression in a dog model of hemophilia. *Gene Ther* 5:40–9.

Muzyczka, N. 1992. Use of adeno–associated virus as a general transduction vector for mammalian cells. *Curr Topics Micro Immunol* 158:97–129.

Nigro, V., Okazaki, Y., Belsito, A., Piluso, G., Matsuda, Y., Politano, L., Nigro, G., Ventura, C., Abbondanza, C., Molinari, A.M., Acampora, D., Nishimura, M., Hayashizaki, Y., and Puca, G.A. 1997. Identification of the Syrian hamster cardiomyopathy gene. *Hum Mol Gen* 6:601–7.

Peel, A.L., Zolotukhin, S., Schrimsher, G.W., Muzyczka, N., and Reier, P.J. 1997. Efficient transduction of green fluorescent protein in spinal cord neurons using adeno–associated virus vectors containing cell type specific promoters. *Gene Ther* 4:16–24.

Ponnazhagan, S., Erikson, D., Kearns, W.G., Zhou, S.Z., Nahreini, P., Wang, X.S., and Srivastava, A. 1997. Lack of site specific integration of the recombinant adeno–associated virus 2 genomes in human cells. *Hum Gene Ther* 8:275–84.

Ponnazhagan, S., Mukherjee, P., Yoder, M.C., Wang, X.S., Zhou, S.Z., Kaplan, J., Wadsworth, S., and Srivastava, A. 1997. Adeno–associated virus 2 mediated gene transfer in vivo: organ-tropism and expression of transduced sequences in mice. *Gene* 190:203–10.

Quantin, B., Perricaudet, L.D., Tajbakhsh, S., and Mandel, J.L. 1992. Adenovirus as an expression vector in muscle cells in vivo. *Proc Natl Acad Sci USA* 89:2581–4.

Ragot, T., Vincent, M., Chafey, P., Vigne, E., Gilgenkrantz, H., Couton, B., Cartaud, J., Briand, B., Kaplan, J.C., Perricaudet, M., and Kahn, A. 1993. Efficient adenovirus mediated transfer of a human mini dystrophin gene to skeletal muscle of mdx mice. *Nature* 361:647–50.

Ryan, J.B., Wheeler, J.H., Hopkinson, W.J., Arciero, R.A., and Kolakowski, K.R. 1991. Quadriceps contusions—West point update. *Am J Sports Med* 19:299–304.

Sakamoto, A., Ono, K., Abe, M., Jasmin, G., Eki, T., Murakami, Y., Masaki, T., Toyo–oka, T., and Hanaoka, F. 1997. Both hypertrophic and dilated cardiomyopathies are caused by mutation of the same gene, δ–sarcoglycan, in hamster: an animal model of disrupted dystrophin associated glycoprotein complex. *Proc Natl Acad Sci USA* 94:13873–8.

Salvatori, G., Ferrari, G., Messogiorno, A., Servidel, S., Colette, M., Tonalli, P., Giarassi, R., Cosso, G., and Mavillo, F. 1993. Retroviral mediated gene transfer into human primary myogenic cells lead to expression in muscle fibers in vivo. *Hum Gene Ther* 713–23.

Samulski, R.J. 1993. Adeno–associated virus: integration at a specific chromosomal locus. *Curr Opin Genet Dev* 3:74–80.

Samulski, R.J., Chang, L.S., and Shenk T. 1989. Helper free stocks of recombinant adeno–associated viruses: normal integration does not require viral gene expression. *J Virol* 63:3822–8.

Samulski, R.J., Zhu, X., Xiao, X., Brook, J.D., Housman, D.E., Epstein, N., and Hunter, L.A. 1991. Targeted integration of adeno–associated virus (AAV) into human chromosome 19 [published erratum appears in 1992 EMBO J 11(3):1228]. *EMBO J* 10:3941–50.

Snyder, R.O., Miao, C.H., Patijn, G.A., Spratt, S.K., Danos, O., Nagy, D., Gown, A.M., Winther, B., Meuse, L., Cohen, L.K., Thompson, A.R., and Kay, M.A. 1997. Persistent and therapeutic concentrations of human factor IX in mice after hepatic gene transfer of recombinant AAV vectors. *Nat Genet* 16:270–6.

Snyder, R.O., Xiao, X., and Samulski, R.J. 1996. Production of recombinant adeno–asociated viral vectors. In N. Dracopoli, J. Haines, B. Krof, D. Moir, C. Seidman, and J.S. Seidman, D. (eds.), *Current protocols in Human Genetics*. 12.1.1–12.3.23, New York: John Wiley & Sons.

Tamayose, K., Hirai, Y., and Shimada, T. 1996. A new strategy for large scale preparation of high titer recombinant adeno–associated virus vectors by using packaging cell lines and sulfonated cellulose column chromatography. *Hum Gene Ther* 7:507–13.

Tripathy, S.K., Black, H.B., Goldwasser, E., and Leiden, J.M. 1996. Immune responses to transgene encoded proteins limit the stability of gene expression after injection of replication defective adenovirus vectors. *Nat Med* 2:545–50.

van Deutekom, J.C.T., Floyd, S.S., Booth, D.K., Oligino, T., Krisky, D., Marconi, P., Glorioso, J.C., and Huard, J. 1998. Implications of maturation for viral gene delivery to skeletal muscle. *Neuromusc Disord* 8:135–48.

van Deutekom, J.C.T., Hoffman, E.P., and Huard, J. 1998. Muscle maturation: implications for gene therapy. *Mol Med Today* 5:214–20.

Verma, I.M. and Somia, N. 1997. Gene therapy—promises, problems and prospects [news]. *Nature* 389:239–42.

Vincent, M., Ragot, T., Gilgenkrantz, H., Couton, D., Chafey, P., Gregoire, A., Briand, P., Kaplan, J.C., Kahn, A., and Perricaudet, M. 1993. Long term correction of mouse dystrophic degeneration by adenovirus mediated transfer of a mini dystrophin gene. *Nat Genet* 5:130–4.

Vitiello, L., Chonn, A., Wasserman, J.D., Duff, C., and Worton, R.G. 1996. Condensation of plamid DNA with polylysine improves liposome mediated gene transfer into established and primary muscle cells. *Gene Ther* 3:369–404.

Walsh, C.E., Liu, J.M., Xiao, X., Young, N.S., Nienhuis, A.W., and Samulski, R.J. 1992. Regulated high level expression of a human gamma–globin gene introduced into erythroid cells by an adeno–associated virus vector. *Proc Natl Acad Sci USA* 89:7257–61.

Weitzman, M.D., Kyostio, S.R., Kotin, R.M., and Owens, R.A. 1994. Adeno–associated virus (AAV) rep proteins mediate complex formation between AAV–DNA and its integration site in human DNA. *Proc Natl Acad Sci USA* 91:5808–12.

Wolff, J.A., Ludtke, J.J., Acsadi, G., Williams, P., and Jani, A. 1992. Long term persistance of plasmid DNA and foreign gene expression in mouse muscle. *Hum Mol Genet* 1:363–9.

Xiao, W., Berta, S.C., Lu, M.M., Tazelaar, J., and Wilson, J.M. 1998. Adeno–associated virus as a vector for liver directed gene therapy. *J Virol* 72:10222–6.

Xiao, X., DeVlaminick, W., and Monahan, J. 1993. Adeno–associated virus (AAV) vectors for gene transfer. *Advanced Drug Delivery Review* 12:201–15.

Xiao, X., Li, J., McCown, T.J., and Samulski, R.J. 1997. Gene transfer by adeno–associated virus vectors into the central nervous system. *Exper Neurol* 144:113–24.

Xiao, X., Li, J., and Samulski, R.J. 1996. Efficient long term gene transfer into muscle tissue of immunocompetent mice by adeno–associated virus vector. *J Virol* 70:8098–108.

Xiao, X., Li, J., and Samulski, R.J. 1998. Production of high titer recombinant adeno–associated virus vectors in the absence of helper adenovirus. *J Virol* 72:2224–32.

Xiao, X., McCown, T.J., Li, J., Breese, G.R., Morrow, A.L., and Samulski, R.J. 1997. Adeno–associated virus (AAV) vector antisense gene transfer in vivo decreases GABA(A) alpha1 containing receptors and increases inferior collicular seizure sensitivity. *Brain Res* 756:76–83.

Xiao, X., Xiao, W., Li, J., and Samulski, R.J. 1997. A novel 165 base–pair terminal repeat sequence is the sole cis requirement for the adeno–associated virus life cycle. *J Virol* 71:941–8.

Yang, Y., Li, Q., Ertl, H.C., and Wilson, J.M. 1995. Cellular and humoral immune responses to viral antigens create barriers to lung directed gene therapy with recombinant adenoviruses. *J Virol* 69:2004–15.

Yang, Y. Nunes, F.A., Berencsi, K., Furth, E.E., Gonczol, E., and Wilson, J.M. 1994. Cellular immunity to viral antigens limits E1 deleted adenoviruses for gene therapy. *Proc Natl Acad Sci USA* 91:4407–11.

Yang, Y. and Wilson, J.M. 1995. Clearance of adenovirus infected hepatocytes by MHC class I restricted CD4+ CTLs in vivo. *J Immunology* 155:2564–70.

Yurchenco, P.D. 1990. Assembly of basement membranes. *Ann New York Acad Sci* 580:195–213.

Index

Acidic fibroblast growth factor
(aFGF), 201
Adeno-associated virus (AAV) vectors,
246
biology of, 262–265
for gene delivery, 19–20, 25
to joints, 253–254
to muscle, 80–81, 201
musculoskeletal, 260–273
immune reaction to, 236
recombinant, 44–45
Adenosine deaminase deficiency
syndrome, 4, 6
Adenovirus, 206, 261
advantages of, for nucleus pulposus
infection, 70
immune response induced by, 90
tacrolimus administered in
conjunction with, 238–239
Adenovirus vectors, 18–19
for gene delivery, 204
to an intervertebral disc, 60–62
direct, of IGF to a cartilage defect,
51
for bone nonunion treatment, 46,
48
to ligaments, 186–187
to skeletal muscle, 80–82, 201
to tendon, 130
for gene transfer, 23, 44
to joints, 249–251
immune response induced by,
236
Alkaline phosphatase, production
by myogenic cells, 203–204

Allograft
bone, 203
rejection of, 144
Annulus fibrosis, nitric oxide
production in cells from, 134
Anterior cruciate ligament (ACL)
delivering genes to, 186, 206–207
healing of, 8, 28–29, 81, 174, 178
nitric oxide involvement in, 131
reconstruction of, surgical, 204
synthesis of fibroblasts of, 181
Antibodies
against dystrophin, in myoblast
transplantation, 236
immunosuppression with, 239
treatment with, to develop tolerance,
240
Antigens, 3
minor, immune responses against,
234
Antisense therapy, for suppressing a
mutant allele, 117–118
Apoptosis, of large cells, 165–166
Arthritis, endstage, preventing, 49. *See
also* Osteoarthritis; Rheumatoid
arthritis
Arthroplasty, abrasion, 213
Arthroscopic techniques, 7
Articular cartilage. *See* Cartilage,
articular
Articular defect repair, protocol for,
rabbit model, 91
Autogenous cells, advantages of using,
198–199
Autologous cells, transplantation of, 241